Sustainable Environmental Economics and Management

Sustainable Environmental Economics and Management

Principles and Practice

Edited by
R. Kerry Turner

JOHN WILEY & SONS
Chichester · New York · Brisbane · Toronto · Singapore

First published in 1993 by Belhaven Press

Published June 1995 by John Wiley & Sons Ltd,
Baffins Lane, Chichester,
West Sussex PO19 1UD, England
National 01243 779777
International +44 1243 779777

Other Wiley Editorial Offices

John Wiley & Sons, Inc., 605 Third Avenue,
New York, NY 10158-0012, USA

Jacaranda Wiley Ltd, 33 Park Road, Milton,
Queensland 4064, Australia

John Wiley & Sons (Canada) Ltd, 22 Worcester Road,
Rexdale, Ontario M9W 1L1, Canada

John Wiley & Sons (SEA) Pte Ltd, 37 Jalan Pemimpin #05-04,
Block B, Union Industrial Building, Singapore 2057

ISBN 0–471–94781–4

Typeset by Mayhew Typesetting, Rhayader, Powys
Printed and bound in Great Britain by
Biddles Ltd, Guildford and King's Lynn

Contents

List of contributors

N. Adger Senior Research Associate, Centre for Social and Economic Research on the Global Environment (CSERGE), University of East Anglia and University College London.

E.B. Barbier Director of the London Environmental Economics Centre (LEEC) at the International Institute for Environment and Development (IIED).

I.J. Bateman School of Environmental Studies, University of East Anglia, Norwich, and Research Fellow, Centre for Social and Economic Research on the Global Environment (CSERGE), University of East Anglia and University College London.

J.F. Benson Department of Town and Country Planning, University of Newcastle upon Tyne.

N. Hanley Department of Economics, University of Stirling.

N. Henderson School of Environmental Sciences, University of East Anglia, Norwich.

I. Moffatt North Australian Research Unit, Australian National University.

T. O'Riordan Associate Director, Centre for Social and Economic Research on the Global Environment (CSERGE), University of East Anglia and University College London.

D.W. Pearce Director, Centre for Social and Economic Research on the Global Environment (CSERGE), University of East Anglia and University College London.

M. Redclift Environment Section, Wye College, University of London.

R.K. Turner Executive Director, Centre for Social and Economic Research on the Global Environment (CSERGE), University of East Anglia and University College London.

K.G. Willis Department of Town and Country Planning, University of Newcastle upon Tyne.

Preface

Sustainable Environmental Management (1988) turned out to be a very successful monograph which sought to analyse the meaning and practical significance of the then emerging concept of sustainable development. This volume represents both an update and extension of the basic principles and practice of sustainability first presented in the 1988 volume. Many of the original contributors have again agreed to write chapters for this new volume. They have been joined by researchers from the ESRC-sponsored Centre for Social and Economic Research on the Global Environment (CSERGE), University of East Anglia and University College London, and from the London Environmental Economics Centre (LEEC), as well as by two environmental economists from the University of East Anglia and the University of Sterling.

Part 1
Principles

Chapter 1
Sustainability: Principles and Practice
R.K. Turner

Principles: Introduction

A large and diverse literature has emerged in recent years concerned with the notion of sustainable development (SD).[1] Many definitions (often incompatible) have been suggested and debated, thereby exposing a range of approaches linked to different world views (Pearce *et al.* 1989; Pearce and Turner 1990).

From the ecocentric perspective, the extreme deep ecologists seem to come close to rejecting even the sustainable utilisation of nature's assets. According to some of their writings, the environment ought not to be conceived of as a collection of goods and services for human use at all (Rolston 1988). From the opposite technocentric perspective, other analysts argue that the concept of sustainability contributes little new to mainstream (largely neo-classical economic) approaches to intertemporal choice. Given this world view, the maintenance of a sustainable growth economy over the long run depends on the adequacy of investment expenditure. First and foremost, it is investment in physical and human capital (i.e. buildings, machines, etc., plus the stock of knowledge) that counts and only to a much lesser degree investment in natural capital (i.e. the stock of non-renewable and renewable resources provided by the biosphere and the opportunities for solar-powered recycling) (Nordhaus 1992). A key assumption of this position is that there will continue to be a very high degree of substitutability between all forms of capital resources.

In a typology, outlined in a later sub-section of this chapter, these two polar positions have been labelled respectively the 'very strong sustainability' (VSS) and the 'very weak sustainability' (VWS) positions (Turner 1992).[2] But defining sustainable development is not the only, and probably not the most important, problem. If the sustainability goal is accepted then a fundamental requirement is a set of sustainability principles that can give some concrete form to a sustainable development strategy. This strategy will necessarily have to encompass multiple and interrelated goals (reflecting the several dimensions of the sustainability concept) – social/cultural, economic, political, environmental and moral – and will have to deploy a package of enabling policy instruments. In Chapter 2 of this volume O'Riordan takes a closer look at the complexities of this wider 'politics' of the sustainability idea and process.

The most publicised definition of sustainability is that of the World Commission on Environment and Development (WCED) (the Brundtland Commission). The Commission defined SD as 'development that meets the needs of the present without compromising the ability of future generations to meet their own needs' (WCED 1987, 43). Both an equity dimension (intragenerational and intergenerational) and a social/psychological dimension (i.e. the term 'needs' is used rather than the economic term 'wants', which is tied into the concept of consumer sovereignty) are clearly highlighted by this definition. If a society accepts the desirability of the goal of SD then it must develop economically and socially in such a way that it minimises the effects of its activities, the costs of which are borne by future generations. In cases where the activities and significant effects are unavoidable, future generations must be compensated for any costs they incur.[3]

The Brundtland Commission also highlighted 'the essential needs of the world's poor, to which overriding priority should be given', and 'the idea of limitations imposed by the state of technology and social organisation on the environment's ability to meet present and future'. The central rationale for SD is therefore to increase people's standard of living (broadly defined) and, in particular, the well-being of the least advantaged people in societies, while at the same time avoiding uncompensated future costs.

So how does the environment fit this requirement? According to Norton (1992), while the environment is mentioned it is given a passive role in the Commission's analysis; it 'does not impose any non-negotiable limits on sustainable use, independent of limitations on the abilities of humans to control it'. Economic growth and environmental protection are therefore at least potentially compatible objectives. In Chapter 3 of this volume, Pearce analyses the prospects for the decoupling of economic growth from environmental impact and the use of incentives to aid the decoupling process. The Brundtland Commission and similar viewpoints are anchored to a greater or lesser extent to the technocentric world view and therefore advocate some version of a weak sustainability-type approach.

Advocates of the strong sustainability approach (linked to or at least influenced by elements of ecocentrism) would take issue (to varying degrees) with the assumption of almost infinite substitutability of resources and that projections of economic growth and levels of social well-being can be forecast without accounting for the *scale* of human activities and its implications for the health and integrity of ecosystems.[4] In the 1950s the Dutch economist Jan Tinbergen formulated what has become known as the 'Tinbergen rule'. Simply stated, the rule lays down that for every independent policy goal there must be a complementary independent enabling policy instrument (Tinbergen 1952). One of the best-known advocates of the 'strong sustainability' viewpoint, Herman Daly, has introduced this 'targets and instruments' rule into the SD debate.

For Daly (1992) any environmental economic analysis of development is fatally flawed unless resource allocation and distribution objectives (which are supported by actual policy instruments) are augmented by a consideration of the 'scale' question.[5] 'Scale' is defined in a 'materials-balance sense'[6] in terms of the throughput of matter/energy in an economic system. It is the product of population times per-capita resource use. Because of the laws of thermodynamics, 'useful' resource inputs (matter/energy) are drawn into the economy and at a later stage reintroduced into the environment as relatively 'useless' waste products. The scale of economic activity should be related to the natural capacities of ecosystems to regenerate resource inputs for the economy and to assimilate (absorb and/or store without long-term damage) the waste flows from the economy. A desirable scale for economic activity would be one that does not erode the environmental carrying capacity over time. What is missing is the appropriate set of policy instruments to regulate the scale of economies. Severance taxes linked to natural capital depletion, and, in cases where uncertainty is great, assurance bonds tied to resource developments, have been suggested (Page 1977; Costanza and Perrings 1990; Costanza and Daly 1992).

Sustainable development, it is generally agreed, is therefore economic development that endures over the long run.[7] Economic development could be narrowly defined in traditional terms as real gross national product (GNP) per capita, or real consumption per capita. Alternatively, the traditional measures can be modified and extended to include a more comprehensive set of welfare indicators – education, health, quality of life, etc. Failure adequately to account for natural capital and the contribution it makes to economic welfare will also lead to misperceptions about how well an economy is really performing (Costanza and Daly 1992). A framework to reflect the use of natural resources at the national level is in the process of being agreed by the United Nations Statistical Office, but there is much debate as to the feasibility of the proposed alterations and other alternative amendments (Bartelmus *et al.* 1991; Bryant and Cook 1992; Daly and Cobb 1989; Nordhaus 1992). Both Barbier (Chapter 9) and Adger (Chapter 10) examine the potential advantages of a modified set of national income accounts with particular reference to developing countries. The economies of these countries are the ones that are most directly dependent on stocks of primary resources such as forests, rangelands, wetlands and mineral deposits.

SD now becomes fairly simply defined as, at least, non-declining consumption per capita, or per unit of GNP or some alternative agreed welfare indicator(s). This is how SD has come to be interpreted by a majority of economists addressing the issue (Pearce *et al.* 1990; Pezzey 1989; Mäler 1991a). Nevertheless, determining the necessary and sufficient conditions for achieving SD is an altogether more complicated task, as we have hinted in previous paragraphs. The so-called 'London School' of environmental economists, among others, has argued that a non-declining stock of natural capital over time is a necessary condition for sustainability, because of substitutability limits in production processes as well as other factors (Pearce *et al.* 1990; Pearce and Turner 1990). This is a strong sustainability (SS) position, with natural capital fulfilling the role of the fair/just compensatory bequest to future generations. To meet the requirements of SD the future must actually be compensated for any impairment of well-being caused by actions and activities engaged in by the current generation. The mechanism by which the current generation can ensure that the future is compensated, and is not therefore worse off, is through the transfer of capital bequests. Capital provides the opportunity and capability to generate well-being through the creation of material goods and amenity and other services that give human life its meaning.[8]

The weak sustainability position (WS) does not single out the environment for special treatment, it is simply another form of capital (natural capital). Therefore, what is required under SD is the transfer of an aggregate capital stock no less than the one that exists now. WS is, as we pointed out earlier, based on a very strong principle of perfect substitutability between the different forms of capital.[9]

Some salient elements in the SD debate

Intragenerational and intergenerational equity

SD is future-orientated in that it seeks to ensure that future generations are at least as well off, on a welfare basis, as current generations. It is therefore in economic terms a matter of intergenerational equity and not just efficiency. The distribution of rights and assets across generations determines whether the efficient allocation of resources sustains welfare across human generations (Howarth and Norgaard 1992). The ethical argument is that future generations have the right to expect an inheritance sufficient to allow them the capacity to generate for themselves a level of welfare no less than that enjoyed by the current generation. What is required, then, is some sort of intergenerational social contract.

SD, as we have seen, also has a poverty focus, which in one sense is an extension of the intergenerational concern. Daly and Cobb (1989) have argued that families endure over intergenerational time. To the extent that any given individual is concerned about the welfare of his/her descendants, he/she should also be concerned about the welfare of all those in the present generation from whom the descendant will inherit. Accordingly, a concern for future generations should reinforce and not weaken the concern for current fairness. Ethical consistency demands (despite the trade-offs involved) that if future generations are to be left the means to secure equal or rising per-capita welfare, the means to maintain and improve the well-being of today's poor must also be provided. Collective rather than individual action is required in order to effect these socially desirable intra- and intergenerational transfers. In any case, it seems to us that no nation that neglects the most vulnerable in society ought to be labelled 'developed' or 'developing'.

The process of industrialisation has created an international economic system out of which the developed countries of the North have gained the lion's share of the economic benefits. The scale of this global economic activity now means that some of its impacts can be classified as global environmental change effects (GECs) – acid rain, ozone layer depletion, global warming, deforestation, etc.[10] The gap between the rich countries and the poorer developing countries (excluding the small number of 'newly industrialising' states such as Taiwan, South Korea and Singapore) has not been closing, and may be growing larger over time. It is the view of some elements in the United Nations that poor countries are growing poorer because they are 'exploited' by the global trading system (United Nations Development Programme 1992, ch.4). It is certainly true that a number of poor countries face extra problems because of the debt burden they are carrying from past attempts at 'development'. Equally, domestic factors including corruption as well as market and intervention failures (i.e. inappropriately priced resources and inefficient and/or uncoordinated domestic economic policies, lack of land tenure reform, etc.) are also to blame for the poverty of ordinary citizens of the South and their harsh environmental conditions.

Arguably, however, the moral lead ought to be taken by the rich countries (if only because of the national 'resources' they have available) to make the international system more beneficial (more just) to poor countries. Aid flows, for example, could be better targeted and made more environmentally sensitive (Pearce *et al.* 1991, ch.9). There is also a sense in which it is in the richer countries' own interests to undertake such reforms, since both rich and poor have a common interest in the health of the biosphere ('global environmental interdependency'). This form of interdependency thesis cannot be pushed too far, but evidence is accumulating and it seems to indicate that GECs are increasing in significance. From a simple equity point of view, it is the rich countries' 'development fallout' that is causing much of the GECs and so it is they who should be carrying the bulk of the abatement cost burden.

Equally, the greater wealth of some nations could be defended if it improved the expectations of the least advantaged countries and did not impair the long-term development of these countries by seriously damaging their environments (Penn 1990). During the early 1980s the Brandt Commission reports (in 1980 and 1983) came close to justifying the international trading system (albeit with some reforms) on this basis. The Commission put forward an economic interdependency thesis buttressed by the supposed evidence of mutual and/or common interests between North and South. The mutual interests argument is, however, not a particularly strong one and if the prospects of the least fortunate countries remain at best uncertain, the global inequality cannot be justified.

The problem still remains as to how to get all rich countries and then rich and poor countries to cooperate over the 'global commons'. The threat to global sustainability will have to be a lot more severe before an enduring ethic of international environmental cooperation can become established. We examine the benefits and limitations of global environmental agreements in a later sub-section of this chapter.

The degree of concern, as expressed by the rate of time discount (DR) attached to the welfare of future generations, that is ethically required of the current generation is another controversial matter. Six positions seem possible: moral obligations to the future exist, but the welfare of the future is less important than present welfare ($0 < DR < \infty$); moral obligations to the future exist and the future's welfare is almost as important as present welfare (social time preference rate = $STPDR$; ($0 < STPDR < DR$); discounting procedure is only acceptable after the imposition of pre-emptive constraints on some forms of economic development; obligations to the future exist and the future is assigned more weight than the present ($DR < 0$); the rights and interests of future people are exactly the same as those of contemporary people ($DR = 0$); there is no obligation at all on the present to care about the future ($DR = \infty$).

Specification of the sustainability inheritance asset portfolio

Victor (1991) has recently remarked that one of the contributions that economists have made to the SD debate has been the idea that the depletion of environmental resources (source and sink resources) in pursuit of economic growth is akin to living off capital rather than income. SD is then defined as the maximum development that can be achieved without running down the capital assets of the nation, which are its resource base. The base is interpreted widely to encompass man-made capital K_m, natural capital K_n, human capital K_h and moral (ethical) and cultural capital K_c. Victor identifies four 'schools' of thought on the 'environmental as capital' issue: the mainstream neoclassical school; the London School (after Pearce, Barbier, Markandya and Turner); the post-Keynsian school; and the thermodynamic school (after Boulding, Georgescu-Roegen and Daly). Roughly speaking, this spectrum of views moves from a position we can label 'very weak sustainability' through to one we call 'very strong sustainability' (see also Klassen and Opschoor 1990). Table 1.1 formalises in simplistic fashion these various sustainability paradigms, which in practice are less clearly defined and are overlapping.

Very weak sustainability (Solow sustainability)
This VWS rule merely requires that the overall stock of capital assets ($K_m + K_n + K_h$) should remain constant over time. The rule is, however, consistent with any one asset being reduced as long as another capital asset is increased to compensate. This approach to sustainability is based on a Hicksian definition of income, the principle of constant consumption (buttressed by a Rawlsian maximin justice rule operating intergenerationally),[11] on production functions with complete substitution properties, and the Hartwick rule governing the reinvestment of resource rents (Common and Perrings 1992).

Thus, following Hicks, income is the maximum real consumption expenditure that leaves society as well endowed at the end of a period as at the start. The definition, therefore, presupposes the deduction of expenditures to compensate for the depreciation or degradation of the total capital asset base that is the source of the income generations, i.e. conservation of the value of the asset base. Assuming a homogeneous capital stock (perfect substitution possibilities) the Hartwick rule states that consumption may be held constant in the face of exhaustible resources if and only if rents deriving from the intertemporally efficient use of those resources are reinvested in reproducible capital.

It is now possible to derive an intuitive weak sustainability measure or indicator (in value terms) for determining whether a country is on or off

Table 1.1 Sustainability rules and indicators

	No critical natural capital	Critical natural capital
Very weak sustainability	$s/y - \delta k/y > 0$	Perfect substitution All K_n and K_m
		Growth economy
Weak sustainability	$s/y - \delta m/y - \delta n/y = WSI$ $WSI > 0$ $\lambda > h$ $n > Z$	$WSI > 0$ $\lambda > h$ $n > Z$ $\delta n^* \leq 0$
Strong sustainability	$\delta n \leq 0$ $WSI > 0$	$WSI > 0$ $\delta n \leq 0$ $\delta n^* \leq 0$ $\delta K_c \leq 0$
Very strong sustainability	Perfect Complementarity All K_n and K_m Stationary-state economy	$WSI > 0$ $\delta n \leq 0$ $\delta n^* \leq$ $n \leq 0$ $\delta k_c \leq 0$ $\delta k_e \leq 0$

Notes:
K = total capital assets
s = savings
δm = depreciation on man-made capital
δn = depreciation on natural capital
λ = technical change
h = rate of population growth
n^* = critical natural capital (no substitutes)
K_c = cultural capital
K_e = moral/ethical capital
Z = lower bound stock limit (determined by SMS) to ensure ecosystem stability

a sustainable development path (Pearce and Atkinson 1992). Thus a nation cannot be said to be sustainable if it fails to save enough to offset the depreciation of its capital assets. That is,

$$WSI > 0 \quad \text{if } S > \delta K \tag{1.1}$$

where *WSI* is a sustainability index, *S* is savings and δK is depreciation on capital. Dividing through by income (Y) we have

$$WSI > 0 \quad \text{if } (S/Y) > (\delta K/Y) \tag{1.2}$$

or

$$WSI > 0 \quad \text{if } (S/Y) > [\delta m/Y + (\delta n/Y)] \tag{1.3}$$

where δm is depreciation on man-made capital, δn is depreciation on natural capital and K_n and K_m are substitutable.

Weak sustainability (modified Solow sustainability)
Perrings (1991) and Common and Perrings (1992) highlight the fact that the technological assumptions (substitution possibilities) of the weak sustainability approach violate scientific understanding of the evolution of thermodynamic systems, and ecological thinking about the complementarity of resources in system structure and the importance of diversity in system resilience.

The London School has also modified the VWS approach by introducing into the analysis an upper bound on the assimilative capacity assumption, as well as a lower bound on the level of K_n stocks necessary to support SD assumption, into the analysis (Barbier and Markandya 1989; Pearce and Turner 1990; Klassen and Opschoor 1990). The concept of critical natural capital (e.g., keystone species and keystone processes) has also been introduced to account for the non-substitutability of certain types of natural capital (K_n) – such as environmental support services – and man-made capital (K_m). Thus the requirement for the conservation of the value of the capital stock has been buttressed by constraints aimed at the preservation of some proportion and/or components of K_n stock in physical terms.

The implications of this modified Solow sustainability thinking seems to be the formulation of a sustainability constraint which will impose some degree of restriction on resource-using economic activities. The constraint will be required to maintain populations/resource stocks within bounds thought to be consistent with ecosystem stability and resilience. To maintain the instrumental value (benefits) humans obtain from healthy ecosystems, the concern is not preservation of specific attributes of the ecological community but rather the management of the system to meet human needs, support species and genetic diversity, and enable the system to adapt (resilience) to changing conditions.

A set of physical indicators will be required in order to monitor and measure biodiversity and ecosystem resilience. As yet there is no scientific consensus over how biodiversity should be measured. Measuring genetic

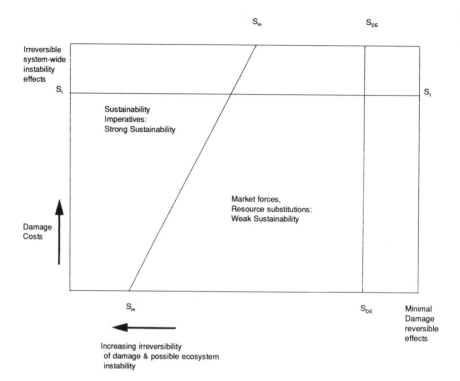

Figure 1.1   Safe minimum standards approach to sustainability

Source: Adapted from B. Norton, Georgia Institute of Technology, quoted in Toman (1992)

diversity presents the least difficulty, but measuring species diversity is more problematic – measures of species richness, taxonomic richness, richness of genera or families have all be investigated and none are without difficulties. Problems associated with measuring biodiversity at the community level are even greater but such measures would be very useful to policy-makers if the aim is conservation at the ecosystem level. Community classification schemes can be developed from the global level, through biogeographic provinces, down to regions within a country. Ecoregion classification is based primarily on the physical environment. A minimum set of 22 indicators (including species richness, species risk index, community diversity, etc.) has been proposed for wild and domesticated species (Reid *et al.* 1992).

For some commentators sustainability constraints of this type should be seen as expressions of the precautionary principle (O'Riordan 1992) and one that is akin to the safe minimum standards (SMS) concept (Bishop 1978). Toman (1992), quoting the work of Norton, has recently suggested that the SMS concept is a way of giving shape to the inter-generational social contract – see Figure 1.1. Given irreversibility and uncertainty about the impact of economic activities on ecosystem performance, SMS posits a socially determined dividing line between moral sustainability imperatives and the freeplay of resource trade-offs (e.g. S_mS_m in Figure 1.1). To satisfy the intergenerational social contract, the current generation might rule out in advance (depending on the social opportunity costs involved) actions that could result in damage impacts beyond a certain threshold of cost and irreversibility. Social and not individual preference values will be part of the SMS setting process (Turner 1988a). Supporters of the technocentric paradigm might favour such a line as S_tS_t, while ecocentrics such as the deep ecologists might favour a line such as $S_{DE}S_{DE}$ (Pearce and Turner 1990).

Strong sustainability (ecological economics approach)
As we have seen, the weaker versions of sustainability are consistent with a declining level of environmental quality and natural resource availability as long as other forms of capital are substituted for K_n, or the imposition of SMS is judged to impose too high a social opportunity cost given the inevitable uncertainties involved in conservation benefits forecasting.

A number of analysts, from a variety of disciplines, have drawn attention to the 'missing elements' in the economic calculus that underlies the weak sustainability rules. Many ecosystem functions and services can be adequately valued in economic terms; others may not be amenable to meaningful monetary valuation. Critics of conventional economics have argued that the full contribution of component species and processes to the aggregate life-support service provided by ecosystems has not been captured in economic values (Ehrlich and Ehrlich 1992). Nor has the prior value of the aggregate ecosystem structure (life-support capacity) been taken into account in economic calculations; indeed, it is probably not fully measurable in value terms at all. Bateman and Turner turn to this concept of what we call 'ecosystem primary value' in more detail in Chapter 5. There is the risk, therefore, that as environmental degradation occurs, some life-support processes and functions will be systematically eroded (because they are undervalued), increasing the vulnerability (reduced stability and resilience) of the ecosystem to further shocks and stress.

On this SS view, it is not sufficient to just protect the overall level of capital, rather K_n must also be protected, because at least some of K_n

(critical K_n) is non-substitutable. Thus the SS rule requires that K_n be constant, and the rule would be monitored and measured via physical indicators. The case for this 'strong' view (linked to the precautionary principle) is based on the combination of a number of factors: presence of uncertainty about ecosystem functions and their total service value; presence of irreversibility in the context of some environmental resource degradation and/or loss; the loss aversion felt by many individuals when environmental degradation processes are at work; and the critically (non-substitutability) of some components of K_n.

While akin to SMS, the SS rule is not the same since what is stressed in the latter approach is the combination of factors, irreversibility, uncertainty, etc., not their presence in isolation. Further, SMS says conserve unless the benefits forgone (social opportunity costs) are very large. SS says, whatever the benefits forgone, K_n losses are unacceptable (i.e. constant 'aggregate' K_n, not constant K_n for each asset, with the exception of 'critical' components of K_n).

SS need not imply a steady-state, stationary economy, but rather changing economic resource allocations over time which are not sufficient to affect the overall ecosystem parameters significantly, i.e. beyond the point where the stability (resilience) of the system or key components of that system are threatened. A certain degree of 'decoupling' of the economy from the environment should, therefore, be possible through technical change and environmental restoration investment in a 'moderated' growth scenario.

Very strong sustainability (stationary state sustainability)
The VSS perspective concentrates on what Daly (1991; 1992) has termed the 'scale effect', i.e. the scale of human impact relative to global carrying capacity. For him the greenhouse effect, ozone layer depletion and acid rain all constitute evidence that we have already gone beyond a prudent 'plimsoll' line for the scale of the macroeconomy. The VSS approach reduces to a call for a steady-state economic system based on thermodynamic limits and the constraints they impose on the overall scale of the macroeconomy. The rate of matter and energy throughput in the economy should be minimised. The second law of thermodynamics implies that 100% recycling is impossible (even if it were socially desirable) and the limited influx of solar energy poses an additional constraint on the sustainable level of production in an economy (the solar influx potential is a matter of some dispute). Zero economic growth and zero population growth are required for a zero increase in the 'scale' of the macroeconomy. Supporters of the steady-state paradigm would, however, emphasise that 'development' is not precluded and that social preferences, community-regarding values and generalised obligations to

future generations can all find full expression in the steady-state economy as it evolves (i.e. conservation of the moral capital (K_e) on which economic activity eventually depends (Hirsch 1976; Daly and Cobb 1989).

A systems and coevolutionary perspective for sustainability

The adoption of a systems perspective serves to re-emphasise the obvious but fundamental point that economic systems are underpinned by ecological systems and not vice versa. There is a dynamic inter-dependency between economy and ecosystem. The properties of bio-physical systems are part of the constraints set which bound economic activity. The constraints set has its own internal dynamics which reacts to economic activity exploiting environmental assets (extraction, harvesting, waste disposal, non-consumptive uses). Feedbacks then occur which influence economic and social relationships. The evolution of the economy and the evolution of the constraints set are interdependent; 'coevolution' is thus a crucial concept (Common and Perrings 1992).

Norton and Ulanowicz (1992) advocate a hierarchical approach to natural systems (which assumes that smaller sub-systems change accord-ing to a faster dynamic than do larger encompassing systems) as a way of conceptualizing problems of scale in determining biodiversity policy. For them, the goal of sustaining biological diversity over several human generations can only be achieved if biodiversity policy is operated at the landscape level. The value of individual species, then, is mainly in their contribution to a larger dynamic, and significant financial expenditure may not always be justified to save ecologically marginal species. A central aim of policy should be to protect as many species as possible, but not all.

Ecosystem health (stability and resilience or creativity), interpreted in terms of an intuitive guide, is useful in that it helps focus attention on the larger systems in nature and away from the special interests of individuals and groups (Norton and Ulanowicz 1992). The full range of public and private instrumental and non-instrumental values all depend on protection of the processes that support the health of larger-scale ecological systems. Thus when a wetland, for example, is disturbed or degraded, we need to look at the impacts of the disturbance on the larger level of the landscape. A successful policy will encourage a patchy land-scape.

Environmental resource valuation

In particular, from the weak sustainability perspective there is an essential link between sustainable development and monetary valuations of the environment in terms of willingness to pay (WTP). The lack of meaningful monetary valuations of environmental assets would greatly circumscribe the weak sustainability case, based as it is on substitution possibilities between K_n and K_m.

According to conventional economic theory the value of environmental assets can be estimated by reference to the preferences for or against conservation of those assets, as displayed by individuals. Individuals possess a number of internal held values which result in objects being given various assigned values. In order to obtain as full an estimate of value as is feasible (known as 'total economic value') economists seek to quantify both use and non-use values. Monetary estimates of use values are relatively straightforward to obtain. Taking the example of a wetland ecosystem, humans derive *direct use values* in the form of outputs such as fish and needs, or service value such as recreation. The wetland ecosystem can also be utilised *indirectly* in the sense that the functions it can perform – floodwater storage and flood protection, effluent storage, storm buffering, etc. – are of benefit to humans. A combination of market prices and more indirect valuation methods can be used to estimate such use values (Turner and Jones 1991).

The valuation of some of the indirect use values are more difficult to derive, but many can be measured via the damage avoidance method (for example, flood damage avoided because of the wetland's capacity to retain excess water); and/or the substitute services (replacement cost) method (the costs of a sewage treatment works to substitute for the natural capacity of a large wetland to retain such effluent, up to some limit).

Individuals may also express a WTP for an option to use the environment some time in the future (option value), or display a willingness to pay to preserve the environment for their children and grandchildren (bequest value).

Non-use values are more complex and are not associated with any current or intended future use of an asset. Individuals may value just the very existence of certain species or whole ecosystems. In principle, many of the use and non-use values can be estimated by methods which infer WTP (or willingness to be compensated for a loss) such as the travel cost, hedonic price and contingent valuation methods. The advantages and limitations of these monetary valuation methods are examined by Bateman and Turner in Chapter 5 and by Bateman in Chapter 6.

Within economics, the debate about these monetary valuation

techniques has basically been about technical issues. It has involved the theoretical validity and the reliability of the value estimates as the analyst moves from a consideration of use values in contexts familiar to individuals being studied to non-use values in contexts that individuals have little experience of. Questions about the form and type of information that survey respondents ought to be provided with and the type of aggregation processes required (i.e. to derive regional or national valuations from sample data) are typically debated. But outside economics, both the model of human behaviour that economics has assumed (rational self-interested maximisation), and the normative role of individualism (consumer sovereignty) have also been drawn into the debate and scrutinised critically (see Redclift's analysis in Chapter 4).

Critics have argued that individuals can and do operate as both individual consumers (close to the conception of the preference- or want-dominated rational economic person assumed by conventional economics) and as citizens (a role that is influenced by 'ethical rationality' – duties, obligations, needs – and requires highly informed deliberation) (Sagoff 1988). Economics would play a less significant role in the context of citizens and social preference-dominated deliberations. However, we would not agree with Sagoff (1988) that economics plays almost no role (except for cost-effectiveness assessments of predetermined standards/policies) in such deliberations and goal setting. Opportunity costs are ever present and therefore the choice of any particular goal carries an implicit cost (value) in relation to alternative goals. But the 'dual self-conception' of the individual as consumer and citizen is an important notion, one that has even attracted the attention of a minority of economists (Harsanyi 1955; Sen 1977; Margolis 1982; Lutz and Lux 1988). It has particular relevance for the contingent valuation method (Kahneman and Knetsch 1992). Overall, the axiom of individualism (consumer sovereignty) is weakened if the 'dual self-conception' thesis is both realistic (we believe it is) and significant in resource management contexts (again, we believe it is).

Sustainability, cultural capital depletion and sustainable livelihoods

Coevolution is a local process, so local human subsystems are a significant starting point for the discussion of evolution in ecological economics (Berkes and Folke 1992). Traditional ecological knowledge may be important and therefore cultural diversity and biological diversity may go hand-in-hand as prerequisites to long-term societal survival. Diverse cultures encompass not just diverse environmental adaptation methods and processes, but also a diversity of world views (technocentrism and ecocentrism) that support these adaptations. The conservation of this rapidly diminishing pool of social experience in adaptability (cultural capital, K_c) may be as pressing a problem as maintenance of biodiversity since local cultures may be as important a reservoir of information as the genetic information contained in species currently threatened with extinction (Perrings et al. 1992).

The most convincing body of evidence for human self-organisational ability may be found in the literature of common property resources (Ostrom 1990). Institutions, co-evolution and traditional ecological knowledge are all components of the common property dilemma. There exist a large number of self-regulating regimes governing access to resources in common property, and their scope in limiting the level of economic stress on particular ecological systems is clearly very wide (Bromley 1992).

Any sustainable strategy for the future will have to confront the question of how a vastly greater number of people can gain at least a basic livelihood in a manner which can be sustained; many will have to endure in environments which are fragile, marginal and vulnerable (Chambers and Conway 1992).

A working definition of sustainable livelihoods (SL) would be

a livelihood comprising the capabilities, assets (stores, resources, claims and access) and activities required for a means of living: a livelihood is sustainable which can cope with and recover from stress and shocks, maintain or enhance its capabilities and assets and provide sustainable livelihood opportunities for the next generation: and which contributes net benefits to other livelihoods at the local and global levels in the short and long term. (Chambers and Conway 1992, 27–8)

From the policy viewpoint the aim must be to promote SL security by means of vulnerability reduction. Both public and private action is required for vulnerability reduction – public action to reduce external stress and shocks through, for example, flood protection, prevention and insurance measures; and private action by households which add to their portfolio of assets and repertoire of responses so that they can respond more effectively in terms of loss limitation. Chambers and Conway (1992) offer a range of monetary value and physical indicators for SL monitoring and assessment of trends – for example, migration (and off-season opportunities), rights and access to resources, and net asset position of households.

Sustainability and ethics

For many commentators traditional ethical reasoning is faced with a number of challenges in the context of the sustainable development debate. Ecological economists would argue that the systems perspective demands an approach that privileges the requirements of the system above those of the individual. This will involve ethical judgements about the role and rights of present individual humans as against the system's survival and therefore the welfare of future generations. We have already argued that the poverty focus of sustainability highlights the issue of intragenerational fairness and equity.

So 'concern for others' is an important issue in the debate. Given that individuals are, to a greater or lesser extent, self-interested and greedy, sustainability analysts explore the extent to which such behaviour could be modified and how to achieve the modification (Turner 1988b; Pearce 1992). Some argue that a stewardship ethic (weak anthropocentrism; Norton 1987) is sufficient for sustainability, i.e. people should be less greedy because other people (including the world's poor and future generations) matter and greed imposes costs on these other people. Bioethicists would argue that people should also be less greedy because other living things matter and greed imposes costs on these other non-human species and things. This would be stewardship on behalf of the plant itself (Gaianism) in various forms up to 'deep ecology' (Naess 1973; Turner 1988a; Wallace and Norton 1992).

The degree of intervention in the functioning of the economic system deemed necessary and sufficient for sustainable development also varies across the spectrum of viewpoints. Supporters of the steady-state economy (extensive intervention) would argue that at the core of the market system is the problem of 'corrosive self-interest'. Self-interest is seen as corroding the very moral context of community that is presupposed by the market. The market depends on a community that shares such values as honesty, freedom, initiative, thrift and other virtues whose authority is diminished by the positivistic individualistic philosophy of value (consumer sovereignty) of conventional economics. If all value derives only from the satisfaction of individual wants then there is nothing left over on the basis of which self-interested, individualistic want satisfaction can be restrained (Daly and Cobb 1989).

Depletion of moral capital (K_e) may be more costly than the depletion of other components of the total capital stock (Hirsh 1976). The market does not accumulate moral capital, it depletes it. Consequently, the market depends on the wide system (community) to regenerate K_e, just as much as it depends on the ecosystem for K_n.

Individual wants (preferences) have to be distinguished from needs.

For humanistic and institutional economists, individuals do not face choices over a flat plane of substitutable wants, but a hierarchy of needs. This hierarchy of needs reflects a hierarchy of values which cannot be completely reduced to a single dimension (Swaney 1987). Sustainability imperatives, therefore, represent high-order needs and values.

Sustainable development: operational principles

The practical implications of SD have yet to be properly assessed, but some broad outlines are presented below. The shift from VWS through to VSS positions involves the progressive rejection of the axioms of consumer sovereignty and infinite substitution possibilities (in both utility and production functions) and their replacement with a new set of axioms. Norton (1992) has recently proposed five axioms of ecological management which are relevant to the stronger versions of sustainability:

(i)   'The Axiom of Dynamism' – nature is a set of processes in a continual state of flux, but larger systems change more slowly than smaller systems.
(ii) 'The Axiom of Relatedness' – all processes are interrelated.
(iii) 'The Axiom of Hierarchy' – systems exist within systems.
(iv) 'The Axiom of Creativity' – processes are the basis for all biologically based productivity.
(v)   'The Axiom of Differential Fragility' – ecological systems vary in their capacity to withstand stress and shock.

These axioms are then linked to a broad normative policy principle/objective, the maintenance of the health/integrity of ecosystems (Leopold 1949).

A number of rules (which fall some way short of a blueprint) for the sustainable utilisation of the natural capital stock can now be outlined (roughly ordered to fit the VWS to VSS progression):

(i)   Market and intervention failures related to resource pricing and property rights should be corrected.
(ii) The regenerative capacity of renewable natural capital (RNC) should be maintained – i.e. harvesting rates should not exceed regeneration rates – and excessive pollution which could threaten waste assimilation capacities and life-support systems should be avoided.
(iii) Technological changes should be steered via an indicative planning system such that switches from non-renewable natural capital (NRNC) to RNC are fostered; and efficiency-increasing technical progress should dominate throughput-increasing technology.

(iv) RNC should be exploited, but at a rate equal to the creation of RNC substitutes (including recycling).

(v) The overall scale of economic activity must be limited so that it remains within the carrying capacity of the remaining natural capital. Given the uncertainties present, a precautionary approach should be adopted with a built-in safety margin.

Sustainable development: practice

Figures 1.2 and 1.3 summarise some of the measures and enabling policy instruments that would be involved in any application of a WS or SS strategy. At a fundamental level, if resource prices are set too low, or are absent completely, excessive use will be made of the resource. Overuse will then contribute to environmental degradation. Many elements of the natural capital stock are underpriced or are unpriced and some may also be open-access resources. Under such conditions the market will fail correctly to take account of the full environment costs and benefits (including externalities) of economic activity – 'market failure' (Helm 1991; Pearce and Turner 1990). Taking the example of natural or semi-natural forests, only part of their function and service value output can be captured and allocated by markets or for that matter by non-market action on the part of national governments operating unilaterally. Nationally, both commodity and direct service outputs (such as timber and recreation) from forests, together with some environmental services such as watershed protection, can be valued and conserved through markets or regulations. But other function services, carbon sequestration and biodiversity provision, can only be captured by international cooperation.

But real-world situations are often even more complex, with both institutional framework and market structure also playing significant roles in pollution and resource depletion problems. 'Multiple' market failures can be present. As if these situations were not complicated enough, the failures phenomenon is not restricted to markets. Policy interventions (inefficient policy and/or uncoordinated policy) can sometimes compound the difficulties caused by market failure (Turner and Jones 1991). Resource degradation is then caused by a combination of interrelated failures, lack of information, market and intervention failures. Pearce (Chapter 3), Willis and Benson (Chapter 7) and Barbier (Chapter 8) all highlight the issue of failures (found in developed and developing economies) and environmental degradation.

Uncertainty is a particular problem in the context of global environmental changes. The externalities present at this level are in the

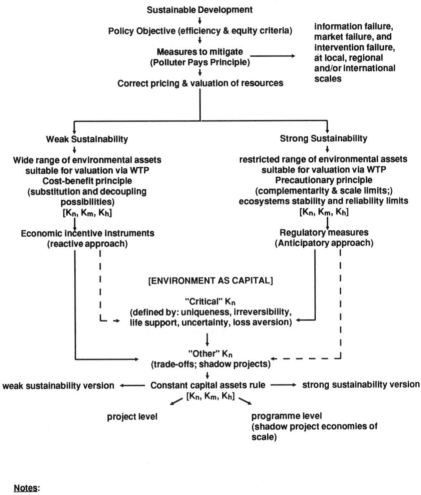

Sustainable Development
↓
Policy Objective (efficiency & equity criteria)
↓
Measures to mitigate ————→ Information failure, market failure, and intervention failure, at local, regional and/or international scales
(Polluter Pays Principle)
↓
Correct pricing & valuation of resources

Weak Sustainability
↓
Wide range of environmental assets
suitable for valuation via WTP
Cost-benefit principle
(substitution and decoupling
possibilities)
$[K_n, K_m, K_h]$
↓
Economic incentive instruments
(reactive approach)

Strong Sustainability
↓
restricted range of environmental assets
suitable for valuation via WTP
Precautionary principle
(complementarity & scale limits;)
ecosystems stability and reliability limits
$[K_n, K_m, K_h]$
↓
Regulatory measures
(Anticipatory approach)

[ENVIRONMENT AS CAPITAL]

"Critical" K_n
(defined by: uniqueness, irreversibility,
life support, uncertainty, loss aversion)
↓
"Other" K_n
(trade-offs; shadow projects)

weak sustainability version ←——— Constant capital assets rule ——→ strong sustainability version
$[K_n, K_m, K_h]$

project level

programme level
(shadow project economies of
scale)

Notes:

TEV	=	Total Economic Value (use + non-use value)
K_n	=	natural capital
K_m	=	man-made capital
K_h	=	human capital
——→	=	strong linkage
– –→	=	weaker linkage

Figure 1.2   Sustainability practice I

Sustainability Mode [overlapping categories]	Management Strategy [as applied to projects, policy or course of action]	Policy Instruments [most favoured]		
		Pollution Control and Waste Management	Raw Materials Policy	Conservation and Amenity Management
VWS	**Conventional Cost-Benefit Approach:** Correction of market and intervention failures via efficiency pricing; potential Pareto criterion (hypothetical compensation); consumer sovereignty; infinite substitution	e.g. pollution taxes, elimination of subsidies, imposition of property rights		
WS	**Modified Cost-Benefit Approach:** extended application of monetary valuation methods; actual compensation, shadow projects etc; systems approach; 'weak' version of safe-minimum standard	e.g. pollution taxes, permits, deposit-refunds; ambient targets		
SS	**Fixed Standards Approach:** Precautionary Principle, Primary and Secondary value of natural capital; constant natural capital rule; dual self-conception, social preference value; 'strong' version of safe minimum standard	e.g. Ambient standards; conservation zoning; process technology-based effluent standards; permits; severance taxes; assurance bonds		
VSS	**Abandonment of Cost-Benefit Analysis:** or severely constrained cost-effectiveness analysis; bioethics	standards and regulations; birth licences		

Figure 1.3 Sustainability practice II

form of global mutual externalities (ozone layer depletion, deforestation, etc.). Trying to correct (internalisation) for global externalities requires imposed valuation of the environmental resources involved and the establishment of property rights to this 'global commons'. Formidable problems are raised in any process of international bargaining over pollution targets or resource usage rates. Thus the informational and transactions costs of facilitating an enduring global environmental policy will be very high. International negotiations (for example, over CO_2 emissions or CFCs) require collusion among many countries (rich and poor, gainers and losers). The so-called 'prisoner's dilemma' analysis indicates that motivation for free-riding (by individual countries) is strong if cheating is hard to detect and the agreement hard to enforce. Each nation would be better off if everyone else cut back emissions (taking a cost in the form of reduced economic growth) while it maintained production (Barrett 1991; Mäler 1991b). Those who gain by cooperation must devise incentives to make those who lose play the game, i.e. sidepayments are required.

In principle, what is required is an international monitoring and enforcement agency. The theory of non-cooperative games does indicate that cooperative outcomes can be optimal outcomes, but in practice formal protocols, treaties and enabling institutions have to be in place. The UNCED process (since Rio 1992) is currently grappling with these very issues. Both a Climate and a Biodiversity Convention have been agreed (see Chapter 12). While an enduring ethic of international environmental cooperation has yet to be established, countries may in certain circumstances be prepared to act counter-preferentially for the common good (on equity or environmental stewardship grounds). It is also the case that, unlike the theory of games, real-world negotiations are repeated over time. Sequential games (played over a long time period) may not face the Prisoner's Dilemma syndrome because of the fear of 'retribution'.

However, we should not be deluded into thinking that all global environmental change problems require macro-scale policy interventions. The materials balance model serves to remind us that individual consumption and production units, operating at the microeconomic level, are the pollution/resource depletion agencies. There is a lot of sense in the maxim 'Think global, act local'. So policy intervention must also focus on this micro level and must serve to change human attitudes and behaviour. A combination of policy instruments, ranging from moral suasion and better 'education' through market incentives and property rights to command-and-control standards and regulations, are available (Opschoor and Vos 1989; Opschoor and Turner 1993).

It has long been the contention of environmental economists that the price mechanism is both an efficient and a powerful incentive mechanism for changing the attitudes and behaviour of economic agents (households and firms). Pollution taxes could be levied, altering prices and costs of production, or subsidies could be given to stimulate so called 'cleaner technology' and to support low-intensity farming methods in environmentally sensitive areas (ESAs). To the extent that taxes rather than subsidies are favoured, then the polluter-pays principle (PPP) has been supported (OECD 1992).

The creation of ESAs (in the European Community) and other forms of conservation zoning are compatible with the SS view of ecological sustainability that has at its core the requirement to protect dynamic creative systems (overall K_n). Since we cannot measure the individual preferences of future people we preserve future options by maintenance of the overall complex system in terms of its relative stability and health. The result should be a patchy landscape. But Henderson (Chapter 8) reminds us that at this landscape level we should not concentrate all our efforts on conservation landscapes in public conservation reserves. For him semi-natural landscapes are educational tools whose main objective should be the altering of man's behaviour outside of conservation landscapes. He identifies what he calls 'Heritage Landscape' which is privately controlled and inhabited and examines the opportunities and difficulties involved in a wider conservation policy. The 'native prairie' grasslands of North America are his case-study example.

Quantity incentives offer an alternative way of supporting moves towards sustainable development. Tradable/marketable permits (in the forms of emission quotas and/or natural capital protection obligations) combine the 'certainty' of some fixed target (ambient quality state or stock of biodiversity) with efficiency in compliance costs. Permits are also compatible with the PPP. The theory of permits was first developed with water pollution in mind, but real-world applications have been restricted to the problems of air pollution in the USA (Tietenberg 1991). Sedjö (1992) has argued that permits could play a role at the global scale in the context of forest conservation. Forest protection obligations would be set and distributed among countries via some international forum or agency. These obligations (like permits) would be tradable. Countries could fulfil them directly or induce another country to assume them, in return for payment.

Poor countries with a large amount of forest stock would assume obligations to protect less forested land than actually exists within their boundaries. Richer countries would assume obligations to protect a greater forest stock than exists within their national boundaries. They would fulfil their 'surplus' obligations by protecting forests in poorer countries (through the medium of payments tied to performance targets).

The potential for using the tradable permit as a sustainability incentive does seem to be substantial. It does, however, require imaginative administration and the system itself will be quite complicated (increasing transactions costs) as it adjusts to reality. Thus the forest conservation systems would need to incorporate the fact that forests differ in type and in condition and management. This results in a differential capacity to produce function and service outputs. So obligations to protect tropical forests, for example, would be calibrated differently (smaller acreage) than that for boreal forests. In a different global environmental change context, carbon dioxide emission reduction, it is the initial allocation of permits that is troublesome. If emission targets are allocated via a grandfathering process, developing countries will acquire a fairly low initial allocation if rights are based on base-year emissions, and will be unable to afford traded permits. One solution is to bias the initial allocation to developing countries according to population (Grubb 1989).

Environmental administrators have been far less enthusiastic than economists about the potential for economic incentive instruments like taxes/charges. They have had a relative preference for command-and-control instruments such as standards and, to a limited extent in the USA, for permits. The reasons for this appear to be that the environmental effectiveness of regulations is more predictable. Price-based instruments like charges offer less absolute protection against environmental catastrophes in an uncertain world (Baumol and Oates 1988; Opschoor and Turner 1993). Regulations appear to provide a better grip on polluters. Thus while technology or performance-based emission regulations provide no incentive for regulated polluters to exceed their prescribed level of abatement they have still proved attractive to control agencies. What they do provide is a measure of environmental quality 'certainty' (as long as there is adequate monitoring and enforcement).

But there may also be an ethics-based explanation for the limited application of economic incentive instruments (Frey 1992). We noted earlier the critique of the simplistic 'rational economic person' view of human nature assumed in conventional economic analysis. Critics have offered alternative conceptions based on 'dual self-conception, social/citizen preferences and ethical rationality'. If such intrinsic motivation exists and is significant then price-based economic instruments for controlling pollution may reduce, or crowd out, environmental ethics. Individuals may feel that their moral stance is unnecessary when they are induced by the price system (extrinsic motivation) to reduce pollution. Pricing may crowd out environmental ethics in the sector in which pricing is applied, which may also damage environmental ethics in areas in which pricing is not applied (Frey 1992). Thus the use of pricing could under certain conditions lead to an increase in pollution in the non-pricing sector, or may increase pollution overall.

Regulatory standards fare better in this ethics-based explanation because while they, too, can cause a substitution of environmental ethics, this effect is likely to be less severe because such regulations are also accompanied by an explicit disapproval of polluting activity (in contrast to pricing, which some have interpreted as merely a 'licence to pollute'). Subsidies (linked, for example, to cleaner technologies) tend to support environmental ethics both because individual self-determination is maintained and because preserving nature for intrinsic reasons is acknowledged (Frey 1992).

Hanley (Chapter 11) examines a range of practical problems associated with the application of market incentive instruments to environmental quality management. He uses the example of water pollution to demonstrate the arguments.

The *valuation* of the environmental damage caused by failure phenomena forms another important part of sustainable development policy design. We examined the theoretical advantages and limitations of the various monetary valuation methods earlier in this chapter (see also Chapters 5 and 6), but the difficulties involved in the estimation of precise values for the majority of environmental goods and services should not be allowed to dismiss monetary valuation *per se*. Willis and Benson (in Chapter 7) argue that the lack of such precise values should not come as any great surprise. Asset values are conditional, depending on institutions, demand, supply and expectations about these conditions. Price at any point in time reflects all the information available at that point in time. Price is conditional upon information unique to that point in time, which is unlikely to be replicated, so over time values will vary.

Pearce (Chapter 3), Willis and Benson (Chapter 7) and Barbier (Chapter 9) all present the results of actual valuation studies (utilising all the main methods) for developed and developing countries. While the results are not precise numbers, their existence still leaves us in a much better position (in decision-making terms) than we would be in if such value information did not exist or was totally ignored. The limits to meaningful monetary valuation of natural capital are still in our view an open research question, but that is an altogether different position from one which completely dismisses such valuation, or severely circumscribes the process (i.e. limiting it to market or near-market situations with some prices data, complemented by opportunity cost-based valuation).

The travel cost (TC) recreation demand model is the longest-established indirect revealed preference approach, and a great deal of empirical data has been accumulated using this approach. US practioners seem to be generally agreed that the TC methodology is successful and has produced meaningful value estimates. Smith (1992) concludes that the estimates made using TC uphold rudimentary predictions of consumer theory (i.e. quantity negatively related to their own price). Further, when TC is applied to comparable sites value estimates are broadly consistent, and when TC is applied to different types of recreation site plausible differences in value have been revealed. The estimation of the value individuals place on changes in the recreation sites' quality features has, however, proved to be more problematic.

A somewhat different picture emerges if UK experiences are reviewed. After a short-lived renewed interest in the TCM in the mid-1980s, there has since been a pronounced loss of confidence in TC in the UK and in the zonal variant (ZTCM) in particular (Turner et al. 1992). As far as the majority of UK valuation practioners are concerned, ZTCM has almost reached the stage of complete rejection in favour of the 'individual' version (Garrod and Willis 1991). A number of significant TCM problems, first raised in the UK literature in the early to mid-1970s, have not been satisfactorily addressed and continue to lie at the core of the critique of TCM.

First, there is what has become known as the 'endogenous price' problem. If potential visitors to a recreation site change their residence in order to be near to the site, the assumption that all zones have the same distribution of tastes collapses. If this type of behaviour does exist across a significant spectrum of the visitor population then the TCM (Clawson model) would substantially underestimate recreation site benefits. The price of a trip to the recreation site has become endogenous, and if corrections are not made then the estimated slope of the conventional travel cost demand curve will be too flat. The estimated consumer surplus for access value of the site will therefore be too small. Parson (1991) has argued that endogeneity may be eliminated using an instrumental variables approach (e.g., place of work, job characteristics) in the questionnaire design and data analysis. Future TC surveys could include a question covering the importance of proximity to recreation sites in choosing place of residence. Split data sets could then be used to test for bias.

Second, a site may be only one of a number visited during a single trip. In this case it is not certain how much of an overestimation of benefits there would be.

Third, some visitors ('meanderers') may derive utility from the journey to a site. A complex behaviour pattern may be present, again making it difficult to discern whether benefits are under- or overestimated.

Fourth, the inclusion of a value for time spent travelling to a site, while theoretically correct, has nevertheless served to open a long debate on which empirical approach should be used.

During the late 1980s the TCM was applied in a study of coastal recreation in the UK (Green *et al.* 1990). Analysts sought to test the assumption that the value of enjoyment must be higher for visitors who travel further (incurring higher costs) to visit a particular site. Visitors to six seaside locations in England were questioned (using CVM) about their valuation of the enjoyment experienced on a day visit. The survey results were then compared with their travel costs. These costs were estimated by using a computer program which calculated the cheapest, quickest and shortest routes between the origin of the trip and the coastal site.

Results of this study did not support an unquestioning and extensive use of TCM in the UK. Only two sites (Dunwich and Spurn Head) produced data indicating a rising enjoyment value as travel cost and travel time and distance increased. For another site (Clacton) enjoyed value increased as trips were made from increasing distances, but only up to a threshold of about one hour's travel time. In the case of two further sites (Frinton and Scarborough) the value of the day's enjoyment seemed to be negatively related to distance travelled. Finally, in the case of Filey, no association between the value of enjoyment and distance travelled was found.

Household production function (HPF) (including hedonic pricing) models also rely on individual actions in order to isolate features of their values. Empirical results for HPF and hedonic models applied to housing markets (and linked to air pollution or water quality) are not as extensive as that of TC models. US researchers believe that good-quality hedonic pricing models are capable of providing credible evidence of a negative and significant relationship between air pollution and property value. Estimation of the representative household's marginal value of reductions in air pollution is less certain (Smith 1992).

There is now a large body of research which elicits individuals' valuations of changes in some environmental resources using the CVM (contingent valuation method), in which individuals express their preferences by answering questions about hypothetical choices. The CVM has been subject to criticism, particularly as a result of theoretical and experimental research by psychologists and economists into the problems of eliciting responses. Supporters of CVM are currently attempting to address both reliability and validity questions.

A basic question for the implementation of the CVM is whether willingness to pay for benefits (WTP) or willingness to accept compensation for disbenefits (WTA) is the most appropriate indicator of value in a given situation. For cost–benefit analysis based on the Hicks–Kaldor compensation test, WTP would seem to be the appropriate measure for gainers from some resource-allocation decision and WTA the proper measure for losers from that same reallocation. But Harris and Brown (1992) have pointed out that it is often not easy conclusively to identify gainers and losers, since this judgement is itself influenced by the valuer's own perspective.

Contingent valuation (CV) studies have become much more sophisticated over time. In terms of reliability, sample sizes have been too small in a number of studies. The issue of temporal bias (for seasonal intervals or longer periods of time) also requires further investigation. But it is the so-called 'part–whole', 'mental account' or 'embedding' bias problem that has recently caused most controversy. If it is the case that individuals possess several mental accounts (i.e. divisions and subdivisions of expenditure related to disposal income), then, when they respond to CV questionnaires (trying to estimate just one environmental benefit, say recreation), they may omit to consider their relevant mental account in relation to all the other accounts. Instead the CV survey may pick up a 'warm glow' effect (i.e. individuals deriving satisfaction from 'doing the right thing', supporting environmental conservation by providing a high WTP answer).

In conclusion, in an uncertain world the deployment of policy instruments in pursuit of the sustainability objective must be conditioned by a recognition of the different points of leverage (local, regional and international) available, and by the requirement for a phased implementation. It makes sense to target 'no regret' interventions first, i.e. policies which make economic sense even if the environmental benefits derived are not necessarily large and/or globally significant in themselves. Thus Pearce (in Chapter 3) suggests the following sequence of priorities (with special reference to developing countries):

(i)   short term – private cost pricing policy (market failure correction) and reform of existing tax policy (intervention failure correction);
(ii)   medium term – resource rights and land tenure reform (establishment of property rights);
(iii)   long term – full social cost pricing of resources.

In the long term the international scale will become dominant. It is most unlikely that an individual economy could guarantee sustainability to its citizens (Pezzey 1992). The nation state cannot easily insulate itself from global resource prices, from global climate change, or from potential social unrest stimulated by access to the international mass media and advertising industries. International environmental agreements will become increasingly important, underpinned by the evolution of an international environmental cooperation ethic.

Notes

1. A number of early issues and arguments encompassed by the sustainability debate were analysed in Turner (1988b). This present book seeks to update and extend that analysis.

2. In economic terms, the VSS position treats the economy and the environment as perfect complements; while the VWS position assumes almost perfect substitution between physical and human capital and natural capital. In fact, VWS assumes smooth production functions (i.e. functions describing the transformation process as raw material, labour and other inputs are transformed into outputs of goods and services and non-productive outputs which often become waste flows) with perfect inputs substitution properties; and smooth welfare (utility) functions (i.e. functions describing economic well-being derived from the consumption of goods and services and ambient environmental conditions) in which all consumer wants are assumed to be substitutable.

3. SD therefore rejects the neo-classical 'potential Pareto criterion' (PPC) which underpins conventional economic cost–benefit analysis as applied to projects, policies or courses of action. The PPC sanctions activities (on efficiency grounds) if the gainers gain enough benefits hypothetically to compensate all the losers while still remaining better off than they were in the status-quo situation. An SD policy would mandate actual not hypothetical compensation for all losers (equity and efficiency trade-offs).

4. Following Leopold's (1949) 'land ethic' (much quoted by ecocentrically influenced writers), sustainable activities are activities that do not destabilise the large-scale, biotic and abiotic systems on which future generations will depend. 'Strong sustainability' (SS) resource management will therefore include a commitment (non-negotiable constraints or fixed standards) to protect the health and integrity of ecological systems (Norton 1992). The extensiveness of the commitment does, however, vary across the range of SS views and positions.

5. 'Allocation' refers to the relative division of the resource flow among alternative product uses. An efficient allocation (desirable) is conditioned by individual human preferences as weighted by the ability of the individual to pay (and therefore value). Broadly speaking, relative prices determined in a competitive market (or incentives to correct for market failures) are the relevant policy instrument for the allocative efficiency objective (target). 'Distribution' refers to the relative division of final goods and services (embodying resources) among current humans and future generations. Some relatively egalitarian distribution is considered fair or just and therefore socially acceptable. Taxes and welfare payments represent the relevant policy instrument.

6. The materials balance model is based on an appreciation of the first and second laws of thermodynamics which, put simplistically, say that matter and energy cannot be destroyed (only transformed) by human activity; and that there is a universal law of entropy which lays down that relatively ordered

(low-entropy) and useful resource inputs are transformed by the economy (as a set of irreversible processes) into less well-ordered and less useful (high-entropy) non-product outputs which, if they are not recycled, re-enter the environment as waste flows (Ayres and Kneese 1989). The waste substances contaminate the environmental media into which they are introduced and can potentially cause pollution damage impacts. A non-zero level of contamination is impossible because 100% recycling is thermo-dynamically impossible. Pollution damage may or may not be economically significant, depending upon the effects on human welfare. Zero 'economic' pollution is probably technologically impossible and is in any case prohibitively costly (in clean-up or control cost terms) and therefore socially undesirable.

An optimal scale for the economy is one that is sustainable and at which humans have not yet sacrificed essential ecosystem functions and services that are at present worth more at the margin than the production benefits derived from further growth in the scale of resource use (Daly 1992).

7. Sustainability interpreted as non-declining utility of a representative member of society for millennia into the future (Pezzey 1992).

8. See, in particular, Page (1982) for an exposition of the 'justice as opportunity' thesis.

9. There is one important caveat that should be added because of the existence of non-renewable resources. The depletion of such resources must be accompanied by investment in substitute resources – for example, investment in renewable energy as a substitute for fossil fuels.

10. The significance of these GEC effects is, however, shrouded in both social and scientific uncertainty, and some analysts doubt that they deserve the amount of attention and scarce investment resources that they now seem to be attracting; see, for example, Beckerman (1992) and Nordhaus (1992).

11. The philosopher John Rawls has used contractarian philosophy in order to formulate principles of justice chosen by rational and risk-averse individual representatives from contemporary society in an 'original position' (the negotiations) and generating from behind what he calls a 'veil of ignorance' (i.e. individuals are assumed not to know to which stratum of society they themselves belong). The contractarian approach is therefore based on actual or hypothetical negotiations which are said to be capable of yielding mutually agreeable principles of conduct, which are also binding upon all parties. One of the principles that Rawls derives is known as the 'differential principle' or maximin criterion. What it boils down to is the guarantee of an acceptable standard of living for the least well-off in contemporary society (Rawls 1972). Other writers have sought to place the maximin criterion in an intergenerational context (Page 1982; Norton 1989). In this context the rule becomes one of passing on over time an 'intact' resource base. The constant capital assets rule (VWS) or the constant natural capital assets rule (SS) would be relevant to this intergenerational equity case (Pearce et al. 1991, ch.11).

References

Ayres, R.U. and Kneese, A.V. (1989). Externalities: Economics and thermo-dynamics. In Archibugi, F. and Nijkamp, P. (eds), *Economy and Ecology: Towards Sustainable Development*. Kluwer, Dordrecht.

Barbier, E.B. and Markandya, A. (1989). *The Conditions for Achieving Environmentally Sustainable Economic Development*, LEEC Paper 89-01, London Environmental Economics Centre, London.

Barrett, S. (1991). The problems of global environmental protection. In Helm, D. (ed.), *Economic Policy towards the Environment*. Blackwell, Oxford.

Baumol, W.J. and Oates, W.E. (1988). *The Theory of Environmental Policy*, 2nd edn. Cambridge University Press, Cambridge.

Bartelmus, P. Stahmer, C. and van Tongeren, J. (1991). Integrated environmental and economic accounting: framework for a SNA satellite system. *Review of Income and Wealth* 37, 111–48.

Beckerman, W. (1992). Economic growth and the environment: Whose growth? Whose environment? *World Development* 20, 481–96.

Berkes, F. and Folke, C. (1992). A systems perspective on the interrelations between nature, human-made and cultural capital. *Ecological Economics* 5, 1–8.

Bishop, R.C. (1978). Economics of endangered species. *American Journal of Agricultural Economics* 60, 10–18.

Brandt Commission (1980). *North–South: A Programme for Survival*. Pan, London; and Brandt Commission (1983) *Common Crisis*. Pan, London.

Bromley, D.W. (1992). The commons, common property and environmental policy. *Environmental and Resource Economics* 2, 1–18.

Bryant, C. and Cook, P. (1992). Environmental issues and the national accounts. *Economic Trends* 469, 99–122.

Buckley, G.P. (ed.) (1989). *Biological Habitat Reconstruction*. Belhaven Press, London.

Chambers, R. and Conway, G. (1992). *Sustainable Rural Livelihoods: Practical Concepts for the 21st Century*, Discussion Paper 296. Institute of Development Studies, Sussex University, Sussex.

Common, M. and Perrings, C. (1992). Towards an ecological economics and sustainability. *Ecological Economics* 6, 7–34.

Costanza, R. and Daly, H.E. (1992). Natural capital and sustainable development. *Conservation Biology* 6, 37–46.

Costanza, R. and Perrings, C. (1990). A flexible assurance bonding system for improved environmental management. *Ecological Economics* 2, 57–76.

Daly, H.E. (1991). Towards an environmental macroeconomics. *Land Economics* 67, 255–59.

Daly, H.E. (1992). Allocation, distribution and scale: towards an economics that is efficient, just, and sustainable. *Ecological Economics* 6, 185–94.

Daly, H.E. and Cobb, J.B. (1989). *For the Common Good: Redirecting the Economy Towards Community, the Environment and a Sustainable Future*. Beacon Press, Boston.

Ehrlich, P.R. and Ehrlich, A. (1992). The value of biodiversity. *Ambio* 21, 219–26.

Frey, B. (1992). Pricing and regulation affect environmental ethics. *Environmental and Resource Economics* 2, 399–414.

Garrod, G.D. and Willis, K.G. (1991). *Some Empirical Estimates of Forest Amenity Value*, Countryside Change Working Paper 13. Countryside Change Unit, University of Newcastle upon Tyne.

Green, C.H. *et al.* (1990). The economic evaluation of environmental goods. *Project Appraisal* 5, 70–82.

Grubb, M. (1989). *The Greenhouse Effect: Negotiating Targets*. Royal Institute of International Affairs, London.

Harris, C.C. and Brown, G. (1992). Gain, loss and personal responsibility: The role of motivation in resource valuation decision-making. *Ecological Economics* 5, 73–92.

Harsanyi, J.C. (1955). Cardinal welfare, individualistic ethics and interpersonal comparisons of utility. *Journal of Political Economy* 61, 309–21.

Helm, D. (ed.) (1991). *Economic Policy towards the Environment*. Blackwell, Oxford.

Hirsch, F. (1976). *Social Limits to Growth*. Routledge, London.

Howarth, R.B. and Norgaard, R.B. (1992). Economics of sustainability or the sustainability of economics: Different paradigms. *Ecological Economics* 4, 93–116.

Kahneman, D. and Knetch, J. (1992). Valuing public goods: The purchase of moral satisfaction. *Journal of Environmental Economics and Management* 22, 57–70.

Klassen, G.K. and Opschoor, J.B. (1990). Economics of sustainability or the sustainability of economics: Different paradigms. *Ecological Economics* 4, 93–116.

Leopold, A. (1949). *Sand County Almanac*. Oxford University Press, Oxford.

Lutz, M.A. and Lux, K. (1988). *Humanistic Economics: The New Challenge*. Bootstrap Press, New York.

Mäler, K.G. (1991a). National accounts and environmental resources. *Environmental and Resource Economics* 1, 1–17.

Mäler, K.G. (1991b). International environmental problems. In Helm, D. (ed.), *Economic Policy towards the Environment*. Blackwell, Oxford.

Margolis, H. (1982). *Selfishness, Altrusim and Rationality: A Theory of Social Choice*. Cambridge University Press, Cambridge.

Maslow, A. (1970). *Motivation and Personality*. Harper and Row, New York.

Naess, A. (1973). The shallow and the deep, long range ecology movement: a summary. *Inquiry* 16, 95–100.

Nordhaus, W. (1992). Is growth sustainable? Reflections on the concept of sustainable economic growth. Paper presented to International Economic Association Meeting, Varenna, Italy, October.

Norton, B.G. (1987). *Why Preserve Natural Variety?* Princeton University Press, Princeton, NJ.

Norton, B.G. (1989). International equity and environmental decisions: A model using Rawls' veil of ignorance. *Ecological Economics* 1, 137–59.

Norton, B.G. (1992). Sustainability, human, welfare and ecosystem health. *Environmental Values* 1, 97–111.

Norton, B.G. and Ulanowicz, R.E. (1992). Scale and biodiversity policy: A hierarchical approach. *Ambio* 21, 244–9.

OECD (1992). *Environmental and Economics: A Survey of OECD Work.* OECD, Paris.

Opschoor, J.B. and Turner, R.K. (eds) (1993). *Economic Incentives and Environmental Policies: Principles and Practice.* Kluwer, Dordrecht, forthcoming.

Opschoor, J.B. and Vos, H.B. (1989). *The Application of Economic Instruments for Environmental Protection in OECD Countries.* OECD, Paris.

O'Riordan, T. (1992). *The Precaution Principle in Environmental Management.* CSERGE GEC Working Paper 92–02. CSERGE, UEA, Norwich and UCL, London.

Ostrom, E. (1990). *Governing the Commons: The Evolution of Institutions for Collective Action.* Cambridge University Press, Cambridge.

Page, T. (1977). *Conservation and Economic Efficiency.* Johns Hopkins University Press, Baltimore, MD.

Page, T. (1982). Intergenerational Justice as Opportunity. In Maclean, D. and Brown, P. (eds) *Energy and The Future.* Rowman and Littlefield, Totowa.

Parson, G.R. (1991). A note on choice of residential location in travel cost demand models. *Land Economics* 67, 360–4.

Pearce, D.W. (1992). Green economics. *Environmental Values* 1, 3–13.

Pearce, D.W. and Atkinson, G.D. (1992). *Are National Economics Sustainable? Measuring Sustainable Development.* CSERGE GEC Working Paper 92-11. CSERGE, UEA, Norwich and UCL, London.

Pearce, D.W. and Turner, R.K. (1990). *Economics of Natural Resources and the Environment.* Harvester Wheatsheaf, Hemel Hempstead.

Pearce, D.W., Markandya, A. and Barbier, E.B. (1989). *Blueprint for a Green Economy.* Earthscan, London.

Pearce, D.W., Markandya, A. and Barbier, E.B. (1990). *Sustainable Development.* Earthscan, London.

Pearce et al. (1991). *Blueprint 2.* Earthscan, London.

Penn, J. (1990). Towards an ecologically-based society: A Rawlsian perspective. *Ecological Economics* 2, 225–42.

Perrings, C. (1991). The preservation of natural capital and environmental control. Paper presented at the Annual Conference of EARE, Stockholm.

Perrings, C. et al. (1992). The ecology and economics of biodiversity loss. *Ambio* 21, 201–11.

Pezzey, J. (1989). *Economic Analysts of Sustainable Growth and Sustainable Development.* Environment Department Working Paper No. 15. World Bank, Washington, DC.

Pezzey, J. (1992) Sustainability: An interdisciplinary guide. *Environmental Values* 1, 321–62.

Rawls, J. (1972). *A Theory of Justice.* Oxford University Press, Oxford.

Rolston III, H. (1988). *Environmental Ethics.* Temple University Press, Philadelphia.

Sagoff, M. (1988). Some problems with environmental economics. *Environmental Ethics* 10.

Sedjö, R.A. (1992). A global forestry initiative. *Resources* 109, 16–18.

Sen, A.K. (1977). Rational fools: A critique of the behavioral foundations of economic theory. *Philosophy and Public Affairs* 16, 317–44.

Smith, V.K. (1992). Non market valuation of environmental resources: An interpretative appraisal. Draft copy of unpublished paper.

Swaney, J. (1987). Elements of a neo-institutional environmental economics. *Journal of Environmental Issues* 21, 1739–79.

Tietenberg, T.H. (1991). Economic instruments for environmental regulation. In Helm, D. (ed.), *Economic Policy towards the Environment*. Blackwell, Oxford.

Tinbergen, J. (1952). *On the Theory of Economic Policy*. North Holland, Amsterdam.

Toman, M.A. (1992). The difficulty of defining sustainability. *Resources* 106, 3–6.

Turner, R.K. (1988a). Wetland conservation: Economics and Ethics. In Collard, D. *et al.* (eds), *Economics Growth and Sustainable Environments*. Macmillan, London.

Turner, R.K. (ed.) (1988b). *Sustainable Environmental Management: Principles and Practice*. Belhaven Press, London.

Turner, R.K. (1992). *Speculations on Weak and Strong Sustainability*, CSERGE Working Paper, GEC 92-26. University of East Anglia, Norwich and University College, London.

Turner, R.K. and Jones, T. (eds) (1991). *Wetlands: Market and Intervention Failure*. Earthscan, London.

Turner, R.K., Bateman, I. and Pearce, D.W. (1992). United Kingdom. In Navrud, S. (ed.), *Valuing the Environment: The European Experience*. Scandinavian University Press, Oslo.

United Nations Development Programme (1992). *Human Development Report*. Oxford University Press, New York.

Victor, P.A. (1991). Indications of sustainable development: Some lessons from capital theory. *Ecological Economics* 4, 191–213.

Wallace, R.R. and Norton, B.G. (1992). Policy implications of Gaian theory. *Ecological Economics* 6, 103–18.

World Commission on Environment and Development (1987). *Our Common Future*. Oxford University Press, Oxford.

Willig, R.D. (1976). Consumer's surplus without apology. *American Economic Review* 66, 587–97.

Chapter 2
The Politics of Sustainability
Timothy O'Riordan

Framing the concept

It is tempting to dismiss the term 'sustainable development' as an impossible ideal that serves to mask the continuation of the exploitation and brutality that have characterised much of human endeavour over millennia. In the first edition of this book I argued (O'Riordan 1988) that sustainability was being used as a mediating term to bridge the widening gulf between 'developers' and 'environmentalists'. I also contended that the concept was deliberately vague and inherently self-contradictory so that endless streams of academics and diplomats could spend many comfortable hours trying to define it without success.

Five years on, the phrase has stuck. There is no doubt that the report of the World Commission on Environment and Development (Brundt-land 1987) ennobled the idea in international political circles. Dozens of reports and books have now been written on the matter, and almost every major international organisation has addressed it in some form or another. It was the central theme in the UN Conference on Environment and Development (UNCED) held in Rio de Janeiro in June 1992. Like it or not, 'sustainable development' is with us for all time.

The staying power of the concept is understandable, if not forgivable. No public figure or private corporation can afford to speak any other language. The greening of society may still be essentially cosmetic but it does have political force and powerful public relations and marketing implications. The growth of discriminating consumer power can weaken even a major company. Witness the successful campaigns against Burger King for allegedly buying meat from former Brazilian rainforested areas and against Sun Kist tuna of the Kellogg Corporation for not controlling the incidental destruction of dolphin in its overaggressive tuna fishing. Nowadays major tuna processors can only market 'dolphin friendly' brands (Elkington *et al.* 1991, 143). National Westminster is one of a number of British banks that now employs ethical investment advisers at a senior level, just as most transnational corporations are committed to environmental audits, published codes of practice, and wide-ranging compensation schemes in case of environmental damage (see Business and Society Review 1990; Adams *et al.* 1991; Elkington *et al.* 1991). There are practical considerations here. Environmental auditing is a management tool as much as it is an environmental account. Used wisely, it should improve product efficiency, streamline management, increase competitiveness, pinpoint product niches, and promote marketing strategies. Many a sound environmental impact assessment (EIA) has warned a company, or an aid organisation, of probable disaster due to environmental ignorance or political naivety. EIAs often save developers a lot of money and even more embarrassment.

Secondly, there is a political reality to this new consciousness about sustainability. Regulations, both domestically and internationally, have tightened. Regulatory agencies are tougher, more independent, more professionally competent and more transparent in their dealings with their clients. This spirit of openness is by no means universal, nor complete, but in relative terms regulatory activities are not only better handled but also much more politically accountable. The European Community directives that mandate the principle of best available techniques not entailing excessive costs (notably the Large Combustion Plant Directive and the so-called 'Red List' Directive) require member states to insist on the best practice for a process irrespective of the financial implications for a company, so long as the application of these techniques can be shown to improve environmental quality. Similarly, the forthcoming EC directive on civil liability for hazardous wastes will require companies to follow the very best practices in waste minimisation. Yet the directive will also ensure that such companies will still be legally and financially liable should some unforeseen calamity afflict individuals or whole communities, even when best practice is followed and all regulations are adhered to.

Thirdly, a wish for a more sustainable future is latent in most emancipated individuals – including those who have to make decisions that run at cross-purposes with such an ideal aim. Consider the conflict of conscience in a modern major chemical corporation faced with an expanding market in chlorofluorocarbon products in flourishing east Asia. This is a region where no countries have yet signed the Montreal protocol controlling the use of these ozone-depleting and atmospheric warming gases. There are massive profits to be made of stocks that cannot be sold in Europe or North America. Yet can a company, in all conscience, deliberately sell this legacy of future global damage? In public relations terms if for no other, the answer must be 'no'. Sustainability, for better or for worse, is definable in the conscience: it remains to be seen if it is definable in the account book as well.

The growing sense of unease over global change is built on a combination of moral guilt and pragmatic fear. It is more evident in personal values, in social communication as well as in employer–employee contracts than ever before. It may still be very subdued, and often very repressed, but it is not eliminated. We have started to create an environmentally sensitised democracy yet have not discovered the means of delivering the goods. This tension between conflicting personal and societal values and emerging self-awareness, or emancipation, which is widespread in modern society, is playing its part in promoting the longevity of the sustainable development paradigm.

The trouble is that sustainable development needs much more than shirtsleeve greenery and the latent guilt of what are charmingly referred to as 'couch potatoes' for its survival. It confronts modern society at the very heart of its purpose. Humankind is still a colonising species. It has no institutional or intellectual capacity for equilibrium. Like it or not, true sustainability requires five conditions that are still far from any political consensus. These conditions include:

(i)   a form of democracy that transcends the nation-state and the next election, namely that alters the meaning of 'self-interest' and 'sacrifice';

(ii)   guarantees of civil rights and social justice to oppressed peoples the world over, so that they are allowed to consume resources in an equilibrium manner, and appreciate the intrinsic rights of nature;

(iii)   commitments of resources, notably technology, intellectual property generally, and cash, to impoverished and environmental vulnerable regimes, many of which are run by politically unstable and inherently corrupt governments (as indeed are many rich nations);

(iv)   elimination of debt where debt is induced by unfair terms of trade and a historical legacy of exploitation;

(v)   establishment of a variety of public–private non-governmental mechanisms for delivering resources, training and management techniques to areas and communities in need, in such a way as to be socially acceptable and democratising.

These are well-worn topics. They have been promoted in official and quasi-official reports for generations. Major non-governmental organisations such as the development charities and Amnesty International have fought for years to bring about a closer relationship between justice, personal opportunity, civil rights and environmental survival. Enormous progress has been made, and one wonders how much worse the world would be had such organisations not existed. Yet the fact remains that the world is still in a desperately sorry state and is growing worse.

Table 2.1   Official development
assistance from OECD countries ($US
billions, 1986 prices)

1980	26.2
1983	24.4
1986	24.2
1988	25.4

Source: World Bank (1989, 10).

Table 2.2   Net financial transfers to developing
countries ($US billions, 1986 prices)

1980	19.6	1985	− 36.2
1981	22.7	1986	− 38.2
1982	6.4	1987	− 40.3
1983	− 2.4	1988	− 52.0
1984	− 21.9	1989	− 51.6 (est)

Source: World Bank (1989, 9).

This is not just the sober conclusion of the environmentalist think
tanks such as World Resources Institute (1991). It is shared by serious
commentators (for example, Holdgate 1991) and in the recent report by
environmental consultants to the OECD (1991). It is also implicit in the
views of distinguished commentators on the idea, expressed in a series of
Cambridge lectures in 1989 (Angell *et al.* 1990). Just to rub salt in the
wound, Tables 2.1 and 2.2 reveal how aid in real terms to the developing
world is falling, when adverse terms of trade mean that twice the North
to South aid disbursement is being returned to the rich North in the form
of transfer payments.

It is not that developers seek to exploit the sustainability idea to justify
almost any investment so long as the trappings of environmental concern
are evident. It is not even that environmental groups try to demand
safeguards   and   compensatory   investments   that   are   not   always
economically efficient or socially just. It is much more that the root
causes of non-sustainability lie in profoundly powerful systems of
exploitation and degradation that are fostered by ignorance, greed,
injustice and oppression.

These are abstract points. A flavour of the issues can be found in the
moving account of the plight of children in the Third World by
Timberlake and Thomas (1990, 114). They cite the example of Pavitra,

an 11-year-old Indian girl, rising at 5 a.m. to draw polluted water from a hand pump because the aquifers of fresh water are overdrawn and have run dry. She has to gather fuel from thorny bushes, with her bare hands, 1.5 kilometres away and cook the breakfast because her mother is too sick from the emissions and toxic dust of the detergent factory where she earns 60 pence per day. Pavitra then works in a granite quarry where she earns £5 in four days of backbreaking work.

This is not her family's choice. They come from Rajasthan but the drought there forced them, a proud people, to migrate to the Delhi region seeking any work that was remotely respectable. The granite quarry spews dust throughout the village. Despite a ruling from the Indian Supreme Court, that required the company to control its pollution, the dust continues to contaminate lungs and food. As Timberlake and Thomas put it (1990, 114):

It seems obscene to say that Pavitra is suffering from 'environmental problems'. She suffers from all the ills inherent in India's social, political and economic organisation, and from India's place in the modern world. But these have been translated, for her and her family, into virtually all the world's environmental ills.

Pavitra's case can be multiplied many million times. In a non-sustainable world her future is almost certainly doomed. Can the politics of sustainable development, as it is currently being played out, save her and her long-suffering sisters and brothers?

Non-sustainability in microcosm: a case study

I began this essay commenting on the tensions between job and conscience, power relations and emancipation. All too often the first of each of these dilemmas wins – hence the continuation of non-sustainable development amid the welter of rhetoric. A micro-case study from the UK provides some clues.

In March 1991 the Council for the Protection of Rural England, one of the most sophisticated non-governmental organisations in the UK, was given a brown paper envelope, presumably by somebody whose private conscience was at odds with his or her occupational loyalties. This document contained the official forecasts for the demand for aggregates – sand and gravel, building stone, road surface material and industrial requirements – to the year 2011. These forecasts were produced in secret by regional aggregate working parties composed of representatives of government, local authorities and the aggregates industry. No non-governmental environmental group is represented, nor are the methodologies which purport to justify the forecasts ever made public – even when demanded by opponents fighting a proposed quarry or mineral working at a statutory public enquiry (Adams 1991).

The basis for the forecasts is demand-led, and in that sense it is similar to the bases for forecasting electricity and, until recently, water in the UK. The effect of both a demand-led forecasting methodology and non-accountable secrecy is to maintain a cartel of prices for the industry so that real prices are less than they were 20 years ago. This tends to discriminate against any incentive to recycle or reuse waste aggregate materials. Furthermore, the forecasts are based on assumptions of economic growth and unquestioned aggregates demand linked to the putative growth. The outcome is allegedly a need for a near-doubling of capacity by 2011. The key variable is that aggregate intensity – the demand for raw material per unit of growth – will increase. Furthermore, assumptions about environmental protection cause the forecasters to increase even more their forecasts on the basis of constructing more sewage works, reservoirs and sea defence projects. There was no serious analysis of what a more sustainable scenario might look like for the aggregates industry – in terms of price- or target-directed demand, prices or incentives to promote the use of recycled materials, and price intervention to reflect the true cost of extraction, transportation and general social nuisance.

At the heart of this dispute lie four issues. The first is the insensitivity of the forecasting methodologies to a wide range of feasible options for reducing aggregate demand, even for a prospective period of real growth. The second is the secret and cabal-like character of the forecasts and incestuous linkages between the government, the industry, and the regulators (the local authority minerals and planning authorities). Third, there is complete lack of interest in any strategy for reusing or recycling existing aggregate material, notably in the road building programme and for land reclamation, through pricing or institutional redesign. Finally, there are the insidious connections with the road industry, the car lobby and the whole paraphernalia of promoting new motorspace at the expense of less mobility, more effective public transport and an energy conservation priority.

What we witness here is the insensitivity of power, lobbying and closed thinking. There is no interest in manipulating market forces to upgrade the secondary aggregate recycling industry, yet the technology is almost there to promote it. This would require government support, yet that is palpably not forthcoming. Nor is there any indication of special protection of scenic countryside or already heavily quarried areas to safeguard against future planning applications. Almost any area is vulnerable, including the granite-rich outer islands of Scotland, ideally suited to supply the road building materials for the southern English motorway programme. As the monthly magazine *ENDS* (Environmental Data Services 1991, 8) put it:

> The ultimate goal is a decoupling of demand for aggregates and economic growth, in much the same way that the previous close relationship between energy demand and economic activity is being changed with progressive policies on energy efficiency and demand management. The scale of the challenge can be seen in the chain which leads from the Department of Transport's road traffic forecasts in which environmental constraints are assumed – to its roads programme – prepared without environmental assessment – to the DoE's aggregates forecasts – also based on the assumption that extraction is unconstrained by environmental considerations.

This is a micro-example. It is repeated in almost every resource sector in the developed and developing world – in water, forests, soil fertility, fisheries and waste management. At the heart of this micro-story is not mendacity or even ignorance and stupidity. It is simply that all our institutions, whether they be extractive, administrative, regulatory or economic (in the broadest sense), are effectively geared to depletion, to passing on third-party costs and, above all, to the comfort of not having to think and act sustainably. When cocooned in a world of throughput, it is hard to balance the environmental books. It goes against the institutional grain and the dominant attitude of mind. People of conscience fall by the wayside, or leak memoranda.

Interpretations of 'sustainability'

The notion of 'sustainability' applies most conveniently to the replenishable use of renewable resources. The aim is to benefit from the advantages provided by such resources to the point where the rate of 'take' equals the rate of renewal, restoration or replenishment. So in agriculture the farmer derives fertility from the soil equal to the ability of the soil to supply nutrition. Similarly, the woodsman removes trees or tree products at a rate equal to tree regeneration. The fisherman catches marine resources in amounts that are equivalent to their refurbishment. This begs the question of whether inherent rates of renewability can be enhanced through scientific management. Even under those conditions, however, the basic principles of sustainability apply.

Implicit in this narrow definition are four seemingly justifiable precepts:

(i) *knowability*: the amount, rate and other characteristics of renewability are knowable and calculable;

(ii) *homoeostasis*: renewable resource systems operate broadly around equilibria or can be manipulated to approximate steady states following human intervention – homoeostasis is a preferential state of nature;

(iii) *internal bioethics*: the act of drawing upon a renewable resource, even below some threshold of take, has implications only for the tightly confined ecosystem that is that resource;

(iv) *external bioethics*: utilising a renewable resource up to the point of sustainable yield is morally justifiable even though that resource, below the threshold of optimal take, may have other ecological values and functions.

We shall see below that none of these principles or conditions is realistic, practicable or justifiable. Yet, if sustainability in this narrow sense is to be operational, then these principles must at least set boundaries for acceptable action.

Sustainability within Gaian laws

The concept of sustainability has a long historical lineage. The very brief historical account that follows is written to illustrate that sustainability has been a desirable objective throughout human history. Interpretations, however, have differed noticeably, depending upon the character of the man–land relationship.

As a specific notion, sustainability probably appeared first in the Greek vision of 'Ge' or 'Gaia' as the Goddess of the Earth, the mother figure of natural replenishment. The historian, Donald Hughes (1983, 55) summarises the Gaian perspective:

[Gaia] nourishes and cares for all creatures as her own children. From her all things spring; to her return all things that die. Her creative womb bore all that is, including the first of all the sky and all that it contains . . .

So important was the practice of sustainability to the Greeks that provincial governors were rewarded or punished according to the look of the land. Signs of erosion or other features of environmental damage led to admonishment or even exile, whereas a healthy-looking land, regardless of the real well-being of its people, would be accorded approval.

Hughes takes the Gaian interpretation of sustainability one stage further by showing that linked to Gaia was her daughter, Themis, the goddess of law or justice. As Hughes (1983, 56) comments:

It is because the Earth has her own law, a natural law in the original sense of these words, deeper than human enactments and beyond repeal . . . Who treats her well receives blessings; who treats her ill suffers privation, for she gives with evenhanded measure. Earth forgives, *but only to a certain point, only until the balance tips and then it is too late* . . . [emphasis added]

There is far more to the Greek concept of Gaia, and its more scientific contemporary interpretation attributed to Lovelock (1987), than this rather confined version of sustainability. Nevertheless, the principles of adherence to natural laws, or what scientists prosaically term 'fundamental environmental processes' and the evenhandedness of retribution (at least in terms of consequence, if not in terms of final impact which relates to poverty and technological adaptability), are pertinent in the original, simple formulation.

Sustainable utilisation as democratic resource management

We can move almost two thousand years to see the notion of sustainability taken into the modern era. The principal advocates were the new elite of scientific environmentalists operating under the Progressive political ideologies found in the United States at the turn of the century. Their champion was President Theodore Roosevelt, but their intellectual leader was Gifford Pinchot, Roosevelt's chief forester and mentor.

Pinchot regarded sustainable utilisation as the application of 'common sense'. To Pinchot, conservation and sustainability were as one:

Conservation advocates the use of foresight, prudence, thrift and intelligence in dealing with public matters. It proclaims the right and duty of the people to act for the benefit of the people. Conservation demands the application of common sense to the common problems for the common good . . . [written in 1910, quoted in Nash 1968, 61]

Conservation is foresighted utilisation, preservation and/or renewal of forests, waters, land and minerals for the greatest good of the greatest number for the longest time . . . [ibid., 61]

Pinchot was a forester. During the latter part of the nineteenth century, foresters had developed a series of management principles around the application of sustained yield silviculture. Sustained yield management was very tightly defined, referring to the replanting of commercial species over logged areas so that productive biomass was sustained. Nothing was said about concomitant ecological losses.

However, foresters had also linked their profession to soil conservation. Indeed, the early British forest reserves in India were known as 'conservancies' as their prime role was to protect the soil and control runoff, rather than to provide timber (for an excellent review of the history of conservation, see Grove 1990). It is from this derivation that Pinchot extended the meaning of conservation, and ecologists later coined the phrase 'sustainable utilisation'. Pinchot and his associates were primarily interested in efficiency of economic transactions over natural resource utilisation. This meant blocking the monopoly power of the major resource-owning monoliths, and extending access to the benefits of natural resources to as many people as possible. Enlightened regulation was supposed to apply the brakes on selfishness or the inherent tendency towards corporatist aggrandisement. Grafted on to the fundamentals of replenishment were the more contemporary themes of efficiency, i.e. minimisation of wastage, achieved by appropriate pricing and/or regulation. Sustainability therefore cannot be divorced from the mechanisms for determining resource allocation, and hence is of central concern to contemporary environmental politics.

Samuel Hays (1959, 266), the eminent student of the history of American conservation, was probably the first to reveal how the principles of sustainability, if vaguely defined and drawn from insecure scientific principles, could be exploited for political ends:

The conservation movement did not involve a reaction against large scale corporate business, but, in fact, shared its views in a mutual revulsion against unrestrained competition and undirected economic development. Both [developers and conservationists] placed a premium on large scale capital organisation, technology and industry-wide cooperation and planning to abolish the uncertainties and waste of competitive resource use.

Hays's analysis goes further. Roosevelt used the Progressive conservation rhetoric to bypass Congress and to outflank established economic institutions and major interest lobbies in his quest to democratise resource use decision-making and management. He exploited the 'common sense' interpretations of Pinchot and his scientific colleagues in his belief that a decentralised society would be more enlightened and frugal in resource utilisation, and that democratic accountability could be extended into the nascent realms of environmental regulation.

Here we begin to see the stretching of the concept of sustainability to embrace institutions and devices capable of reducing waste and ensuring adequate surveillance. Sustainability is only promoted if it is seen to be achievable.

Sustainability and protective buffers

The first economist to appreciate that the 'knowability' and 'equilibrium' precepts that were built into the early interpretations of sustainability were not practicable was Ciriacy-Wantrup (1963). Wantrup was alive to the economic dangers of the irreversibility – or what he termed 'the critical zone' – and argued for a buffer to protect managers from folly, annoyance or ignorance. He termed this buffer, or sub-threshold optimum for resource draw, the 'safe minimum standard'. But even Wantrup erred on the side of a technical analysis of the economic–ecological relationship. He gave no indication of an interest in or the relevance of any eco-morality issues associated with the 'sub-threshold' take, nor in the possible ecological side-effects as renewable resources are drawn. This is how he interprets the safe minimum standard:

In *many* practical situations, maintenance of a safe minimum standard does not involve any use [forgone]; rather it involves a change in the technical ways (*not the quantities*) of utilisation . . . with proper timing and choice of tools . . . costs of maintaining the safe minimum standard are not only small in absolute amount, but very small relative to the loss which is being guarded against, a decrease of flexibility in the continuing development of a society. [quoted in Burton and Kates, 1965, 57; second emphasis added]

For Ciriacy-Wantrup sustainability was regarded primarily as a technical matter, involving appropriate economic incentives for correcting resource mismanagement. The buffer against misuse was determined through what he termed 'economic rationale', not ecological morality.

One should compare this with the powerful analysis offered by Charles Frankel (1976, 111–12). Frankel was considering what are the 'messages' or 'rights' of nature which should provide maxims for sustainability. He concluded that nature ought to provide a moral brake on arrogant prescriptions about knowledge of environmental processes.

Indeed the appeal to 'Nature' may well be a useful reminder that human purposes fade, and that the sacred truths of an era are usually only collective follies. It also reminds us that, although there are laws, presumably, that explain what happens in human life, we do not know these laws and, from our partial point of view, we must accept Nature as in part random, unpredictable, mysterious. *So it is that the experts must be wrong, are destined to be wrong, unless they make explicit provision for reversing their plans and hedging their bets* . . . Perhaps when people say 'Nature has rights' they mean only to say that we ought to have institutional protection against being carried away by temporary enthusiasms. [emphasis added]

This is a far cry from Ciriacy-Wantrup, but a useful bridge into the contemporary politics of sustainability. Frankel was seeking to establish a set of rules for sustainability that transcended short-term practicalities. He provided no coherent set of new guides to assist those responsible for allocating environmental resources. Supporters of the strong and very strong versions of sustainability, in particular (see Turner, Chapter 1), recognise that economic transactions now have to contain and/or be constrained by an ecological element if full valuation of resource use is to be undertaken. Very strong sustainability advocates also admit that it is extraordinarily difficult to do this under present practices. The time has come to go beyond rhetoric. The contemporary politics of sustainability attempts to deal with this problem.

The evolution of the sustainability concept

It all began with a series of African-based conferences in the mid-1960s. The aim was in part to safeguard the habitat of African wildlife, regarded as the epitome of conservation at that time. But it was also recognised that African rural development had to confront the limits imposed by soil, climate and water availability if it was to retain a physical basis for self-development (see McCormick, 1989, 43–6). As Africa emerged from colonialism, so European conservationists felt it was essential to encourage African governments publicly to associate themselves with wildlife conservation, and to agree in principle to reconcile conservation with national development plans. Various waves of visiting European environmental consultants to Africa recognised that this could only be achieved if non-game reserve land management was aimed at providing self-sustaining food output. Otherwise there would be a great temptation for farmers to cross into game parks in order to graze livestock, plant crops and attack wild game for food or for protection (Hillaby 1961).

There was an element of Euro-orientated self-serving morality about this analysis. Wildlife management outside and adjacent to national parks and protected areas depended upon the needs and ways of life of local rural communities, whose understanding and cooperation were vital if wildlife management projects were to succeed. As McCormick (1989, 44–5) put it:

Traditional African bonds with nature, which had been undermined by European economic development models of the sort that had caused the destruction of nature in Europe, needed to be revitalised . . . Preservation alone was not the answer, rational management based on clear objectives was needed.

This resulted in a convention, agreed through the auspices of the International Union for the Conservation of Nature (IUCN) and signed by 33 Organisation of African Unity states in 1969, that conservation management must be based on a combination of prevention *and* cure, and that conservation should involve

the management (which includes survey, research, administration, preservation, utilization) of air, water, soil, minerals and living species including man, so as to achieve the highest sustainable quality of life. [quoted in McCormick 1989, 46]

Wildlife protection was essentially regarded as a barometer or marker of humanity's commitment to a relatively unchanged world. Healthy habitat was also seen as a monitor of environmental stress as other habitats became degraded from the protected 'optima' of the jewel-in-the-crown reserves.

This coupling of wildlife protection to the maintenance of life support services, such as air and water quality and biological diversity, became the linchpin of the IUCN's World Conservation Strategy (1980), where the concept of sustainable utilisation of resources received a more formal airing. Arguably this is the first global statement about sustainable development, for it required considered response by all national governments and the development of a plan of action (Adams 1990, 42–65).

Attention is given here to the World Conservation Strategy process because the original document has been revised by the IUCN (1989; 1991). In part, also, the new versions lay much greater emphasis on the protection of cultures, and indigenous ways of managing resources, than on the resources themselves. This process of updating is important because both the revised document and the subsequent national response will seek to take on board all relevant developments since 1980, including the growing scientific awareness of global environmental change generally (for a comprehensive review, see *Scientific American* 1989). These developments will be linked to the changing geopolitics of the globe, given a friendlier but potentially unstable Commonwealth of Independent States (the former Soviet Union), the new democracies of eastern Europe which will be forced to develop in an environmentally friendlier manner, and the continued concern over population–resource imbalances in the underdeveloped world, especially Africa (World Resources Institute 1990).

The World Conservation Strategy of 1980 identified three vital ingredients to sustainable development. These continue to form the basis of modern thinking on the topic.

(i)   the maintenance of essential ecological processes, of oxygen, carbon, sulphur and nitrogen fluxes governed or strongly moderated by ecosystems, and providing the primary basis for food production, health and other aspects of human survival;

(ii) the preservation of genetic diversity, both in the species varieties of local food production as well as for species in the wild to act as an insurance against disease or climatic stress, and an investment for future food varieties and genetically modified organisms;

(iii) the sustainable development of species and ecosystems, particularly fisheries and other wild species which are cropped, forests and timber resources, and grazing land.

Table 2.3 summarises the IUCN position in its first major report. Much of the thinking of the early 1960s has continued into this statement of conservation requirements. Adams (1990, 46–8) points out that the morality implicit in this influential document reflects both the ecological determinism and the bio-ethics of the early 1970s. Environmental degradation would set the limits to population growth, but in the process a lot of people and innocent wildlife would be lost. Wildlife is important partly for utilitarian reasons. We never know what species or genetic strain might become useful or of economic value, if it is eliminated without being surveyed and recorded.

Partly, too, there was an ethical argument. Wildlife was there before human beings, so it has a 'right' to exist in a way in which humans should not interfere. The solution, according to the IUCN, is to give conservation a high priority in all aspects of development – in agriculture, industry, health, education and regional redevelopment – so that the two become one and the same.

The World Conservation Strategy exercise fell foul of three crucial problems linked to the sustainable development argument:

(i)   It tried to look both ways – to retain conservation principles yet permit only very modest changes in traditional forms of development to proceed. No practical examples were offered as to how this unstable relationship could be set right.

(ii) It ignored the neo-Marxist political interpretations of science, conservation and development, namely that the generation of new wealth was an imperative of power, and that those who buttressed that power made sure that they benefited from the fruits of any new wealth provided (Verhelst 1990).

(iii) It appeared also to be unaware of the structures of power and their supporting institutions that promote inequality, poverty and desperation in modern society. As Adams (1990, 51) put it, the

Table 2.3 Priority requirements of the World Conservation Strategy

A. Ecological processes
1. Reserve good cropland for crops (para 5.1).
2. Manage cropland to high, ecologically sound standards (para 5.3).
3. Ensure that the principal management goal for watershed forests and pastures is protection of the watershed.
4. Ensure that the principal management goal for coastal wetlands is the maintenance of the processes on which the fisheries depend (para 5.6).
5. Control the discharge of pollutants (para 5.8).

B. Genetic diversity
1. Prevent the extinction of species (para 6.1).
2. Preserve as many kinds as possible of crop plants, forage plants, timber trees, livestock, animals for aquaculture, microbes and other domestic organisms and their wild relatives (para 6.4).
3. Ensure on-site preservation programmes protect:
   - the wild relatives of economically valuable and other useful plants and animals and their habitats;
   - the habitats of threatened and unique species;
   - unique ecosystems;
   - representative samples of ecosystem types (para 6.8).
4. Determine the size, distribution and management of protected areas on the basis of the needs of the ecosystems and the plant and animal communities they are intended to protect (para 6.10).
5. Coordinate national and international protected area programmes (para 6.12).

C. Sustainable utilization
1. Determine the productive capacities of exploited species and ecosystems and ensure that utilization does not exceed those capacities (para 7.1).
2. Adopt conservation management objectives for the utilization of species and ecosystems (para 7.2).
3. Ensure that access to a resource does not exceed the resource's capacity to sustain exploitation (para 7.3).
4. Reduce excessive yields to sustainable levels (para 7.4).
5. Reduce incidental take as much as possible (para 7.5).
6. Equip subsistence communities to utilise resources sustainably (para 7.6).
7. Maintain the habitats of resource species (para 7.7).
8. Regulate international trade in wild animals and plants (para 7.8).
9. Allocate timber concessions with care and manage them to high standards (para 7.9).
10. Limit firewood consumption to sustainable levels (para 7.10).
11. Regulate the stocking of grazing lands to maintain the long-term productivity of plants and animals (para 7.11).
12. Utilise indigenous wild herbivores, alone or with livestock, where domestic stock alone would degrade the environment.

Source: IUCN (1980)

WCS was 'pious, liberal and benign, inevitably ideological and disastrously naive'.

Ecodevelopment

One antidote to the almost neo-Malthusian approach, enshrined in the World Conservation Strategy, of resource scarcity requiring economic sacrifice for longer-term population reduction and ultimate economic gains, was the emergence of *ecodevelopment*. This was first advocated by Maurice Strong in the 1972 Stockholm Conference on the Human Environment. It is regarded as more of a 'geographical' notion, i.e. it is sensitive to people, the land, history and culture. The intertwining of all these provides ecodevelopment with a set of ideas that remain important for the modern notion of sustainability.

The essential principles of ecodevelopment are as follows:

(i)   *The provision of basic needs*, starting with the needs of the poorest. This is not new (see McHale and McHale 1979) but it provides an essential foundation to sustainable development that has yet to be met, so cannot be tested as true or false until the necessary experiments are tried out.

(ii) *Participation for the community itself.* Again this is a well-established idea (see, for example, Sachs 1979; Riddell 1981). But in the ecodevelopment mode it took on a new political meaning of power sharing, access to resources as of right, and greater equality of opportunity for all peoples, especially women and children (see, for example, Galtung 1984; Glaeser 1984; and, generally, Adams 1990).

(iii) *The use of appropriate or intermediate technology.* Once again this idea had developed in the early 1970s with the writing of Schumacher (1989). Here the concept is coupled with the practice of participation, the experience of local people in understanding and defining their resource needs, and the scope for aid and regional development policies that specifically aimed at introducing such technologies in a form that was ecologically and culturally appropriate. This meant training and new forms of social organisation, not just the application of the equipment and techniques.

All these ideas consolidated around the concept of sustainable development in the late 1970s and early 1980s. In every case, however, rhetoric was strong but practical implication was weak or non-existent. None of the studies seriously addressed the complicating issues of inappropriate economic incentives (i.e. taxes and subsidies which distort the use of resources in a non-sustainable manner). None looked at the pattern of political power, and patronage, the reinforcing the position of elites whose support maintained governments and armies but whose resource use was not sustainable. None canvassed the realities of inequality of opportunity which repressed large numbers of people, some systematically because of their gender or race or tribal affiliation, in positions of poverty and destitution (see, for example, Repetto and Gillis 1988; Rau 1991; Shiva 1991). Such views were always implicit, but rarely incorporated into the practice of sustainable development (Redclift 1987; Adams 1990).

Table 2.4   Requirements for sustainable development

1. A political system that secures effective citizen participation in decision-making.

2. An economic system that is able to generate surpluses and technical knowledge on a self-reliant and self-sustained basis.

3. A social system that provides for solutions for the tensions arising from disharmonious development.

4. A production system that respects the obligation to preserve the ecological basis for development.

5. A technological system that can search continuously for new solutions.

6. An international system that fosters sustainable patterns of trade and finance.

7. An administrative system that is flexible and has the capacity for self-correction.

Source: Brundtland (1987, 65).

The publication of the Brundtland Report did little to change this. It did, however, put the issue of sustainable development more firmly on the international political agenda. Table 2.4 shows how, at least in principle, the Brundtland Commission saw beyond technicalities and vague rhetoric and paved the way for UNCED (see de la Court 1990). It was the intention that this particular meeting would focus on the practice and the continuation of meaningful and identifiable sustainable development schemes well beyond the Conference itself. Many of the critiques of the sustainability concept and its application, as summarised here, were extensively aired at that meeting. The main point to emphasise is that UNCED was designed to create a workable plan for action (Agenda 21) and will seek the necessary commitments on financing, training and experimentation to ensure that at least some genuinely practical outcomes will ensue. In this post-UNCED era it remains to be seen how extensively the practical achievements (through the Global Environmental Facility, etc.) will turn out to be.

Impediments to the application of the sustainability principle

Both the interpretation and enactment of sustainability embrace political values and the exercise of power. For sustainability to become implanted in the political culture, the character of developmental and resource management institutions will require reform and remodelling. Opposition to such reforms is ideological and structural. Sustainability is not regarded seriously by those who really count, namely those at the top of political structures and those who control the flows of national and international capital. The promotion of sustainability implies a reorganisation of agency alignments and priorities which cuts across the prejudices of those at the top. The full application of sustainability also demands new arrangements for budget-sharing and cross-organisational responsibilities that are deemed unacceptable or unworkable by those who benefit and operate through existing arrangements. By contrast, sustainable development, especially where it can be shown that thoughtless mismanagement of ecological systems can result in economic disaster and unprofitable capital investment, is now visualised as an appropriate way forward. But the pursuit of sustainable development is simply not enough.

By way of illustration, let us look at three levels at which sustainability principles might operate:

(i)   environmental development in poor Third World countries;
(ii) putting into effect the concept of best practicable environmental option;
(iii) trading environmental losses for environmental gains in project evaluation.

These three themes are chosen to reflect important arenas where the ideas contained in other chapters of this volume apply most directly.

The Third World dilemma
The Third World is a diverse collection of countries facing a myriad of environmental, social and economic problems with an equally varied mix of developmental histories and political-military governments. What

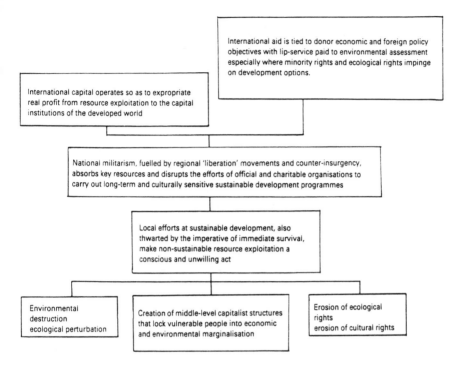

Figure 2.1   Non-sustainable resource draw and international capitalism

follows is a generic analysis of the difficulties of putting sustainability into practice. The arguments cannot be applied neatly to any Third World country.

We start with Figure 2.1. This illustrates the relationship between the pressures for non-sustainable resource draw and the imperatives of international capitalism, international aid, national militarism and counter-insurgency, and cultural conflict which render sustainable development impossible.

Making this more specific Figure 2.2 illustrates the links between capital movement, the seed–fertilizer–pesticide axis of major multinational chemical corporations, desertification and indebtedness, all leading to marginalisation of the vulnerable who intensify the degradation of the soil in their desperate efforts to survive. The solution requires coordination between agriculture, soil management, forestry, aid schemes and industrial regeneration departments, integrating their efforts at a regional and local level through extensive use of *animateurs*, locally based extension agents acting as catalysts for local self-help schemes.

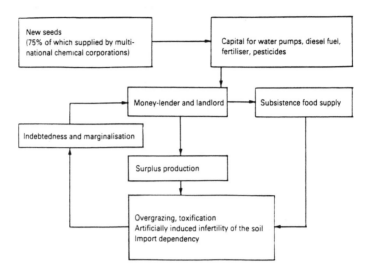

Figure 2.2   Constraints on sustainable development

These diagrams expose the central dilemma. Putting sustainability into practice requires restructuring institutional arrangements and transforming economic thinking as well as undertaking politically formidable decisions regarding North to South economic and technology transfers. Much of what follows in this book illustrates the economic dilemmas, notably the wasteful subsidies that support the already rich to exploit even more resources at the expense of the poor, and the failure to take into account third party ecological-economic effects of resource extraction and waste disposal where real costs are being imposed on the functioning capability of environmental systems and the scope for future generations to live sustainably even if they want to.

Examples abound, with a ready reference in Pearce *et al.* (1990). Take the case of charcoal burning in Sudan. This fuel accounts for some two-thirds of all forest scrub removal in Sudan, exposing friable soils to nutrient loss and punishing erosion (Pearce *et al.* 1990, 145). The fuel is in effect subsidised for the relatively rich Khartoum populations, who certainly do not pay for the prospect of displacing 250,000 poor farmers barely on subsistence. Up to 29 per cent of this demand could be reduced by improved design for burners, but there is no agency charged with enforcing such a transition. Price rises to force improved consumption would have to be fairly draconian to be effective, but it is unlikely that there is the political will to take on the restless and influential urban elite.

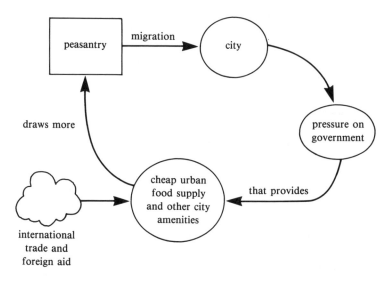

Figure 2.3

Figure 2.3, reproduced from a recent article by Nathan Keyfitz, illustrates the politics of this problem in an unsimplified but revealing form. Urban concentrations become political impediments to fundamental economic reform. The larger the urban population, according to Keyfitz (1991, 66) the more governments have to make concessions in setting prices and maintaining employment. As urban areas are favoured, so more immigrants are attracted, reinforcing the process that may distort natural resource prices and inhibit fundamental institutional reform.

In a similar vein, biotechnological developments in Europe and the USA have found substitutes for traditional African export commodities such as cocoa and gum arabic. Indeed, with the prospect of transfers of indigenous knowledge about natural products in food or in medicine, it is entirely likely that the 'designer gene' technology of computerised evolution could eventually create a marketable and patented product from almost any natural substance. Remove the income support and land protective roles of cash crops that can be given in tandem with subsistence agriculture or forestry extraction, and one can see just how the politics of biotechnology could, if left unregulated for its sustainability implications, create a potentially devastating social and economic effect on impoverished peoples, many of whom are already environmentally bereft of subsistence.

Non-sustainability and environmental security
In 1988 there were 22 wars in progress, all of which passed the definition of being associated with over 1000 battlefield deaths. War disrupts the prospects for sustainability in three ways. First, it creates damage, often wanton damage, of resources and livelihoods. This is self-evident, but the suffering can be immense, and may lead to massive translocations of desperate peoples into marginal lands. Such is the case in Sudan, Ethiopia and Somalia, with as many as 6 million people on the move across national borders and a further 9 million as a result of armed conflict internally (for a summary, see *The Guardian*, 14 June 1991, 21–5).

Second, wars absorb precious capital in armaments and military pay and divert that money and manpower from sustainable activities. As much as one-third of foreign earnings may be sequestered in the military of a poor developing country, with much of its able-bodied elite being seconded to military activity rather than restoring damaged ecosystems. As yet it is difficult to prove that conflicts over resource availability lead to wars, though it is arguable that the recent Gulf conflict was in part an oil war, and that some at least of current Middle East tensions are a result of disputes over water (see Clarke, 1991). Renner (1989, 30) provided Table 2.5, which shows the potential flashpoints over possible water conflict.

Third, modern wars may actually use ecological catastrophe as part of its weaponry of destruction. The practice is ancient, but has been brought to a new level of sophistication by the use of defoliant chemicals in Vietnam and Nicaragua, and by the spilling and burning of oil in the Arabian Gulf and Kuwait. No longer is it possible to avoid the use of ecological destruction as part of the trappings of contemporary conflict. The implications for sustainability are potentially severe, especially because of the incidental effect of ecological refugees.

The ecological-economic devastation of deliberate destruction is not confined to any one nation or ideology. Tragically, it appears universal. Consider the effect of the Soviet invasion of Afghanistan in the 1980s (Rupert 1989, 772).

the Soviets' technological rape was primarily accomplished with mines, perhaps a million of which remain buried or scattered in the countryside. Simply making the land safe to walk on will take many years and lives, not to mention foreign assistance. Considerable resources will likewise be required to rebuild the numerous villages' irrigation networks that were destroyed by the Soviets as part of their policy of depopulating strategic areas. Many of these fragile irrigation networks have taken centuries to build, and allowed Afghans to use marginal lands that were once, and are again, unsuitable for food production.

Table 2.5   Unresolved international water issues, mid-1980

Rivers	Countries involved in dispute	Subject of dispute
Nile	Egypt, Ethiopia, Sudan	Siltation flooding, water flow/diversion
Euphrates, Tigris	Iraq, Syria, Turkey	Reduced water flow, salination (constraints on irrigation & hydropower)
Jordan, Yarmuk, Litani, West Bank aquifer	Israel, Jordan, Syria, Lebanon	Water flow/diversion
Indus, Sutlei	India, Pakistan	Irrigation
Brahmaputra, Ganges	Bangladesh, India	Siltation, flooding, water flow
Salween/Nu Jiang	Burma, China	Siltation, flooding
Mekong	Kampuchea, Laos, Thailand, Vietnam	Water flow, flooding
Paraná	Argentina, Brazil	Dam, land inundation
Lauca	Bolivia, Chile	Dam, salination
Rio Grande, Colorado	Mexico, United States	Salination, water flow agrochemical pollution
Great Lakes	Canada, United States	Water diversion
Rhine	France, Netherlands, West Germany, Switzerland	Industrial pollution
Maas, Schedle	Belgium, Netherlands	Salination, industrial pollution
Elbe	Czechoslovakia, East and West Germany	Industrial pollution
Werra/Weser	East and West Germany	Industrial pollution
Szamos	Hungary, Romania	Industrial pollution

Sources (as cited by Renner 1989, 32): *Environment and Conflict*, Earthscan Briefing Document 40 (International Institute for Environment and Development, London, 1984); Norman Myers, *Not Far Afield: U.S. Interests and the Global Environment* (Washington, D.C.: World Resources Institute, 1988); Joyce R. Starr and Daniel C. Stoll (eds.), *The Politics of Scarcity: Water in the Middle East* (Boulder and London: Westview Press, 1988); P.C. Mayer-Tasch, *Die Verseuchte Landkarte* (Munich: C.H. Beck, 1987); K. Bouwer, Geographische Aspekte grenzüberschreitender Umweltprobleme, in Kurt Tudyka (ed.), *Umweltpolitik in Ost- und Westeuropa* (Opladen: Leske & Budrich, 1988); Where Dams Can Cause Wars, *Economist*, 18 July 1987; and other sources.

In a similar vein, military governments suppress any non-governmental organisations that could protect workers and the general population from environmental abuse. Thus trade unions virtually disappeared in Nicaragua in the 1970s, enabling all kinds of mischief over the illegal use of pesticides (Karliner 1989, 792–3).

The link between military activity and non-sustainability is apparent but not easily proved. Like the alleged connection between population growth and lack of access to basic survival needs such as water, fuelwood and fertile land, requiring children as beasts of burden and vital agricultural or slave labour, non-sustainability cannot easily be pinned down as creating the very causes of its continuation. It is difficult not to draw the conclusion that only by transforming swords into sustainable ploughshares, military personnel into green developers, and development assistance into survivable livelihoods for impoverished people, can the vicious spiral be broken and sustainable development begin its path to salvation.

The critical paradox remains that the concept of sustainable development is more encompassing than the tools it has at its command to promote its cause. The scale of rethinking, restructuring and depoliticising institutions, economies, peoples and international organisations is formidable. One way forward is to embark on a series of sustainable livelihood experiments, attempted in various continents and countries, to see just how possible it is to move towards a way of living that approximates equilibrium in a politically, culturally and institutionally acceptable manner. To make such experiments successful requires some kind of international backing launched through UNCED, and financed by the UN or one of its leading institutions, coupled with the international non-governmental organisations (notably the IUCN) and the World Bank.

This proposal may not be possible to implement in time, but one bold research move would be to establish a network of experimental projects in both physical and cultural sustainability in, ideally, every continent. The point here is that no two schemes would be alike, but each would share three common objectives.

(i)   To explore the scope for identifying and planning a coordinated approach to comprehensive physical, social and economic agro-development, coupling agricultural sustainability to programmes of fuelwood and other energy provision, health care, family planning, small-scale job creation and pollution control.

(ii) To try out a series of experimental approaches in appropriate resource pricing, appropriate technology, socially acceptable community development skills, and training programmes to discover what are the impediments, what might form the ways of overcoming these impediments, and what might be the markers to evaluating success. The impediments may be structural (because of the global considerations already mentioned). They may be organisational, or administrative, locked in the separation of agency powers and the inability of regional or national governments to coordinate policy across administrative agencies, or they may be attitudinal, a function of power relationships, custom, or other social factors that imprison knowledge and communication in the programme.

(iii)   To devise a variety of measures for launching and maintaining such schemes. This may be through existing or newly created non-governmental organisations, involving economy (e.g., communal credit), ecology (e.g., common land holding trusts), society (e.g., rural women's cooperatives), or politics (e.g., new political parties aimed at promoting cultural sustainability). Another vehicle might be non-governmental aid channelled either through the major rural development charities or through an international agency such as the IUCN. This body is intergovernmental but non-governmental. It is committed to the interests of people as well as the land. It is embarking on a new programme of sustainable agricultural development and it has a network of conservation officers on the ground who could help coordinate the programme.

Such schemes should start small, they should be based on established success, and they should explicitly involve the local people in ways that are familiar and beckoning. It may be necessary to begin with a few schemes in one country, then build on good experience. The main point is to coordinate effort, bridging the field sciences and local experience, so as to provide a viable learning laboratory.

Training and evaluation programmes   None of these proposals will be possible unless the whole effort is backed up by substantial investment in appropriate training, coupled with a wider and more practical educational programme for children, linking the health of the land to the heart of the person and the family. Again research is needed to find ways of ensuring that local environmental knowledge is built upon, not lost, and that the combined knowledge of tradition and reform are coupled via the experimental networks. Once again international aid capital needs to be channelled through suitable institutions to allow this to be explored in as free a way as possible.

The politics of UNCED
In the run-up to the Rio Conference, it was always thought that the key

players would be those economies with sufficient surplus, or capacity for surplus, as to be the leading contributors to a global environmental trust fund, or some equivalent. The obvious candidates were the USA, Japan, Germany and to some extent Saudi Arabia.

In fact this is not the focal point of post-Rio environmental politics. The new Clinton administration in the USA gives the impression of wishing to put its own environmental house in order before turning to the rest of the world. In addition, it has let it be known that investment in environmentally friendly technology, supported by tax credits, will be the prime concern, as this should create new jobs and industrial opportunities both at home and abroad. The Japanese, despite brave promises of spare yen for environmental well-being, have actually come up with very little – less than $2 billion over ten years – and are beleaguered by economic crises and political scandal. The Germans are unexpectedly hampered by the costly reconstruction of the beleaguered eastern provinces, and by a crisis over the non-acceptance of an open immigration policy, again most especially in the impoverished east. Saudi Arabia is deeply concerned about some form of carbon taxation, and the prospects that oil demand will be controlled by measures to reduce CO_2 emissions. There is no particular love for sustainable development in Saudi Arabia.

The post-Rio picture is still confused. The original expectations that 179 countries would somehow miraculously embrace the notion of sustainable development were hopelessly naive, despite all the promises following the publication of the Brundtland Report. Rio showed that international diplomacy is inadequately structured to handle the crisscrossing complexities of linking the 'big' issues of development and environment across so many countries and cultures. Yet the modest success of the framework conventions on climate change and biodiversity were no mean achievements in international legal diplomacy. The aim was to get agreement in principle on key concepts of acceptable international action via the utilisation of 'soft law'. This is the process by which promises, declarations and broad targets can be steadily shaped into committing national action by a mechanism of moral and political exhortation and example rather than by the threatening legalities of 'hard law'. Both conventions contain principles such as ecological tolerance, common but differentiated responsibility, precaution and burden sharing that tie to arrangements aimed at ensuring good scientific assessments and transfers of resources and technical assistance. These novel arrangements will become clearer and more binding as negotiations evolve and trust develops between parties that at present are very suspicious.

In the post-Rio age there are no specific key indicators of success or failure. Granted, nothing like enough money was promised to assist the developing world to limit damage that is in everyone's interest to control. Probably only about a tenth of what is required was actually promised. Even then, virtually no money has no strings attached. So in the foreseeable future something alarmingly close to the status quo remains. At the same time, deep suspicion remains embedded between both the North and the South towards each other's motives. The richer countries of the North are not yet entirely convinced of the true extent of the economic negatives associated with non-sustainable development. In addition, they cannot readily identify ways in which money and know-how can be transferred without some form of corruption or mischief intervening between intent and practice.

The impoverished nations of the South remain very dubious about the full meaning of sustainable development when they have never tasted wealth, and when their elites know that to achieve such an objective must necessarily mean that they give up some of their privileges and opulence in favour of groups whom they believe cannot be trusted not to become highly aggressive and dangerous when offered a larger slice of the cake. Likewise at a national level, there is very great suspicion that the demand for biodiversity and lower emissions of greenhouse gases and ozone-depleting substances will be made at the expense of any hope for them to gain access to any of the resources or technology that should be transferred to them to allow for legitimate changes in property rights and sovereign choices over patterns of technological advance.

What can be said at least is that these notions were at least aired and debated in Rio, and will continue to be the focus of attention in the many post-Rio gatherings to come. Obviously one target of attention will be the role and exuberance of the new Sustainable Development Commission set up by the Rio participants to receive proposals for sustainable development by all countries, to set guidelines for good practice, and to monitor performance. It is far too early to double-guess the likely success of this body. Much depends on the emergence of the UN generally as a global coordinator and interventionist body in the cause of humanitarianism and environmental well-being. So far the signs are promising. In a post-cold war era, when the old stabilities have been replaced by much more ethnic and interregional conflict with no established order of control via military or economic 'blocks' of power, the UN has begun to show that it may be able to find acceptable forms of intervention and negotiating structures that promote the cause of humanitarianism and human dignity. These are desperately early days, so it would be foolhardy to be too optimistic. Nevertheless, if these trends do continue, then it is possible that the new Sustainable Development Commission can be set in a wider pattern of international diplomacy that should give it real influence.

Obviously, for the Commission to succeed, there must be a will to promote the cause of sustainable development, backed up by mechanisms and legal-diplomatic institutions. Here is where Agenda 21 fits in. Agenda 21 is the list of activities that ought to be followed to give sustainable development a sporting chance across the world. At the time of writing, it is little more than a list of good intentions and estimated budgets with no commitments. No country is prepared to take the full extent of Agenda 21 very seriously. So for it to succeed, not only will the Sustainable Development Commission have to be given teeth, but also the national plans for sustainable development will have to include measures to contribute to poorer countries to help them switch to a tolerable path of equitable social and economic transition.

One device for assisting this process is the Global Environment Facility (GEF), established by the World Bank in conjunction with the UN Development Programme and the UN Environment Programme. There is deep suspicion about the connection between the GEF and the World Bank among non-governmental organisations and the 'group of 77' nations of the South. Again it would be dangerous to jump to conclusions. Like it or not, the GEF is 'the only game in town' through which major international cash and technical transfers can take place in the cause of environmental well-being. The alternative is bilateral aid, with all its complications of tied commitments and national security implications. GEF, at least in principle, can bypass such impediments and channel resources directly into schemes for safeguarding the global commons via investments in CO_2 offsets, water management schemes, deforestation alternatives and sustainable development projects. Again the signs are promising that GEF is more 'good' than 'bad', but the scope for it to fail to meet its highly adventurous objectives is very considerable. So one must be prepared for much acrimony in the foreseeable future as the GEF gathers experience and seeks to stand its ground over the forces of manipulation and conception. Of vital importance is a regular independent audit of GEF's activities and investments. Until that is provided and published via the new Commission there will be no risk from the critique that the GEF is a rich nations' plaything to buy off the charges of inaction over major and significant international resource transfers (Jordan 1993).

Another post-UNCED issue is that of ensuring meaningful self-reliance for all nations, but especially the developing nations. This means decorrupting governments, guaranteeing civil rights, providing access to basic needs, and reducing debt. This is a very tall order, especially if such demands are imposed on Third World governments by nations whose past environmental and civil rights records are not exemplary, and who seek to use the threat of environmental conditionality as a means of assuring their own survival under the guise of promoting global sustainability. One cannot see this matter being resolved easily and without pain.

Waiting in the wings are even more radical proposals to change UN institutions so as to ensure fair treatment of all nations and adequate sanctions against those who violate an earth charter. Proposals include the extension of the existing Trusteeship Council to a Trusteeship Council for the Planet as an adjunct to the Security Council (Tickell 1990, 177). Another idea is to create an International Environmental Protection Agency, along the lines of the International Atomic Energy Agency, to oversee scientific information, provide basic rules of conduct, supervise training and generate codes of conduct for national and bilateral environmental regulatory agencies. A third proposition is to create some form of Environmental Security and Cooperation Council to ensure that countries that do not comply with international conventions are forced to do so in the interests of global well-being. This would have to be backed by reciprocal arrangements of technology transfer and training schemes to enable countries to comply through appropriate means.

Finally, there is a suggestion for an Office of Environmental Administration in the Bureau of the Secretary General (Brown-Weiss, 1989). This would be a form of global environmental ombudsman with powers to investigate the failure of a country to meet its obligations to sustainable development under any UN earth charter or convention. The notification of failure to comply could come from an aggrieved citizen or group. The very fact that such an arrangement existed should encourage most nations to try to play by the rules.

All of these ideas require considerable thought and refinement. Reforming the UN is sometimes counterproductive in its own right, so one should be very careful about suggesting changes simply because they appear fashionable. There is no doubt, however, that global problems will require global solutions, and that existing patterns of diplomacy, the law, economics and administration will have continually to evolve to deal with the challenge as it unfolds.

Sustainable development may be a chimera. It may mark all kinds of contradictions. It may be ambiguously interpreted by all manner of people for all manner of reasons. But as an ideal it is nowadays as persistent a political concept as are democracy, justice and liberty. Indeed, it cannot be disconnected from these three other ideals. If the prospect of an earth uninhabitable for many millions of people does not encourage the fusion of these great verities of human existence, this speaks much for the failure of the human family to manage its earthly household.

Bibliography

Note: This is a longer reference list than just the compilation of citations in the text. This is deliberate to help the reader look at sources that widen the debate on sustainable development in many interesting ways.

Abel, N. and Blaikie, P.M. 1986. Elephants, people, parks and development: the case of the Luangwa Valley Zambia. *Environmental Management* 10(6), 735–51.

Adams, J.S. 1991. *Determined to Dig: The Role of Aggregates Demand Forecasting in National Minerals Planning Guidance.* Council for the Protection of Rural England.

Adams, T., Carruthes, J., and Hamil, S. 1991. *Changing Corporate Values.* New Consumer, London.

Adams, W.M. 1990. *Green Development: Environment and Sustainability in the Third World.* Routledge, London.

Ahmad, T.J., El Serafy, S. and Lutz, E. 1989. *Environmental Accounting for Sustainable Development.* World Bank, Washington, DC.

Anderson, D.M. and Grove, A.H. (eds) 1987. *Conservation in Africa: People, Policies and Practice.* Cambridge University Press, Cambridge.

Angell, D.J.R., Comer, J.D. and Wilkinson, M.L.N. (eds) 1990. *Sustaining Earth: Response to Environmental Threats.* Macmillan, Basingstoke.

Archibugi, F. and Nijkamp, P. (eds) 1989. *Economy and Ecology: Towards Sustainable Development.* Kluwer Academic, Dordrecht, Netherlands.

Barraclough, S. 1991. *The Social Origins of Poverty and Food Strategies.* Zed Books, London.

Batisse, M. 1982. The biosphere reserve: a tool for environmental conservation and management. *Environmental Conservation* 9, 101–11.

Biot, Y. 1988. Modelling productivity losses by erosion. School of Development Studies, University of East Anglia, Norwich.

Blaikie, P.M. 1989. Environment and access to resources in Africa. *Africa* 51(9), 18–40.

Blaikie, P. and Brookfield, H. 1987. *Land Degradation and Society.* Methuen, London.

Brookfield, H. 1991. Environmental sustainability with development: what prospects for a research agenda? *European Journal of Development Research* 3, 41–67.

Brown-Weiss, E. 1989. *In Fairness to Future Generations: International Law, Common Patrimony and Intergenerational Equity.* United Nations University and Transnational Publishers, New York.

Brundtland, G.H. (chair) 1987. *Our Common Future.* Report of the World Commission on Environment and Development. Oxford University Press, Oxford.

Business and Society Review, 1990. *Greening or Preening?* No.75, 96 pp.

Burton, I. and Kates, R.W. (eds) 1965. *Readings in Resource Management and Conservation.* University of Chicago Press, Chicago.

Ciriacy-Wantrup, S.V. 1963. *Resource Conservation*. University of California Press, Berkeley.

Clark, W.C. and Munn, R.E. (eds) 1986. *Sustainable Development of the Biosphere*. Cambridge University Press, Cambridge.

Clarke. R. 1991. *Water: The International Crisis*. Earthscan, London.

Cheru, F. 1990. *The Silent Revolution in Africa: Debt, Development and Democracy*. Zed Books, London.

Coomer, J.C. 1979. The nature and quest for a sustainable society. In Coomer, J.C. (ed.), *Quest for a Sustainable Society*. Pergamon Press, New York, 7–14.

de la Court, T. 1990. *Beyond Brundtland: Green Development in the 1990s*. Zed Books, London.

Elkington, J., Knight, P. and Hales, J. 1991. *The Green Business Guide*. Victor Gollancz, London.

Environmental Data Services 1991. *Demand Forecasts for Aggregates: Nightmare Scenarios or Symbol of Prosperity?* Report 196. ENDS, London, pp 14–18.

Engel, J.R. and Engel, J.G. (eds) 1989. *Ethics of Environment and Development*. Belhaven Press, London.

Food and Agriculture Organisation 1986. *African Agriculture: The Next 25 Years*. FAO, Rome.

Frankel, C. 1976. The rights of nature. In Tribe, L.H., Schelling, L.S. and Voss, J. (eds), *When Values Conflict*. Wiley, New York, 93–114.

Galtung, J. 1984. Perspectives on environmental politics in overdeveloped and underdeveloped countries. In B. Glaeser (ed.), *Ecodevelopment: Concepts, Projects, Strategies*. Pergamon Press, Oxford, 9–21.

Ghabbour, S.I. 1982. Definitions, issues and perspectives. In UN Environmental Programme, *Basic Needs in an Arab Region: Environmental Aspects Technologies and Policies*. UNEP, Nairobi, 19–46.

Glaeser, B. (ed.) 1984. *Ecodevelopment: Concepts, Projects, Strategies*. Pergamon Press, Oxford.

Grove, P. 1990. Threatened islands, threatened earth: early professional science and the historical origins of global environmental concerns. In Angell, D.J.R., Coomer, J.D. and Wilkinson, M.L.N. (eds), *Sustaining Earth: Response to Environmental Threats*. Macmillan, Basingstoke, 15–32.

Hammond, A., Rodenberg, E. and Moormaw, W.R. 1991. Calculating national accountability for climate change. *Environment* 33(1), 10–15, 33–5.

Harrison, P. 1987. *The Greening of Africa*. Penguin Books, Harmondsworth, Middlesex.

Hays, S.P. 1959. *Conservation and the Gospel of Efficiency*. Harvard University Press, Cambridge, MA.

Hillaby, J. 1961. Conservation in Africa: a crucial conference. *New Scientist* 31 August, 536–8.

Hughes, D. 1983. Gaia: an ancient view of our planet. *The Ecologist* 13, 2–3.

Holmberg, J., Bass, S. and Timberlake, L. 1991. *Defending the Future: A Guide to Sustainable Development*. International Institute for Environment and Development, London.

Holdgate, M. 1991. The environment of tomorrow. *Environment* 33, 14–20; 40–4.

International Union for the Conservation of Nature, 1980. *World Conservation Strategy.* IUCN, Gland, Switzerland.

International Union for the Conservation of Nature, 1989. *World Conservation Strategy for the 1990s.* First Draft Report. IUCN, Gland, Switzerland.

International Union for the Conservation of Nature, 1991. *Caring for the Earth: A Strategy for Sustainable Living.* IUCN, Gland, Switzerland.

Jazairy, I. 1987. How to make Africa self-sufficient in food. *Journal of the Society of International Development* 2(3), 50–4.

Jordan, A. 1993. *The Global Environment Facility,* Working Paper, GEC 92-37. Centre for Social and Economic Research on the Global Environment, University of East Anglia, Norwich.

Karliner, T. 1989. Central America's other war. *World Policy Journal* 4(4), 788–810.

Keyfitz, N. 1991. Population growth can prevent the development that would slow population growth. In Matthews, J.T. (ed.), *Preserving the Global Environment: The Challenge of Shared Leadership.* Norton, New York, 39–77.

Lovdiyi, D., Nagle, W. and Ofosu-Amauh, 1989. *The African Women's Assembly: Women and Sustainable Development.* World Wide, Washington, DC.

Lovelock, J. 1987. *The Ages of Gaia.* Oxford University Press, Oxford.

Markandya, A. and Pearce, D.W. 1988. Natural environments and the social rate of discount. *Project Appraisal* 3(1), 2–12.

McCormick, J. 1989. *The Global Environmental Movement.* Belhaven Press, London.

McHale, J. and McHale, J.C. 1979. *Basic Human Needs: A Framework for Action.* Transaction Books, New Brunswick, NJ.

Myers, N. 1989. Environment and security. *Foreign Affairs* 74, 23–41.

Nash, R. (ed.) 1968. *Americans and Environment.* DC Heath, Lexington, MA.

OECD 1991. *The State of the Environment Report.* Organisation for Economic Cooperative and Development, Paris.

O'Riordan, T. 1988. The politics of sustainability. In R.K. Turner (ed.), *Sustainable Environmental Management: Principles and Practice.* Belhaven Press, London, 29–50.

Omari, C.K. 1989. Traditional African land ethics. In J.R. Engel and S.G. Engel (eds), *Ethics of Environment and Development.* Belhaven Press, London, 168–75.

Pearce, D.W. and Turner, R.K. 1990. *Economics of Natural Resources and the Environment.* Harvester Wheatsheaf, Hemel Hempstead.

Pearce, D.W., Markandya, A. and Barbier, E. 1989. *Blueprint for a Green Economy.* Earthscan, London.

Pearce, D.W., Barbier, E. and Markandya, A. 1990. *Sustainable Development: Economics and Environment in the Third World.* Earthscan, London.

Pezzey, J. 1989. *Definitions of Sustainability,* Working Paper No.9. UK Centre for Environment and Development, London.

Pinchot, G. 1910. *The Fight for Conservation.* Garden City, New York.

Rau, W. 1991. *From Feast to Famine: Official Cures and Grassroots Remedies to Africa's Food Crisis.* Zed Books, London.

Redclift, M. 1987. *Sustainable Development: Exploring the Contradictions*. Methuen, London.

Reij, C. 1990. Strategies for initiating sustainable moisture conservation programs in semi-arid West Africa. In Trolldalen, J.M. *Professional Development Workshop on Dryland Management*, Environment Working Paper No.33. World Bank, Washington, DC.

Renner, M. 1989. *National Security: The Economic and Environmental Dimensions*. Worldwatch Paper No.89. Worldwatch Institute Washington, DC.

Repetto, R. 1989. *Losing Ground: A Study of Soil Erosion in Indonesia*. World Resources Institute, Washington, DC.

Repetto, R. and Gillis, M. (eds) 1988. *Public Policies and the Misuse of Forest Resources*. Cambridge University Press, Cambridge, UK.

Riddell, R. 1981. *Ecodevelopment*. Gower Publishing, Aldershot.

Rupert, J. 1989. Afghanistan's slide towards civil war. *World Policy Journal* 4(4), 759–85.

Sachs, I. 1979. Ecodevelopment: a definition. *Ambio* 8(2/3), 113.

Schumacher, E. 1989. *Small is Beautiful: Economics as if People Really Mattered*. Harper and Row, New York.

Scientific American, 1989. *Managing Planet Earth*. September issue.

Sen, A., and Dreze, J. 1990. *Hunger and Public Action*. Clarendon Press, Oxford.

Shiva, V. 1989. *Staying Alive: Women Ecology and Development*. Zed Books, London.

Shiva, V. 1991. *The Violence of the Green Revolution: Ecological Degradation and Political Conflict*. Zed Books, London.

Stocking, M.A. 1987. Measuring land degradation. In P.M. Blaikie and H.C. Brookfield (eds), *Land Degradation and Society*. Methuen, London, 49–63.

Svard, R.L. 1989. *World Military and Social Expenditures 1989*. World Priorities Inc., Washington, DC.

Tickell, C. 1990. Diplomacy and sustainable development. In Angell, D.J.R., Coomer, J.D. and Wilkinson, M.L.N. (eds), *Sustaining Earth: Response to Environmental Threats*. Macmillan, Basingstoke, 172–80.

Timberlake, L. and Thomas, L. 1990. *When the Bough Breaks: The Plight of Children in the Third World*. Earthscan, London.

Tonn, B.E. 1988. Philosophical aspects of 500 year planning. *Environment and Planning A* 20, 1064–72.

Turner, R.K. (ed.) 1988. *Sustainable Environmental Management: Principles and Practice*. Belhaven Press, London.

Verhelst, T. 1990. *No Life Without Roots: Culture and Development*. Zed Books, London.

World Bank 1989. *World Development Report*. Oxford University Press, Oxford.

World Resources Institute 1990. *World Resources 1990/91*. Oxford University Press, Oxford.

World Resources Institute 1990. *World Resources 1991/92*. Oxford University Press, Oxford.

Chapter 3

Sustainable Development and Developing Country Economies

David Pearce

Two environmental revolutions

In historical terms, the science of economics has always had something to say about the relationship between economic welfare and the stock of natural assets.[1] It is only in the last twenty years, however, that the building blocks for a comprehensive theory of environment and economic development have emerged. It is hardly surprising, therefore, to find the theory imperfectly developed. The empirical analysis is correspondingly also underdeveloped. None the less, much has been learned from the economics that emerged in the context of the 'first environmental revolution' of the late 1960s and early 1970s – a revolution characterised best by the debate about *environmental quality versus economic growth*. The second revolution of the late 1980s and early 1990s has revisited many of the original concepts and arguments in the context of *sustainable development*. At their most basic, the differences between the two periods can be summarised in the responses to five broad questions.

Are environment and growth complementary or in conflict?

• Many of the 1970s participants viewed growth and environment as incompatible.
• In the 1980s the pendulum moved towards the view that growth and environment are *potentially* compatible.

The environmentalists' prescription for policy in the 1970s was that growth of the economy and of population levels had to be constrained in environmental quality was to be preserved. The most famous publication advocating this view was *Limits to Growth*.[2] The rationale for preserving environments was that even a 'stationary-state' level of

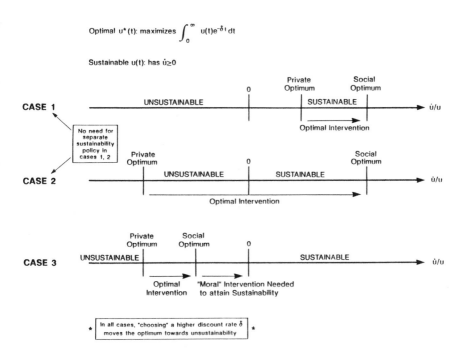

Figure 3.1 Sustainability, optimality and government intervention

Source: adapted from J. Pezzey, *Economic Analysis of Sustainable Growth and Sustainable Development*, World Bank Environment Department Working Paper, No. 15. World Bank, Washington, DC, March 1989.

economic activity could not be sustained unless the environment was conserved. Additionally, there was a moral case for conservation, based on a concern for other living species.

The 'limits to growth' approach was extensively analysed by economists in the 1970s. The main body of the professional environmental economics literature was devoted to economic growth models in which 'resources' were an extra factor of production, along with labour and capital, and sometimes also a source of direct welfare from amenity. This literature was generally concerned with finding the best or *optimal* path of economic growth in the context of assumed fixed stocks of exhaustible resources and stocks of renewable resources.[3] Optimal growth in this literature is typically defined to mean that growth which maximises the 'present value' of future streams of consumption, i.e. the discounted value of future flows of consumption per head.[4] One important finding of this literature is that such optimality may be consistent with *non-*

sustainable growth paths, and this is more likely the higher is the rate at which people discount the future.

The growth models showed that sustainability – interpreted as constant or increasing per-capita welfare levels over time – *could* be achieved through suitable interventions which reduce the rate at which natural assets are depleted. The essence of the interventions is that they should 'drive a wedge' between economic activity and its impacts on the environment (Figure 3.1). Put another way, efforts have to be made to 'decouple' economic growth from environmental impact. The experience of the recent decades shows that decoupling is possible as technological change occurs, as substitution between resources takes place, and as higher real prices for polluting goods lead to their conservation and substitution. It is this finding which underscores the need to design *incentives* to switch from dirty to clean technologies and from polluting to non-polluting products.

What is the value of the environment?

- The 1970s environmentalists assumed that environmental quality mattered, and used some scientific evidence to support the case, but they did not demonstrate the *economic importance* of environmental quality.[5]
- The 1980s have seen the rapid development of techniques and practice in the measurement of economic damage and benefit from environmental change.

In both periods, environmental quality has been regarded as important both in itself as a direct source of human welfare, and as an 'input' to economic activity designed to raise human welfare. It seems fair to say that recent environmental economics has devoted more attention to demonstrating the importance of environment through *valuation* exercises, i.e. by attempting to show that environmental degradation involves not just direct losses of welfare (e.g., through loss of amenity), but also indirect losses through impacts on health and productivity.

Sustainability means that per-capita welfare increases (or is at least constant) over time. Optimality means that the growth path maximises the 'present value' of the future flows of welfare. The concept of present value takes account of the discounting of future welfare gains: they are regarded as less important the further into the future they occur. Private and social optima diverge: the most desirable rate at which to deplete resources from the standpoint of the resource owner is unlikely to be the best rate for society as a whole. In case 1 private and social optima are both sustainable. In case 2 the private optimum is unsustainable so that intervention is needed for *both* sustainability and social optimality. In case 3 both the private and social optima are unsustainable. Achieving the social optimum still does not secure sustainability.

Is environment important in the developing world?

- Most of the 1970s debate about natural environments was confined to the problems of the developed world.
- The 1980s debate expanded to embrace both the developing world and problems of the 'global commons' such as ozone layer depletion and global warming.

These propositions are simplifications, of course. The 1970s witnessed a major debate about the development process in the Third World, but it tended to focus less on natural environments than on the traditional need to reduce poverty through alternative development paths (e.g., basic needs versus the rapid industrialization and export-led growth approaches).[6] In contrast, the 1980s have extended the focus to embrace developing country issues and global problems.

Are we running out of resources?

Many of the 'limits' to growth in the 1970s were presented in terms of 'running out' of exhaustible resources. Fairly simple indicators of exhaustion were used, based on some measure of the consumption rate of the resource relative to its stock. Thus *Limits to Growth* estimated that the remaining 'life' of aluminium, given existing estimates of reserves, was 31 years from the year of estimation, 1972. World gold reserves would last only 9 years (and hence gold production would have stopped in 1981). Longer lifetimes are secured if larger estimates of reserves are made. The most notable exhaustible resource is, of course, fossil energy: coal, oil and gas. The 'energy crisis' of the 1970s, brought on by the substantial price hikes in OPEC oil, tended to give the discussion of resource exhaustion a politically relevant basis, even though the crisis was itself one of price, not physical availability of resources.

In the 1980s the emphasis switched from exhaustible resources to renewable resources. First, it became apparent that 'renewability' did not mean that such resources *would* renew themselves. That depended critically on the actual management regime in place. The experience of ocean fishing and whaling was more than ample to show that overexploitation of a renewable resource could take place. The theoretical literature had also drawn attention to the conditions likely to give rise to exhaustion of a renewable resource: high discounting of the future, a high ratio of resource price to cost of harvesting, and the extent to which the management regime limits access to the resource, the greatest risk being when access is not controlled at all.

Second, the professional literature had tended to focus on selected resources only – fisheries and temperate zone forests were particularly well studied. As scientific knowledge grew, however, and as the ecologists' perspective on things began to permeate environmental economics,[7] it was recognised that ecosystem functions in general behave as renewable resources. Thus the carbon cycle operates as a balanced system in which carbon is emitted from living things and is absorbed by carbon 'sinks', notably oceans and forests. The rate of use of fossil fuels may not occasion much concern from the point of view of fossil fuel scarcity, but the fact that such use generates carbon and methane emissions which act as greenhouse gases, and sulphur and nitrogen oxide emissions which cause 'acid rain', does give rise for concern. What matters is the constraint that 'bites' first, and in the case of fossil fuels it is the limited capacity of the *waste receiving* environments which is likely to be more significant.

Third, as the developing country perspective emerged, so it quickly became apparent that it is the renewable resources that matter most for the immediate livelihoods of most of the world's poor. If water, biomass (trees, crop residues, grass cover) and soil (a mix of exhaustible and renewable characteristics) are overexploited the implication for human welfare can be formidable.

What is the focus of environmental concern?

- The 1970s exhibited a marked concern for global environmental threats, but
- the 'global focus' expanded in the 1980s.

Publications such as *Limits to Growth* were concerned with threats to global well-being, even survival. In the 1980s, however, global concerns grew as increased evidence emerged about the 'internationalisation' of environmental issues. The major development was the recognition that many resources are 'shared' either directly in a physical sense, or as a shared value. Shared physical resources would include the European-Scandinavian airshed, river systems which serve several countries (e.g., the Nile serving Ethiopia, Sudan and Egypt), the oceans outside of exclusive economic zones, and the carbon cycle. By shared values is meant that a resource may be wholly located within one country, but its economic value 'resides' both in that country and elsewhere. Indeed, the share of total value residing outside the country of location may dominate, as many would argue is the case with tropical forests and with many endangered species.

In summary, the modern economic approach to the environment and development issue has four main features:

(i)   greater concern for long-run human welfare, in addition to the traditional concern with the poorest groups of society;
(ii) the extent to which environment has to be 'traded off' against economic growth as traditionally measured (positive changes in real income per capita);
(iii) the measurement and valuation of unmarketed environmental goods and environmental damage to inform the trade-off decision; and
(iv) the use of incentive schemes to 'decouple' economic growth and environmental impact.

Decoupling growth and environmental impact

Simple intuition suggests that sustained growth is possible without damaging the environment if growth can be 'decoupled' from environmental impact. This much is also suggested by the traditional economic growth models in which natural resources and environmental quality are allowed for. The ways in which decoupling can be achieved are considered shortly. But is decoupling possible? The experience of the developed world suggests that it is. Pollutants and economic activity are linked by what is known as the 'materials balance principle'. Basically, the first law of thermodynamics tells us that whatever is taken out of the environment in the form of minerals and energy must reappear somewhere else in the economic system. Matter and energy cannot be destroyed. All that happens is that their form is changed. They will appear as waste products and gases. It is also important to note that waste energy cannot be recycled, whereas waste materials can, up to a point. While this implies that any economic activity must always have *some* impact on the environment, since some waste materials and energy must always be produced, the decoupling argument is that the amount of waste per unit of economic activity can be reduced. Moreover, if it can be reduced at a faster rate than economic activity grows, the aggregate impact of waste on the environment can be reduced.

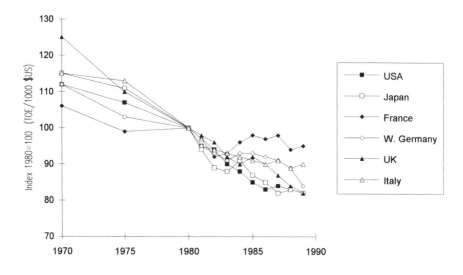

Figure 3.2   The decline in energy requirements per unit of GDP in OECD countries, 1970–87

Source: OECD

Figure 3.2 shows the relationship between gross domestic product (GDP) and the total primary energy consumed in the OECD countries from 1970 to 1987. The trend is systematically downward, meaning that a unit of economic activity in 1987 was using up less energy than in 1970. Specific rates of change for selected individual countries are:

USA	− 27%
France	− 13%
German	− 21%
Italy	− 17%
UK	− 30%

In part, the reductions have come about because of structural change towards less energy-intensive outputs, but they have also occurred because of *energy conservation* induced by real energy price changes. Thus:

Increases in energy prices, especially from 1978 to 1982, have been the most important single factor behind the substantial improvements in energy efficiency over the past ten years, the corresponding reductions in energy intensity and the slowing of energy demand growth.[8]

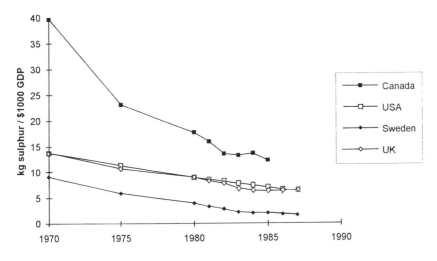

Figure 3.3 Ratio of sulphur oxide emissions to GDP in selected OECD countries, 1970–87

Source: OECD

Reduced energy consumption per unit GDP is environmentally beneficial because energy consumption in the developed world is linked to the emission of a number of air pollutants: sulphur oxides and nitrogen oxides (which contribute to 'acid rain'); carbon dioxide (which contributes to global warming); and particulate matter (which is linked to respiratory illness).

Figure 3.3 shows the relationship between GDP and sulphur oxide (SO_x) emissions for selected OECD countries. Again the trend is downward. This would be expected because of the changes in energy efficiency – SO_x is mainly produced by power stations and industrial boilers. But the change in the ratio also reflects regulatory policies within OECD countries, notably the UN Economic Commission for Europe's Convention on Long Range Transboundary Air Pollution, and the European Community's Large Combustion Plants Directive. These measures seek to reduce the amount of sulphur and nitrogen oxides emitted into the atmosphere.

The data on energy consumption and SO_x emissions suggest that 'decoupling' is possible. Of course, the picture is far more complicated than just analysing ratios of single pollutants to GDP. If GDP grows faster than the ratio declines, total emissions will increase. Reduction in one pollutant may also be secured by increasing another one. Thus,

Table 3.1   Percentage change in commercial energy consumption per unit of GNP, 1970–86

Kenya	− 43	Senegal	+ 23	Nigeria	+ 252
Sudan	− 47	Tunisia	+ 46	Somalia	+ 229
Tanzania	− 19	Zaire	+ 30	Cameroon	+ 118
Chile	− 14	Costa Rica	+ 10	Bolivia	+ 83
Uruguay	− 38	Ecuador	+ 45	Trinidad	+ 59
China	− 14	Turkey	+ 32	Nepal	+ 69
Philippines	− 23	Thailand	+ 39		

Source: World Resources Institute, *World Resources Report 1988/9*. Basic Books, New York, 1989, Table 20.1.

reduced dumping of sewage sludge at sea, for example, may increase pollution problems on the land where the sludge may be tipped, or in the air if it is incinerated. None the less, the evidence is sufficient to suggest that technological change can substantially alter the coefficients between real income and environmental impact. It is this technological change that holds out the greatest promise for sustainable development.

   How far is this finding relevant to the developing world? Pursuing the energy example, changes in *commercial* energy consumption (petroleum products, coal and gas) per unit of GNP over time show marked variability between countries (see Table 3.1). It is not possible to explain the changes in any simple fashion. In some cases, such as Sudan, commercial energy supplies have been constrained because of foreign exchange shortages. Other countries have launched fairly successful energy conservation campaigns: the Philippines and Thailand are examples.[9] In China, energy conservation has been advanced through schemes which set quotas of energy. Usage in excess of quotas can only be secured at higher prices. The central issue, however, is that reducing the ratio of energy consumption to GNP is feasible. Thus:

In most developing countries, the opportunities to reduce the energy bill are substantial. Economically, increasing the efficiency of energy use is normally more attractive than investing additional resources to increase the domestic supply. In addition, energy conservation and demand management can produce results faster than can measures to increase supply.[10]

The costs of environmental degradation

The developed world

Environmental economists have made major advances in recent years in terms of measuring the economic cost of environmental and resource degradation. At the *microeconomic* level, various techniques for 'monetising' the environmental damage or gain that ensues from economic activity have been developed. These have tended to be of three kinds. The first set of approaches uses *surrogate markets* – for example, valuing air pollution impacts according to the effects of air pollution on property values, valuing health hazards by looking at risk premiums in labour markets, and valuing recreational benefits through analysis of travel expenditures. The second set 'creates' a market by direct questioning of willingness to pay for benefits through interviews and questionnaires. This direct questioning approach has special use in the context of valuations of wilderness value or endangered species. The third set involves estimating physical 'dose-response' functions, for example between air pollution and human health, and then valuing the impact with available market prices or other unit values.[11]

Table 3.2   Pollution damage in the Netherlands

	Cumulative damage (1986 US$ billion)	Annual damage	Annual damage/GNP %
Air	1.2–3.0	0.5–0.8	
Water	n.a.	0.1–0.3	
Noise	0.5	0.1	
Total	1.7–3.5	0.7–1.2	0.5–0.8

Source: Adapted from J. Opschoor, A review of monetary estimates of benefits of environmental improvements in the Netherlands. OECD Workshop on the Benefits of Environmental Policy and Decision-Making, Avignon, October 1986.

Macroeconomic valuations of resource degradation and environmental loss are few and far between. Tables 3.2 and 3.3 show estimates for the Netherlands and Germany. Making due allowance for the speculative nature of some of the estimates of damage for the Netherlands and Germany, and for lack of comprehensiveness (e.g., the German data exclude water-based amenity affected by pollution), the analyses reveal some indicative policy conclusions.

First and foremost, if the German data, which are more comprehensive

Table 3.3   Pollution damage in Germany

	1983–5 US$ billions	% GNP
Air pollution		
health	0.8–1.9	
materials damage	0.8	
agriculture	0.1	
forestry losses	0.8–1.0	
forestry recreation	1.0–1.8	
other forestry	0.1–0.2	
disamenity	15.7	
(Total air pollution)	(19.3–21.5)	2.6–2.9
Water pollution		
fishing	0.1	
groundwater	2.9	
(Total water pollution)	(3.0)	0.4
Noise		
Workplace noise	1.1	
House price depreciation	9.8	
Other	0.7	
(Total noise)	(11.6)	1.6
Total	33.9–36.1	4.6–4.9

Source: W. Schulz, A survey on the status of research concerning the benefits of environmental policy in the Federal Republic of Germany. OECD Workshop on the Benefits of Environmental Policy and Decision-Making, Avignon, October 1986.

than the Dutch data, are approximately correct, there are significant net social benefits to be obtained by protecting the environment. On average, industrialised countries spend around 1.5–2.0% of the GNP on pollution control. If the return is close to 5% then the policy is more than justified. More detail on the cost–benefit comparison is considered below for the USA.

Second, air pollution damage tends to dominate both the Dutch and German estimates. This is as expected. As countries industrialise so polluting technologies are adopted. Agriculture diminishes in its share of GNP so that loss of agricultural output from environmental degradation could be expected to be small. The German data are also interesting in suggesting that noise nuisance is valued significantly, a finding in keeping

Table 3.4 The benefits of pollution control in the USA

	Benefits in 1978 ($ billion, 1978)	% of GNP
Air pollution		
health	17.0	
soiling	3.0	
vegetation	0.3	
materials	0.7	
property values	0.7	
(Total air pollution)	(21.7)	1.0
Water pollution		
recreational fishing	1.0	
boating	0.8	
swimming	0.5	
waterfowl hunting	0.1	
non-user benefits	0.6	
commercial fishing	0.4	
diversionary uses	1.4	
(Total water pollution)	(4.8)	0.2
Total	26.5	1.2

Source: A.M. Freeman, *Air and Water Pollution Control: A Benefit–Cost Assessment*. Wiley, New York, 1982.

with attitude surveys where noise is frequently cited as a dominant source of environmental disamenity.

Only one study has considered the balance of costs and benefits from *environmental policy*. An American study estimated the benefits from the US Clean Air Act of 1970 and the Federal Water Pollution Control Act of 1972.[12] Air pollution benefits were evaluated for 1970–8 and water pollution control benefits from 1972 to their expected level in 1985. Table 3.4 shows the results for a single year, 1978. Of the total benefits of $26.5 billion (1978 prices) some 80% came from air pollution control, and 20% from water pollution control. Noise nuisance was not evaluated. This result is broadly consistent with the damage analysis for the Netherlands and Germany. Some $24.3 billion of the $26.5 billion aggregate damage was 'utility-increasing', i.e. showed up in individuals' valuations of the benefits but not in measured GNP. This finding is significant in connection with the debate over the extent to which GNP measures 'capture' environmental losses and gains. The data for the USA suggest they do not. Only 8% of the environmental benefits reduced market costs or increased marketed output. Finally, when compared to the *costs* of pollution control, the data suggested that government policy to control *stationary source* air pollution had been cost-beneficial. Expenditures of some $9 billion per annum were producing benefits of perhaps $21 billion per annum. But for *mobile* source air pollution, expenditures of $7.6 billion per annum were achieving benefits of some $0.3 billion per annum, remarkably cost-inefficient.

The developed country environmental cost and benefit analyses suggest that national damage and benefit estimation is feasible, albeit with significant margins of uncertainty. But they also indicate clearly that the real damages lie, as one would expect, in the pollution problems associated with industrialization, and notably air pollution. The importance of noise pollution reflects the rapid growth of road and air travel, very much a function of income growth. The data do not suggest loss figures associated with loss of habitat, nor do they adequately capture 'non-user' costs, i.e. losses of welfare associated with the loss of species and habitat but unrelated to direct use of the assets in question.

The developing world

Africa

Assessing the costs of resource degradation and environmental pollution in the developing world is in its infancy. Yet, if the theme of the Brundtland Commission and others – that environment matters even in the narrow terms of measured economic growth – is to be substantiated, 'damage assessment' needs to be afforded high priority. Some examples of damage assessment for land degradation in Africa are given in Tables 3.5 and 3.6. Soil erosion effects on agricultural crops in Mali are estimated to cost perhaps 0.2% of GDP per annum, more if allowance is made for losses in subsequent years. Loss of biomass in general, arising from both man-made and climatic factors, could be costing Burkina Faso as much as 9% of its GDP, with crop productivity losses alone standing at perhaps 1.8% of GDP.

Soil erosion is endemic to many developing countries. Soil erodes 'naturally' but lack of investment in conservation, poor extension services, inability to raise credit and insecure land tenure all contribute to poor management of soils. A standard approach to estimating the costs of soil erosion is to estimate soil loss through the Universal Soil Loss Equation (USLE). The USLE estimates soil loss by relating it to rainfall erosivity, R; the 'erodibility' of soils, K; the slope of land, SL; a 'crop factor', C, which measures the ratio of soil loss under a given crop to that from bare soil; and conservation practice, P (so that 'no conservation' is measured as unity). The USLE is then:

$$\text{Soil loss} = R.K.SL.C.P$$

The next step is to link soil loss to crop productivity. In a study of soil loss effects in southern Mali, researchers applied the following equation to estimate the impact.

$$\text{Yield} = C^{-bx}$$

where C is the yield on newly cleared and hence uneroded land, b is a coefficient varying with crop and slope, and x is cumulative soil loss.

Finally, the resulting yield reductions need to be valued. A crude approach is simply to multiply the estimated crop loss by its market price if it is a cash crop. But the impact of yield changes on farm incomes will generally be more complex than this. For example, yield reductions would reduce the requirement for weeding and harvesting. The Mali study allowed for these effects by looking at the total impact on farm budgets with and without erosion.

The procedure described is an example of a 'dose-response', or 'production function' approach to valuation. The 'dose' is soil erosion, the 'response' is crop loss. Another approach would be to look at the costs of replacing the nutrients that are lost with soil erosion. Nutrient losses can be replaced with chemical fertilisers which have explicit market values. The replacement cost approach is helpful, but assumes that all soil loss is undesirable. Since we do not know if it is worth attempting to correct for all soil loss (indeed, it is very unlikely to be worthwhile) replacement cost approaches to valuation need to be used with caution.[13]

Table 3.5 shows the results for Mali using the two approaches. Because soil loss in any one year has effects in subsequent years the data show both an annual loss and a present value loss expressed as a loss in a single year. The authors draw several conclusions from their study:

(i)   Economic losses from soil erosion are high enough to warrant conservation investments in some areas in southern Mali.

(ii)   Investing in additional agricultural output may be less profitable than a simple financial appraisal would suggest. It is necessary to build into the analysis some estimate of expected soil erosion, and this will lower rates of return.

(iii)   Most importantly, it is necessary to ask *why* soil erosion occurs. The authors cite restrictions on access to informal credit and

Table 3.5   The on-site costs of soil erosion in Mali: Farm income losses due to soil erosion, 1988

Based on USLE and farm budgets: Nationwide annual agric. GDP income losses	US$4.6 million	= 0.2%	GDP = 0.6%
Discounted present agric. GDP	US$31.0 million	= 1.5%	GDP = 4.0%
Value of income loss Based on nutrient GDP replacement	US$7.4 million	= 0.4%	GDP = 1.0% Agric.

Source: J. Bishop and J. Allen, *The On-Site Costs of Soil Erosion in Mali*, Environment Department Working Paper No. 21. World Bank, Washington, DC, November 1989.

Table 3.6   Natural resource degradation in Burkina Faso: Crop, livestock and fuelwood losses

Zone	Fuelwood loss (m³)	Livestock loss (UGB*)	Cereal loss (tonnes)
Sahel	0	175,000	19,000
Plateau	900,000	26,000	260,000
Central Sudano-Guinean	1,200,000	0	27,360
Total	2,100,000	201,000	306,360
Prices (CFAF)	22,258	50,000/UGB	50,000/tonne
Values (CFAF billion)	46.7	10.0	15.3
Total damage cost = 72 billion CFAF = 8.8% of GDP			

Source: Adapted with corrections from D. Lallement, *Burkina Faso: Economic Issues in Renewable Natural Resource Management*. Agriculture Operations, Sahelian Department, Africa Region, World Bank, Washington DC, June 1990.
Note: *UGB = Unités de Gros Bétail, a standardised livestock unit measure.

insecure land tenure as important factors. High risks also contribute to high farmer discount rates: measures can be taken to reduce risks.

Where it is not possible to engage in detailed assessment of the costs of resource degradation it is still useful to obtain 'best guess' calculations in order to check that the issue is worth pursuing. Such an exercise has been carried out for Burkina Faso. Estimates were made of the total amount of biomass lost each year in the form of fuelwood and vegetation. The resulting losses show up as forgone household energy (fuelwood) which can be valued at market prices for fuelwood; forgone millet and sorghum crops which can be valued at market prices; and reduced livestock yield due to fodder losses. Table 3.6 shows how the valuations are computed.

In the poorest countries there is heavy reliance on fuelwood for energy. The only feasible substitutes are crop residues and animal dung. Yet both have value as sources of organic and nutrient inputs to the soil which, in turn, is the critical factor in sustainable subsistence agriculture. 'Cycles' of degradation can be observed. As population grows, the harvesting of fuelwood exceeds the regeneration rate of forests and woodlands. The forest is 'mined'. The loss of trees that fix atmospheric nitrogen impairs the fertility of the soil. As fuelwood becomes scarce, crop residues and grass are used for fuel. Further deterioration of soils occurs as these inputs to the soil are lost. Finally, dung is diverted from being a manure to being a fuel. Soil depletion worsens still further.

One of the costs of deforestation, then, is the loss of crop productivity arising from the diversion of dung from the soil to use as a fuel. In a study of Ethiopia, Newcombe showed that some 90% of cattle dung produced in Eritrea, and 60% in Tigrai and Gondar, is used as a fuel.[14] The 'damage cost' of this diversionary use can be estimated in three ways. First, a crop-response function can be estimated and the resulting fall in the yield of crops can be valued at market prices. This is analogous to the 'production function' approach for soil erosion. Second, the dung has an equivalent worth in terms of chemical fertilisers (although it will be an understatement since dung contributes organic matter as well as nutrients). The 'replacement cost' approach can be used to value the dung. Finally, dung is bought and sold on markets, so it can be directly valued at the ruling market price. The resulting valuations of dung, in US dollars per tonne, were: production function approach (grain response), $47–114; replacement cost approach (fertiliser cost), $22; market price approach, $61–91. Newcombe estimated that Ethiopian households burned some 7.9 million tonnes of dung per annum. At the average grain response value ($76) the dung is worth some $600 million. Clearly, the whole $600 million cannot be debited to deforestation. Some dung would have been burned anyway. But if half the dung burned was 'induced' by deforestation, then the cost would be $300 million per annum, or some 6% of Ethiopian GNP in 1983. At the highest implicit values of dung, the cost would be 9% of GNP. As a percentage of

Table 3.7 Deforestation and soil erosion costs in Indonesia

	1975	1980	1984
Forest loss* ($ million, current prices)	994	6262	3054
% GDP	3.6	8.9	3.6
Soil erosion ($ million, current prices)			
On site			315
Siltation of irrigation systems			10
Harbour dredging costs			2
Reservoir sedimentation			46
Total soil erosion			373
% GDP			0.4

* Value of physical depreciation of tree stocks.

Sources: for forestry, see R. Repetto *et al.*, *Wasting Assets: Natural Resources in the National Income Accounts*, World Resources Institute, Washington, DC, 1989; for soil erosion, see W. Magrath and P. Arens, *The Costs of Soil Erosion on Java: a Natural Resource Accounting Approach*, Working Paper No. 18, Environment Department, World Bank, Washington DC, August 1989.

agricultural GNP, these proportions would be roughly doubled.

The Ethiopian study suggests that deforestation costs show up in the form of forgone agricultural output, but this time through the linkage of diverting livestock dung from use as a fertiliser to use as a fuel to substitute for scarce fuelwood. The result cost is perhaps some 6% of total GDP and twice that figure as a proportion of agricultural GDP. Interestingly, this is the same proportion of GDP as that estimated for the direct costs of fuelwood loss in Mali.

What, then, do the damage cost estimates for Africa show? However imperfect the methodologies, and the databases, it seems fairly clear that natural resource degradation imposes severe costs on the economies of the poorest economies. There can be no pretence that the procedures are comprehensive, but they do suggest that land degradation in general is imposing costs of the order of 5% and more of GDP.[15]

Asia

The most detailed study of national environmental damage and resource depletion costs has been carried out for Indonesia.[16] The analysis covers the depreciation of oil and forest assets and the costs of soil erosion.

While oil is self-evidently a natural resource, the interest here focuses on forestry and soil erosion.

A forest can be viewed as a natural capital asset in the same way as we think of man-made capital assets. The asset yields a 'service' over time to which there corresponds an income. An asset can *depreciate* for two reasons: *value* changes and *physical* changes. Value may change quite independently of any physical change, for instance because its price varies with demand. Obviously, value also changes if physical depreciation occurs. There is, in fact, a dispute in the national accounting literature as to the correct way to 'account' for the depletion of environmental assets.[17] The issue is further complicated by the fact that the relevant cost should be that of 'non-optimal' depletion: it cannot be assumed that all depletion is somehow reprehensible.

Table 3.7 shows the results of an exercise to value deforestation and soil erosion in Indonesia. The cost of deforestation estimates are obtained by the depreciation approach (which is described in note 16 at the end of this chapter). The soil erosion damage estimates are based on the 'production function' approach.

Eastern Europe
Political changes in eastern Europe have brought to light serious environmental degradation as one of the by-products of the socialist nations' 'push for growth'. Most detail on the costs of this damage is available for Poland which has a history of costing environmental impacts. In government terms, some 11% of the land area of Poland, supporting 35% of the population, is classified as 'areas of ecological hazard'. Five regions – Upper Silesia, Kraków, Rybnik (all in the south-west of the country), Legnica-Głogów (west-central Poland) and Gdańsk (on the Baltic) – are classified as 'areas of ecological disaster'. All forms of pollution are serious, but air and water pollution and soil contamination give rise to the greatest economic costs. One estimate suggests that 30% of economic costs arise from water pollution, 35% from air pollution and a further 35% from soil contamination.[18]

Table 3.8 shows estimates of the money value of the damage done by all forms of pollution. The figures relate to 1980 damage (but are expressed in 1987 prices) so they almost certainly understate the extent of damage. The estimates also include figures for the cost of 'excessive' resource use, so that the table tries to separate out damage due to pollution from damage due to overuse of resources. The startling result is that pollution damage alone may amount to 5–10% of Polands' GNP or more.

Table 3.8   The economic costs of pollution damage and excess resource use in Poland

Category of damage	Damage (billion zlotys 1987)	% GNP
Agriculture	150	
Forests	50	
Water resources	65	
Corrosion	215	
Excessive use of mineral resources	130	
Raw materials lost in discharges to air and water	50	
Health effects	115	
Total pollution impacts*	595	7.7%
Total environmental losses*	775	10.0%

Source: A. Kassenberg, Zones of ecological threat – a new planning tool, *Kosmos*, 1/86 (in Polish) quoted in S. Kabala, The costs of environmental degradation in Poland, Background Paper for the series *The Environment in Poland*, World Bank, February 1989. See also S. Kabala, The economic effects of sulfur dioxide pollution in Poland, *Ambio* (4), 1989.
* The 10% of GNP figure is widely quoted, but World Bank *World Tables 1989* puts Poland's GNP at 12,700 billion zlotys in 1987 (current prices). If Kassenberg's estimate relates to 1987, then the shares would be 6.1% for total environmental degradation and 4.7% for pollution damage. Note, however, that the damage done relates to 1980 levels of pollution.

Conclusions on the costs of environmental damage

The estimation of the national costs of environmental damage is in its infancy, but the studies to date have some fairly common features. Table 3.9 shows a summary. For the *developed* world the German data suggest damage costs of a little under 5% of GNP. Dutch estimates are lower at 0.5–0.8% but relate to partial impacts only. Damage *avoided* in the USA because of environmental policy was some 1.2% of GNP but the estimates omit noise nuisance and relate to 1978. A range of 1–5% of GNP thus seems fairly reasonable as an estimate of the damage done by environmental degradation (but only for pollution damage) in the developed world.

The expectation would be that damage in the developing world would be higher given the absence of environmental protection legislation and institutions. The estimates that exist tend to bear this out. Soil erosion

Table 3.9 Summary of the social costs of environmental degradation in the developed and developing world

Country	Nature of damage	Year	% of GNP
Netherlands	Some pollution damage	1986	0.5–0.8
Germany	Most pollution damage	1983–5	4.6–4.9
USA	*Avoided* damage due to environmental legislation	1978	1.2
Mali	Soil erosion	1988	0.4
Burkina Faso	Biomass loss	1988	8.8
Ethiopia	Deforestation	1983	6.0–9.0
Indonesia	Deforestation	1984	3.6
Indonesia	Soil erosion	1984	0.4
Poland	Pollution damage	1987	4.7–7.7

in Indonesia and Mali may cost some 0.4% of GNP. Deforestation in Ethiopia would appear to cause at least 6% of GNP losses, while, on a different basis, the cost is 3.6% of GNP in Indonesia. Pollution damage in Poland runs at at least 5% of GNP and may be very much higher. Finally, total biomass loss imposes a cost on the economy of Burkina Faso of just under 9% of GNP. Estimates for other countries, not reported here, tend to support the notion that environmental damage costs developing countries some 5% of their GNP. Moreover, this cost is in the form of lost productive potential, i.e. there are real resource flows associated with these losses. In the developed economy cases, probably the major part of the loss shows up in 'non-GNP' flows, i.e. changes in human welfare that are not captured by the conventional methods of national accounting. In itself this is telling of the need to adjust the national accounts.

Even if the data are subject to significant degrees of uncertainty – as seems likely – they underline a point of major significance: environmental deterioration does damage the economies of both the rich and poor world, and it imposes large costs in particular on the developing world's development capability.

Incentives for sustainable development

Sustainable development stresses the importance of growth and development with the characteristic of 'permanence'. Gains are not short-lived but sustainable. Many past development policies have been based on the idea of a 'rush for growth' on the basis of several prior beliefs about the lessons of history for the development process. Thus, the achievements of the currently developed world reflect a particular path of transition from agriculture to industry to service-orientated economies. That transition was achieved at the cost of environmental losses, many of them in the form of irreversible costs, as with disappearing species. The idea that this development process can be replicated for the currently developing world is what is challenged by the philosophy of sustainable development. Clearly, securing real income gains is a priority, but these gains are not, it is argued, sustainable if they impose heavy costs in the form of environmental damage.

The fact that the developed world sacrificed environmental quality for real income is no rationale for believing that the developing world must do the same. There are at least three reasons for this:

(i)   Sacrifices of renewable resources in temperate zones are likely to cause less loss to human welfare than similar sacrifices in tropical zones where the 'margin of fragility' is much lower. Simply put, small changes in environment have bigger economic effects in most developing countries. The valuation studies discussed previously show the kinds of costs involved.

(ii) The developed world is suffering the consequences of indifference to environmental quality and is now exerting great efforts to repair past damage, salvage what it can and prevent further damage. *Planning* such a policy of reacting to damage once it is done is dangerous. The examples of the ozone layer depletion and global warming are cases in point. Moreover, the developed world has been responsible for much of the global damage done, and that imposes costs on all the nations of the world.

(iii) There is no *need* to engage in the same process of degradation. Growth in real per-capita incomes is achievable without major degradation. The requirement is for environmental impacts to be 'decoupled' from economic growth. Sustainable development is about that decoupling process.

The last reason is the most important. To pursue growth without assessing first whether the same goal can be achieved at less environmental cost is not rational. Disregarding environmental damage also threatens, as we have seen, the permanence of the development effort. Hence, environment matters. How can 'decoupling' be achieved? The clue lies in *incentives* and *information*. Two sets of incentives and information systems are critically important:

(i)   those which reduce *uncertainty* about the future; and
(ii)   those which send out the correct *price and quality signals* in the marketplace.

Two further information systems are important:

(i)   modifying the presentation of *environmental and economic statistics* so that environmental impacts of economic change can be discerned, and the 'services' of the environment highlighted;
(ii)   revising systems of *appraisal* for investments and policies so that they adequately reflect and integrate environmental impacts.

Incentive systems to reduce uncertainty

Most economic decisions are made in the context of uncertainty. Yet uncertainty can be both beneficial and adverse for environmental quality. By choosing crops, crop mixes or rotations that minimise the risk of failure in the event of drought, for example, the farmer is likely to adopt weather- and pest-resistant strains. Many problems with agricultural output arise from choosing the wrong technology or output mix in the face of uncertainty, often because of false beliefs about the potential to correct problems once they have occurred, or because the future is 'discounted' heavily. The technology and production choice can be seen as one of trading off *productivity, stability, equitability* and *sustainability* (Figure 3.4). Highly productive systems may not be sustainable – as with the Mayan tropical forest agriculture systems in Central America. Equitable and sustainable systems may not be as productive – as with the Manorial system of agriculture in medieval Europe. Modern examples include large-scale irrigation systems: high productivity appears incompatible with equitability and sustainability. Biological pest control is likely to be more sustainable but will produce more fluctuating yields (instability) and lower productivity. Uncertainty about the future tends to bias the trade-off towards productive but unsustainable systems.

Ensuring sustainability therefore requires efforts to reduce uncertainty. Some of the most important sources of uncertainty lie in 'resource rights', i.e. the lack of security of tenure over land and/or the resources

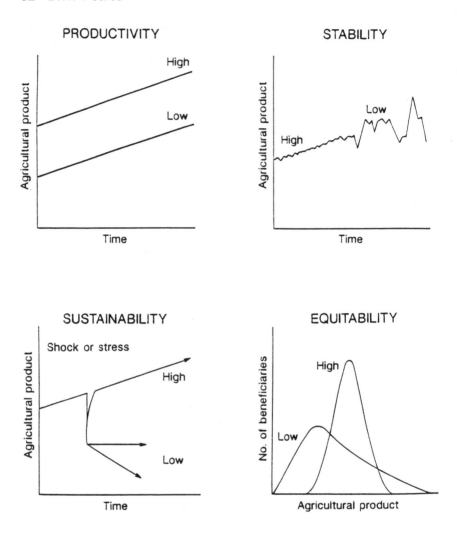

Figure 3.4 Characteristics of agricultural performance: the trade-off between sustainability, productivity, stability and equity

Source: G. Conway and E. Barbier, *After the Green Revolution: Sustainable Agriculture for Development*. Earthscan, London, 1990.

on the land. While the evidence is not conclusive, available studies do suggest that this shows in the capitalization of such gains in land values.[19] Granting tenure of itself may also be insufficient to improve natural resource management. The *way* in which tenure is granted also matters. Thus, in many parts of the developing world tenure is recognised *de facto*, and often *de jure*, only if land is cleared of vegetation. One study in Ecuador shows quite clearly that deforestation is related not only to population pressure but also the desire to establish land rights through land clearance.[20] The influence of uncertainty shows up clearly in this example. Since there are uncertain rights to forested land, but certain (or near-certain) rights to cleared land, the land is cleared.

The Ecuador study tested two hypotheses: first, that rural population pressure is determined by the local demand for agricultural products (measured by the size of the urban population in each area), the availability of soils suitable for agriculture and accessibility (measured by the kilometres of roads in the area):

$$AGPOP = b_0 + b_1.URBPOP + b_2.SOILS + b_3.ROADS \qquad (3.1)$$

and second, that deforestation (*DEFOR*) arises from population pressure (*AGPOP*) and a measure of secure land tenure (*TENSEC*):

$$DEFOR = b_4 + b_5.AGPOP + b_6.TENSEC \qquad (3.2)$$

Using regression techniques, equation (3.1) was shown to have significance and to confirm the view that the pressure to deforest arises from opportunities to capture 'economic rents' through colonising new land areas. Similarly, equation (3.2) showed that both population pressure and the need to confirm land tenure (albeit a tenuous right to the land) are important in explaining deforestation.

Insecurity of tenure may also account for resource destruction by major corporations and government agencies. It is well known from the theory of natural resource economics that uncertainty can accelerate the rate of depletion of an exhaustible resource.[21] An example arises in the context of the logging of tropical forests. The 'stumpage value' (the value of the standing stock of timber) of tropical forests is very large. Governments, however, frequently fail to capture the rents because of policies which effectively enable the exploiter of the forest stock to secure the rent. Economics suggests that taxation policies can be designed to appropriate rents, return them to the public revenues as opposed to private logging companies, and yet not deter investment in the country in question.[22] If rents accumulate in the hands of those responsible for exploiting the resource there are two likely environmental effects. First, the concessionaire is likely to deplete the resource rapidly to capture more of the rent earlier rather than later. This tendency will be reinforced if there is uncertainty about policy towards the concession – for example, doubts about whether it will be renegotiated. Environmental degradation will be made even worse if the concession has a time horizon less than that needed to regenerate the resource, i.e. the rotation period. No new planting is likely to occur under these conditions. Second, the existence of exploitable rents leads to 'rent seeking' whereby other economic agents seek to acquire rights to the resource, again accelerating depletion.[23]

Prices as incentives

Prices are powerful incentives. If resource prices are set too low, excessive use will be made of the resource. The extreme example is zero-priced resources which have no established market – the carbon-fixing functions of the oceans or forests, for example. But the same argument holds for other resources – energy, irrigation water, fertilisers and pesticides, for example. If their prices are set too low they will tend to be overused, and such overuse can readily contribute to environmental degradation. To secure an efficient use of resources, outputs should be priced at their marginal social cost, which comprises the marginal costs of production and the 'external costs' of pollution or resource degradation caused by producing the good. If markets functioned near-perfectly there could be some assurance that prices in the marketplace would reflect their marginal private costs of production. But there are at least two forms of 'market failure'. First, many marketed goods may have prices reflecting private costs of production, but not the social cost. The means of intervening to ensure that social prices are charged are several: regulation by standard-setting, pollution taxes and tradable permits. Second, many goods have no markets at all and hence prices have to be established for them. Assigning property rights to the 'free resource' is one way in which this can be achieved, although it may or may not be associated with full social cost pricing.

In the developing world there is ample evidence that a considerable amount of environmental degradation arises from a failure to price resources and goods at their marginal *private* cost, let alone their marginal social cost. Table 3.10 shows the level of pesticide subsidies in nine countries. Subsidies range from 19% to 90% of the full cost of the pesticides, thus maintaining artificially low prices. Damage from excess use of pesticides shows up in several ways. There may, for example, be

Table 3.10 Pesticide subsidies in eight countries

Country	Subsidy as % of full cost	Total value of subsidy (US$ million)
Senegal	89	4
Egypt	83	207
Ghana	67	20
Honduras	29	12
Colombia	44	69
Ecuador	41	14
Indonesia	82	128
China	19	285

Source: World Resources Institute, *Paying the Price: Pesticide Subsidies in Developing Countries*. World Resources Institute, Washington, DC, 1985.

some 2 million cases of pesticide poisonings per year in the Asia and Pacific region, 40,000 of them probably giving rise to fatality.[24] Exposure is highest among men and death rates among men have risen significantly in communities where insecticides have been introduced on an intensive scale. There is also evidence of health risks from fish caught in pesticide-contaminated ponds, paddies and irrigation channels. New pest biotypes have emerged in response to large applications of some pesticides, increasing crop production instability. Clearly, subsidies to pesticides are not the only factor causing excess use – ignorance of the risks and the use of pesticides that are not permitted in the developed world also contribute. But the subsidies must play a significant role.

What is true of pesticides is also true of irrigation water, energy, fertilisers, lease values for land for mechanised agriculture and other resources.[25] Thus, irrigation charges tend to be a small proportion of the benefits of irrigation water to farmers: 8–21% in Indonesia, 11–26% in Mexico and only 10% in the Philippines, 9% in Thailand and 5% in Nepal. Excess use of irrigation water contributes to waterlogging and salination of soils, let alone wasting the water resource itself. Pricing reform thus has a major role to play in securing sustainable development, with the first priority being to relate prices charged to costs of production (and border prices where the resource is tradable). Later in the process of development, price will need to be related more to the full social costs of production.

But 'getting prices right' is not a simple matter in terms of securing the right balance of environmental quality and shorter-term gains in output. This is readily illustrated in the context of agricultural output prices. It is tempting to think that higher farmgate prices will stimulate an aggregate supply increase, making farmers better off and thus more able to invest in longer-term investments needed for sustainability – for example, soil conservation, tree windbreaks and water conservation. There is certainly evidence to suggest that 'price discrimination' against producers (i.e. keeping farmgate prices low relative to border prices) is associated with lower agricultural growth rates. One study for Africa suggested that countries with low or no discrimination had annual growth rates of 2.9% (1970–81); medium price discrimination was associated with annual growth rates of 1.8%, and high price discrimination produced annual growth rates of only 0.8%.[26] Yet, the impacts of reducing discrimination are not clear. First, price changes must be perceived to be permanent for investment decisions to favour conservation practices. Second, the nature of the response will depend on security of tenure, resource rights generally, and access to credit. Third, if open access resources are available – for example, virgin forest land – the supply response may consist of extensification rather than intensification, i.e. 'new' lands may be cleared rather than existing lands being subject to conservation investments. Fourth, and offsetting the incentive to extensify, higher farm incomes may lower personal discount rates and improve credit 'ratings', even with informal credit markets. These effects should assist conservation. Finally, while there is evidence of supply responses for single crops, price increases may often result in switches between crops rather than an overall increase in output. The nature of the crops matters from the environmental standpoint. Tree crops and perennials, for example, are more likely to be good for soil conditions, whereas many root and grain crops (peanuts, cassava, sorghum, millet) are erosive. Since there is nothing to link price increases with the ecological status of the crop, it is as likely that supply responses will bring increases in erosive crops as not.

As yet, then, the linkages between output price changes and environmental quality in the long run remain to be established. Remarkably, hardly any agricultural supply response studies mention the 'sustainability' of the responses.

Fiscal policies

Since price is instrumental in changing behaviour it follows that taxation policy will also be an important influence on behaviour which affects the environment. The scope for pollution taxes in developing countries is likely to grow in the future, although taxes in the sense of damage-related charges are still a rarity in the developed world.[27] But other

Table 3.11 Rent capture in tropical forests

Country and period	Actual rent $billion	Government rent $billion	Govt share/ actual rent
Indonesia 1979–82	4.4	1.6	37.5%
Sabah 1979–82	2.0	1.7	82.5%
Ghana 1971–74	0.08	0.03	38.0%
Philippines 1979–82	1.0	0.14	14.0%

Source: R. Repetto, Overview. In R. Repetto and M. Gillis (eds), *Public Policies and the Misuse of Forest Resources*. Cambridge University Press, Cambridge, 1988.

taxation policies are capable of adjustment and existing policies frequently discriminate against the environment. As noted previously, governments frequently fail to capture rents from existing valuable resources such as forests. Table 3.11 shows data for four countries and reveals that, for the years shown, government rent capture was very low in Indonesia, Ghana and the Philippines. The scale of the rents is worth noting: in Indonesia, for example, actual rents for the four-year period were $4.4 *billion*, or some $1.1 billion per annum, and of this the government took only $400 million per annum. There is scope, then, for the revision of existing fiscal policies even before attention is paid to the potential for introducing pollution and other resource degradation taxes. In much the same way, user charges – for example, for water connections – have high potential for reducing water wastage which, in turn, has pollution impacts.

Information systems

Information is a major input to sustainable development. At the most 'micro' level individual households need to be informed of the consequences of particular input and output decisions. Looked at from the outsider's standpoint there is also a need to utilise local knowledge and to observe and counteract the constraints that prevent sustainable practices from being employed. Extension systems are clearly important in both these respects. Governments also need to be informed. The most important aspect of this information need is to establish that economic policy impacts on the environment and that the environment impacts on economic welfare. Valuation studies are therefore valuable in the latter respect, and these may be sufficiently formalised to warrant modified presentations of national accounts, as discussed previously. Showing that economic decisions impact on the environment is less easy, but analysis of incentive systems is a useful starting point. Since price policy tends to be fairly easily modified, subject to concerns about the social incidence of price changes and political implications, pricing presents itself as a high-priority candidate for action on sustainable development. As noted above, resource rights and land tenure offer another vital dimension of sustainable development policy.

Other information systems deserve far more emphasis, too. Geographical Information Systems (GISs) have the capability to mix satellite imagery with more standard ground-based information ('ground truthing') which can be used not just for the traditional purposes of mapping and assessing land capability, but also as a database for interpreting environmental change over time. The systematic interaction of socio-economic databases and satellite imagery is still underdeveloped.[28]

Towards sustainable development

What can be learned from this brief overview of issues in the environment-economy connection? First, environmental damage matters not just in the sense of 'psychic' or 'non-economic' welfare, but also in the sense of costs that show up as lost production possibilities. These costs can be large; examples have been given of damage costs that amount to 5% and more of GNP.

Second, given the scale of environmental damage and its nature, there is sufficient evidence to support the view of advocates of sustainable development that a greater priority needs to be given to environmental policy if policies are to be sustainable.

Third, in so far as past development policy has been influenced by the theory of 'optimal growth' – and it clearly has – there is a critical need to analyse the conditions under which optimal growth is also sustainable growth.

Fourth, since raising real per-capita incomes must remain a major objective, though not the only objective, of development policy, the only way in which growth with environmental quality can be achieved is by decoupling growth from its environmental impacts. Where decoupling cannot be achieved, it is essential to understand the nature of the trade-offs between orthodox development goals and environmental deterioration. That can only be achieved through better and more sophisticated attempts to value environmental functions.

Do priority areas emerge from the analysis? Because of the paucity of studies and their uncertainty, hard and fast recommendations are difficult to make. But the evidence does suggest that, for developing countries still at an early stage of development, *deforestation* is likely to be imposing heavy losses on the economy. Even for rapidly industrialising countries there is evidence that forest resources are being mismanaged and that the costs of deforestation are significant. *Soil erosion* is important but perhaps less important than might at first be thought. *Pollution impacts* are important and the experience of eastern Europe and the industrialised countries shows just how important pollution is in terms of economic costs. In the developing world it is likely that *water pollution* is the biggest hazard, primarily because of its health impacts. But costing those impacts still seems very uncertain.[29]

In terms of policy, priority areas for action appear to lie in the following areas:

(i)   short term: private cost pricing policy reform of existing tax policy
(ii)   medium–long term: resources rights and land tenure
(iii) long term: social cost pricing

while a continuous theme has to be information for households, productive units and government.

While some economists have not perceived any change in the nature of the development process arising from the debate about sustainable development, others have urged an even higher policy status for natural resources than that suggested by valuation studies. They stress several features of the workings of ecosystems: the extensive uncertainty surrounding their role in global and local life-support systems; the irreversibility of many of the impacts, in contrast to man-made capital which can be increased or decreased at comparative will; and the non-use values for environmental quality. The last feature relates to the phenomenon of values that lie outside those associated with the 'use' of resources and would explain much of the public concern over tropical deforestation and endangered species, resources which the vast majority of people will never experience. In terms of traditional economic growth models, the 'amenity' value of environmental assets becomes of major significance, along with the environment as productive input. As is well known, combining both features of environmental assets with traditional concerns to secure optimal growth and intergenerational equity tends to support policies of conservation of natural assets. Sustainable development, then, still seems likely to have challenges in store for development economists and environmental economists alike.

Notes

1. For an overview of the history of environmental economics theory, see D.W. Pearce, 'Economics of the Environment', in D. Greenaway, M. Bleaney and I. Stewart, *Economics in Perspective*, Routledge, London, 1991. Pearce highlights four major but non-integrated developments prior to the 1960s. These are: first, the idea that there exist ecological bounds to the scale of economic activity, as in the works of Malthus, Ricardo and Mill, and popularised in such works as H. Daly, *Toward a Steady State Economy*, Freeman, San Francisco, 1973; second, pollution as an 'external effect', first analysed by A. Pigou, *The Economics of Welfare*, Macmillan, London, 1920, and later by K. Kapp, *The Social Costs of Private Enterprise*, Harvard University Press, Cambridge, MA, 1950 and R. Coase, 'The problem of social cost', *Journal of Law and Economics*, 3, 1–44 1960; third, the theory of the optimal rate of depletion of an exhaustible resource, developed by L. Gray, 'Rent under the assumption of exhaustibility', *Quarterly Journal of Economics*, 28, 466–89 1914, and formalised by H. Hotelling, 'The economics of exhaustible resources', *Journal of Political Economy*, 39, 137–75 1931; and fourth, the theory of optimal use of a renewable resource, ranging from M. Faustman's work on forest rotations in 1849 to H.S. Gordon's 'Economic theory of a common property resource: The fishery', *Journal of Political Economy*, 62, 124–42 1954.
2. D. Meadows *et al.*, *Limits to Growth*, Earth Island, New York.
3. This literature is admirably surveyed in J. Pezzey, *Economic Analysis of Sustainable Growth and Sustainable Development*, Working Paper 15, World Bank Environment Department, World Bank, Washington, DC, March 1989.
4. For example, in the simplest 'cake eating' model, per-capita consumption, c, depends on the rate of use of the per-capita stock of a non-renewable resource, s. Thus

$$c = -\mathrm{d}s/\mathrm{d}t.e^{rt}$$

where r is the rate of technological progress and t is time. Utility, u, depends on c, such that

$$u(c) = c^v$$

where $0 < v < 1$. The aim is then to choose $s^*(t)$, i.e. the optimal stock of the resource, to maximise:

$$\int_0^\infty u(c).e^{-kt}.\mathrm{d}t$$

where k is the discount rate.

The solution to such a problem can exhibit *both* properties of optimality and sustainability (in the sense of non-declining utility) *provided* the discount rate k, is less than (or equal to) the rate of technological progress.

If the model is extended to allow for 'environmental productivity' then we can rewrite the equation for c as follows:

$$c = -S^{m/v}.(\ -ds/dt).e^{rt}$$

where S is now the *total stock* of the resource, s is per-capita stock, and $m > 0$ measures the productivity effect of the total stock. Maximisation of the present value of utility in this model yields a number of possible outcomes. If individuals behave 'non-cooperatively', i.e. ignore the environmental value of the resource when planning their own privately optimal path, then sustainability is harder to achieve than in the simple cake-eating model. Significantly, higher discount rates tip the balance towards *non-sustainability*, and in some cases the optimal growth path is itself non-sustainable (recall that optimality is defined in present value terms).

See Pezzey, op.cit.

5. 'Economic importance' here does not mean *financial* importance. Something is economically important if it has a significant impact on human welfare.

6. An early exception was F. Dasmann, *The Conservation Alternative*, Wiley, Chichester, 1975.

7. Although, sadly, far too much of the professional environmental economics theory literature still reveals a marked lack of understanding of ecological functions.

8. International Energy Agency, *Energy Conservation in IEA Countries*, OECD, Paris, 1987, p.44.

9. See J. Camba, D. Caplin and J. Mulckhuyse, *Industrial Energy Rationalization in Developing Countries*, Johns Hopkins University Press, Baltimore, MD, 1986.

10. Ibid., p.3.

11. For surveys of the various methodologies, see D.W. Pearce and A. Markandya, *Environmental Policy Benefits: Monetary Valuation*, OECD, Paris, 1989.

12. See A.M. Freeman, *Air and Water Pollution Control: A Benefit-Cost Assessment*, Wiley, New York, 1982.

13. Using the replacement cost approach, Stocking estimates the following annual losses of nutrients from soil erosion in Zimbabwe: nitrogen 1.6 million tonnes, organic carbon 15.6 Mt and phosphorus 0.24 Mt. In financial terms (1985 prices) the total cost for all Zimbabwe lands, using commercial fertiliser prices to value the replacement nutrients, was $1.5 billion, or a startling 26% of 1985 GDP. See M. Stocking, *The Cost of Soil Erosion in Zimbabwe in Terms of the Loss of Three Major Nutrients*, Soil Conservation Programme, Land and Water Development Division, Food

and Agriculture Organisation, Rome, 1986.

14. K. Newcombe, 'An economic justification for rural afforestation: The case of Ethiopia' in G. Schramm and J. Warford, *Environmental Management and Economic Development*, Johns Hopkins University Press, Baltimore, MD, 1989.

15. Fuelwood output arising from deforestation implies both a *gain* in GDP and, if the harvest is unsustainable, a *loss* as well. Thus household fuelwood production should be *added* to GDP if a value is not already imputed for it. But the net depreciation of forest stocks needs to be deducted. In a study for Tanzania, Peskin estimated the following adjustments to 'conventional' GDP. 137 million person-days were estimated to have been spent in 1980 in Tanzania in the 'production' of fuelwood. Valued at the then minimum wage of 20 shillings per day, this activity was worth 2746 million shillings. The existing imputed value was 207 million shillings, so the adjustment led to a gain of 2746 − 207 = 2539 million shillings. Since around 69% of the wood cut was on a non-sustainable basis, depreciation of (0.69 × 2746) = 1895 million shillings constitutes depreciation which should be deducted from GDP if, as it should be, the interest is in *net* national product. The overall effect of the adjustments is to increase GDP by 6% compared to the standard presentation, and increase NNP by 2%. Not all 'environmental adjustments' therefore diminish GDP. Note, however, that the adjustments made relate *only* to the forests as sources of fuelwood, and nothing else. See H. Peskin, *Accounting for Natural Resource Depletion and Degradation in Developing Countries*, Environment Department Working Paper No.13, World Bank, Washington, DC, January 1989.

16. See R. Repetto, W. Magrath, M. Wells, C. Beer and F. Rossini, *Wasting Assets: Natural Resources in the National Income Accounts*, World Resources Institute, Washington, DC, 1989; and W. Magrath and P. Arens, *The Costs of Soil Erosion on Java: A Natural Resource Accounting Approach*, Environment Department, Working Paper No.18, World Bank, Washington, DC, August 1989.

17. Essentially, this dispute is between the 'depreciation' approach and the 'user cost' approach. The depreciation approach looks at natural asset depreciation in the same was as national accountants look at the depreciation on man-made capital. Net income equals gross income minus depreciation. So, a given stock of a resource is estimated in physical terms and the change in the stock over a year is estimated also in physical terms. These physical changes are then 'valued' at a price. In the case of forests, for example, the price could be that of timber, although the 'true' price may be much higher if non-timber values are also accounted for. The choice of prices is not straightforward. In the Indonesia forest case, for example, the valuation was at the 'rent' of the timber, i.e. the difference between the border price of timber and the costs of harvesting and transport (the 'primary rent' or 'stumpage value'). But such valuations are not relevant to 'secondary' forest, forest which has generated after the primary forest has been removed. This is because the 'best is taken first' and timber from secondary forests is simply not worth as much as timber from primary forests. In the

Indonesia case this 'secondary rent' was put at 0.5 of the primary rent. An example will illustrate how the analysis proceeds. In 1983 the opening physical stock was 24,239 million cubic metres. Natural growth was 51.9 mcm. This growth was in secondary forest and hence was valued at the secondary rent of $17.72 per cubic metre, and is an *addition* to value. Harvesting of primary forest was 15.2 mcm so this is valued at the primary rent of $35.44, i.e. $538.7 million is *deducted* from the value of the forest. Deforestation was of secondary forest, 120 mcm at $17.72 = $2126.4 million; logging damage was 30 mcm at $17.72 = $531.6 million; and fire damage was 153.8 mcm (it was the year of a major fire) valued at an average of $25.1 to account for different mixes of primary and secondary timber, i.e. $3870.9 million. The opening stock in 1983 was valued at $1,020,960 million. By adding the 'growth' value and deducting the various losses to closing stock is then $1,020,960 million, plus $919.7 million minus $7067.6 million = $1,014,812 million. *However*, the rents changed during the year because world prices fell and costs of harvesting rose slightly. So the stock at the end of 1983 needed to be *revalued*. The adjustment for revaluation was basically to multiply the end of year physical stock by the difference in the 1983 and 1982 rentals. Overall, then, the equation is:

End 1983 value = Opening 1983 value − (physical change
                          × 1983 rentals) − (end 1983 physical stock
                          × [1982 rentals − 1983 rentals]).

The *user cost* approach is different. It looks at the revenues from the sale of the resource and splits that revenue into a 'capital' element and a value-added element. The capital element is a 'user cost', i.e. it is the forgone benefit to the future of being without that part of the resource. The value added component, on the other hand, is true income. To estimate user cost one needs to estimate the discounted stream of future benefits forgone, and it is this that is the 'cost' of resource depletion and that should be deducted from GDP. The important point is that the depreciation approach does not affect GDP; whatever is depreciated is offset by the revenues from selling the resource. What it affects is net domestic product. The user cost approach would affect GDP, on the other hand. How is the ratio of true income to user cost determined? It is still a matter of debate, but it has been suggested that the formula:

$$Y/R \; = \; 1 \; - \; \frac{1}{(1 + r)^{n+1}}$$

gives the ratio of true income (Y) to royalties (R = sales value minus costs), where r is the discount rate and n is the number of periods over which the resource is being liquidated. $R - Y$ would then be the user cost. See S. El Serafy and E. Lutz, 'Environmental and resource accounting: an overview', in Y. Ahmad, S. El Serafy and E. Lutz, *Environmental Accounting for Sustainable Development*, World Bank, Washington, DC, 1989.

18. Central School of Planning and Statistics, Warsaw, quoted in S.J. Kabala, *The Costs of Environmental Degradation in Poland*, Background Paper on the Environment in Poland. World Bank, Washington, DC, February 1989.

19. See, for example, G. Feder, 'Land ownership security and farm productivity in rural Thailand', *Journal of Development Studies*, 23, 8–14, 1987; Y. Chalamwong and G. Feder, *Land Ownership Security and Land Values in Rural Thailand*, World Bank Staff Working Paper 790, Washington, DC, 1986. The picture is less clear for Africa: see G. Feder and R. Noronha, 'Land rights systems and agricultural development in sub-Saharan Africa', *World Bank Research Observer*, 2, 21–30 (2), 1987.

20. D. Southgate, R. Sierra and L. Brown, *The Causes of Tropical Deforestation in Ecuador: a Statistical Analysis*, London Environmental Economics Centre, Paper 89-09, London, 1989.

21. See, for example, P. Dasgupta and G. Heal, *Economic Theory and Exhaustible Resources*, Cambridge University Press, Cambridge, 1979, ch.13.

22. This is the theory of resource-rent taxation. See R. Garnaut and A. Clunies-Ross, 'Uncertainty, risk aversion and the taxing of natural resource projects', *Economic Journal*, June 1975.

23. See R. Repetto and M. Gillis (eds), *Public Policies and the Misuse of Forest Resources*, Cambridge University Press, Cambridge, 1988.

24. See R. Repetto, *Economic Policy Reform for Natural Resource Conservation*, Environment Department Working Paper No.4, World Bank, Washington, DC, May 1988.

25. For surveys, see ibid.; and D.W. Pearce and J. Warford, *World Without End: Economics, Environment and Sustainable Development*, Oxford University Press, New York and London.

26. See M. Fones-Sundell, *Role of Price Policy in Stimulating Agricultural Production in Africa*, Swedish University of Agricultural Sciences, International Rural Development Centre, Uppsala, Issue Paper No.2, May 1987.

27. See D. Anderson, *Environmental Policy and the Public Revenue in Developing Countries*, Department of Economics, University College London, April 1990; J. Bernstein, *Alternative Approaches to Pollution Control and Waste Management: Regulatory and Economic Instruments*, Infrastructure and Urban Development Department, World Bank, Washington, DC, May 1990; D.W. Pearce, *Public Policy and Environment in Mexico*, Latin American and Caribbean Country Department, World Bank, Washington, DC, May 1990.

28. For interesting experiments in this respect, looking at the interactions of poverty and environmental degradation, see V. Jaggernathan, *Poverty, Public Policies and the Environment*, Environment Department Working Paper No.24, World Bank, Washington, DC, December 1989.

29. Some idea of the magnitude may be gauged, however. It is estimated that in 1979 something like 360–400 billion working days were lost in Africa, Asia and Latin America because of water-related diseases that prevented work. AT US$0.50 per day, this means that these continents lost some $180–200 billion in that year. GNP in 1979 for these continents was around $370

billion, so that output was below productive potential by perhaps 35% $(200/(200+370))$. On working days lost see J. Walsh and K. Warren, 'Selective primary health care: an interim strategy for disease control in developing countries', *New England Journal of Medicine*, 301, 18, 1979.

Chapter 4

Environmental Economics, Policy Consensus and Political Empowerment

Michael Redclift

The social sciences can no longer afford to look upon the environment as an area of marginal concern, lying outside the parameters of human behaviour, and best addressed by the expertise of the natural scientist. Indeed, social scientists are increasingly called upon to address environmental issues on behalf of society as a whole: they are *expected* to provide both an analysis and prescriptions for environmental problems. The social sciences in the 1990s find themselves, somewhat unwillingly, in the vanguard of expectations over environmental policy, rather as, under the influence of Keynes, Beveridge and Titmuss, they were expected to deliver workable economic and social policies in the 1940s and 1950s. Are they up to the challenge? In particular, is economics – the leading discipline within the environmental policy field – adequately equipped to meet the expectations placed on it in the 1990s?

To answer these questions we need to take a long, hard look at the adequacy of environmental economics, and the extent to which other thinking in other social science disciplines, particularly sociology and anthropology, might lead us to question the ability of economics to deliver on its promises. We need to examine the scientific discourse which surrounds environmental policy issues, and the assumptions that lie behind the neo-classical model. We also need to examine the social and political commitments which, although rarely referred to in the burgeoning professional literature, explain the attraction of this model in advanced industrial societies. Finally, we need to explore the relevance, and limitations, of environmental economics for people in the developing countries of the South. It will be objected, not unreasonably, that the approach I am advocating is 'one-sided', in that it does not subject other social science disciplines to the same degree of scrutiny. This is true, but it should be remembered that, standing at the touchline, few of the other social sciences make strong claims for their own paradigm's ability to contribute to the environmental policy debate. This, of course, is their weakness, as well as their strength.

Environmental knowledge and environmental policy

Environmental issues are not alone in being couched in terms of uncertainty, but uncertainty and indeterminacy clearly play an exceptionally large part in any discussion of the environment. The physical world external to human beings, and often not readily controlled by them, throws up uncertainties of many kinds. These include 'natural' disasters, changes in biophysical systems which cannot be predicted, and the unintended consequences of human behaviour and resource use. Action to prevent or impede environmental changes is necessarily pre-emptive: it embodies an insurance principle, which most of us readily accept. However, this does not prevent confusions surrounding what we know about nature (scientific 'knowledge') being confused with the limitations in our understanding of nature (scientific 'understanding'). Much of the discussion of environmental policy is restricted to what we know, rather than how we come to know it. Consequently, it is usually assumed that we can make up for the deficiencies in our science by knowing more facts, rather than developing new ways of understanding them. There are real problems, which I will come to later, in the way our understanding of the environment (what we know) comes to influence what we do not understand (what we do not know we do not know).

There are two immediate problems here which should receive attention. Is 'science' what we think it is? And what are the costs and benefits of recognising similarities, rather than differences, in environmental knowledge?

The point about our scientific understanding is that we cannot 'know' about nature except through our discourse about it. Our science is part of this social discourse. If the environment is regarded as something which exists separately from us, then this is part of our discourse about it. In practice, we often tend to regard as 'science' the knowledge we possess which our values have had least influence upon, which is least susceptible to human manipulation. Science is broken down into specialist areas, and expertise is distributed by those who demonstrate abilities within these specialisms. This may not be obvious to non-scientists: 'Few laymen [realize] how tightly compartmentalised the scientific community [has] become, a battleship with bulkheads sealed against leaks' (Gleick 1987, 31).

Many social scientists, especially economists, aspire to be scientific in the positivist sense referred to above. That is, they aspire to as objective a view as possible of human activities, in the belief that their objectivity, their value neutrality, will produce more accurate forecasts. There are real problems in this position. First, the natural sciences are less 'positivist' than many scientists and social scientists suppose. For example, quantum theory suggests that the laws of the natural and human worlds are not altogether dissimilar. Sub-atomic particles, for example, are now seen by physicists as connections, rather than 'things'. The second nail in the positivist coffin concerns observation, and the meaning we attach to our observations. Shown the same images of tropical hillsides bare of vegetation, some scientists see 'land cover' (that is, the loss of natural activity) while others see 'land use' (the extension of human activity). We cannot arrive at an objective account of what is happening to eroded hillsides; simply a series of different accounts, based on different scientific perceptions and traditions of thought. What we observe is not given by the object of interest, but the trained eye with which we see it. One of the strengths of both the social sciences, and the natural sciences, is their ability to recognise difference, as well as similarity.

Economics, more than any other social science discipline, has sought to emulate the natural sciences, in subjecting human behaviour to rigorous analysis from the perspective of one paradigm, the neo-classical. Like the natural sciences, economics promises (and often delivers) predictions about outcomes and advice for policy-makers. During the 1980s most of the other social sciences travelled in the opposite direction; making fewer generalising statements, and even fewer predictions, they sought to identify difference rather than similarity in human societies. Postmodernism, whose influence has passed from architecture and art history to literature, anthropology and sociology, is concerned with the inability of any intellectual paradigm to represent the 'truth'. In the social sciences, in particular, postmodernism is concerned with the 'decentring of subjectivity' (Lash and Urry 1987). Taking its intellectual inspiration from Bourdieu (1984) Foucault and others, intellectual enquiry has turned cultural theory into an area of major academic concern (Foster 1984; Harvey 1990). For most of the social sciences today the main focus of attention is not the construction of grand theories of human behaviour, but rather, an enquiry directed at the business of social science itself. It is concerned with what we have invested in our models of human behaviour.

Economics has not been part of this intellectual enquiry. Riding the wave of neo-liberal policies in all the advanced industrial societies, economics has grown towards, rather than away from, the world view of the powerful. This implies clothing oneself in the language of 'objectivity', notably, a culturally grounded view of 'management' as the necessary means to effect desirable changes. In postmodernist terms the economists' view of the environment is a good example of 'displacement'. The environment is actually the product of human culture (our discourse about nature) but it tends to be represented as about science.

In its attempt to maintain its authority to prescribe, economics has chosen to ignore the analysis of difference, in favour of an analysis founded on similarity. It has chosen to understand what we do not know in terms of what we know. This brings us to a second set of issues, arising from our knowledge about the environment. That is the need to explore the human commitments of the models we use, especially those derived from neo-classical economics.

The human commitments underlying environmental models

It is usually assumed that most uncertainties over the environment lie in our knowledge of problems: how they occur, and their possible effects. What is usually ignored is the role of existing human commitments, and routinised human behaviour, in contributing to uncertainty. We need, in other words, to examine the assumptions that lie behind the way we use the environment, before we can really address the problem of uncertainty.

Economic frameworks and methods are founded upon an a priori commitment to a particular model of human nature and social behaviour. The assumption is that the market is an efficient way of allocating goods and services between competitive uses. Few economists believe that in practice the market mechanism works well enough to leave to uninterrupted forces, and government inevitably plays some part in modifying market forces. The premise of neo-classical economics is expressed by Gleick (1987, 181):

Modern economics relies heavily on the efficient market theory. Knowledge is assumed to flow freely from place to place. The people making important decisions are supposed to have access to more or less the same body of information. . . . on the whole, once knowledge is public economists assume that it is known everywhere.

There are two aspects of this model which deserve attention at this stage. First, social interaction is seen as essentially *instrumental* in the neo-classical model. It is designed to achieve the maximisation of utility for each individual. However, the tastes and preferences of individuals are considered a 'given'. It is not part of the business of economics to explain these tastes and preferences, whether they can be altered or prevent wider social objectives from being realised. In essence, individuals are only social beings because they live in societies: there is nothing about societies that makes them individuals.

Second, in the neo-classical model human behaviour is not seen as constituting value in its own right. Although human actions create value they do not possess value – unless it can be traded for an economic return, like a qualification or skill of some sort. This problem with intrinsic values bedevils the way economists look at the environment, since many environmental goods never reach the marketplace, or the values attached to them prove to be unrelated to use. The subjectivity of human beings, which makes them 'subject persons' in Max-Neef's phrase, rather than 'object persons', plays a very considerable role in the way they use the environment. However, if we start with a set of assumptions about human nature, like neo-classical economics, which we use to explain human behaviour, then it soon becomes difficult either fully to understand or to change that behaviour. You become locked into a set of self-fulfilling hypotheses, which are treated more like axioms, while the underlying assumptions of your model are not exposed to close examination. Cultural theorists, such as Michael Thompson (1991) have drawn attention to the dangers attendant on one view of rationality, while economists like Richard Norgaard (1991) have aschewed conventional economics in favour of a more ecumenical, interdisciplinary perspective. However, most of this work does not percolate into economics journals, and criticism that the paradigm used by economists is deficient can usually be countered by claiming that it is not the paradigm which is the problem, but its lack of refinement.

The view of human beings as 'rational individual calculators' is deficient in several ways. First, as I have argued, it is at best a partial view, which in no way captures the full value of the individual, and corresponds with only one mode of action in which the individual is employed. Second, it is ethnocentric, reflecting aspects of advanced industrial societies, whose histories have been interwoven with the discipline of economics. The rational individual calculator is a creation of one cultural heritage, it is not culturally neutral. It is not clear that rational processes are 'universal' in any sense, so we need to be careful (as many governments and their economic advisers fail to be) in not depicting behaviour as 'irrational' because it occurs within a quite distinct cultural context.

Having disposed of the 'rational' and the 'calculator' in the economist's formulation, we should not neglect the 'individual'. We need to ask whether calculations are individually based, and whether the model is able to recognise and represent social or collective calculations. To what extent was a Latvian or an Armenian in the Soviet Union, who was prepared to risk his life in defence of his ethnic identity, acting as an 'individual'? It is no good responding that, ultimately, behaviour lies with individuals whatever their identity. The point is that the calculations of the individual make no sense outside the framework of common cultural identity which informs the individual's actions. This observation, it needs to be stressed, is particularly relevant to the environment, as Garret Hardin's (1968) discussion of the 'tragedy of the commons' made clear.

The economist's model, then, provides a plausible hypothesis, but no more, for examining the individual's behaviour. We would need to know its conditions of validity to test it, and these are rarely specified in practice. The economist's model should not be treated as an axiom by policy-makers, as self-evident 'truth'. If we treat the economist's model as an axiom, then the assumptions that underlie it are not examined, but quickly become treated as normative impositions by policy-makers. We quickly move to the stage where, having forgotten that our analysis is predicated on a model of human behaviour, we look in despair at human attempts to depart from 'model behaviour'. At its best this involves the familiar gap emerging between policies and their implementation, in which people fail to respond as expected. At its worse it means social engineering of fascist or Stalinist proportions, as the state seeks to ensure that people behave according to 'scientific' laws and certainties.

The environment provides numerous examples of ways in which public disquiet has focused upon the assumptions behind policy, and the divergent ways in which the value of policies is assessed. If people are asked how much they are 'willing to pay' to avoid environmental risks, their difficulty is usually with the question, rather than the answer. William Waldegrave, the British Health Secretary, was quoted in December 1990 as saying that 'the language of commerce was more appropriate to supermarkets than to hospitals'. The language of environmental policy suffers from fewer inhibitions, but presents similar problems to many people. The North Sea is sometimes referred to as a 'waste sink resource', but this description, accurate for anybody seeking to dispose of sewage or chemical wastes, is very wide of the mark for fishing communities or (praise the Lord!) seaside holiday-makers. The North Sea regarded as an instrumental resource is treated differently from the North Sea as a supply depot (for fish) or a habitat (for marine life) or an amenity (for tourists). One of the problems of using an economic paradigm, to the exclusion of other approaches, is that the environment loses its multifaceted character. It no longer constitutes value in its own right.

Regulatory policy and the social authority of science

We are unlikely to be able to generate alternative models, and alternative policies to combat environmental problems, unless we can gain a perspective on the models we currently use. The problem with neo-classical economics is that recourse is made to it, as a conceptual framework, without examining thoroughly the assumptions on which it is based, and the weaknesses, as well as strengths, of this approach. Environmental policy is particularly vulnerable to the accusation that without examining the underlying assumptions on which it is based, we are likely to emerge with 'solutions' to a relatively narrow range of problems, which are defined in ways that are fundamentally flawed.

The first assumption is that the kind of behaviour people exhibit towards the environment, based on enlightened self-interest (the 'rational individual calculator') is inherently 'natural', rather than socially constructed. When social behaviour is heavily contested, those who argue for the status quo usually invoke 'naturalness' in their defence. In this way women's roles are often referred to as 'natural', since any other view of them would leave open the question of who should perform them, and how they should be performed. The environment is an analogous domain, in that it is centrally concerned with nature, and is said to obey natural laws, in many instances. Our vocabulary has incorporated this assumption with a vengeance: 'natural disasters', 'natural selection', and so forth. Indeed, it can be contended that the term 'nature' is usually invoked when it is the loss of natural qualities which is at issue – for example, over processed foods, colourings and additives. Recourse to 'naturalness' should send a message to our brains, requesting us to pay closer attention to the claims we are being asked to support. Human behaviour is so diverse, and prompted by so many cultural factors, which themselves change frequently, that the epithet 'natural' only applies under very strictly prescribed conditions. It cannot be taken as read that it is natural to be an 'optimiser' or a 'maximiser' or even a 'rational calculator'.

Second, if human behaviour cannot be considered 'natural', we cannot also assume that the knowledge people possess and utilise *explains the way that they behave*. As Giddens (1981, 19) has expressed it: 'knowledge is always *bounded* by unacknowledged conditions of action on the one side, and unintended consequences of action on the other'. The relationship between knowledge and human behaviour is complex, and 'the stocks of knowledge drawn upon . . . are at the same time the source of accounts . . . which individuals provide of their reasons, motives and purposes' (Giddens 1981, 17). Individual behaviour is the outcome of complex structures of possibility, limited resources of knowledge and mechanisms for self-justification. To reduce behaviour to a set of self-evident principles, such as that people respond to prices or markets in certain discernible ways, is simply to touch the surface of this behaviour. It excludes large areas of behaviour from attention, and reduces behaviour to a series of motives or commitments that are never explored systematically. This is extremely important in considering environmental policy, as we shall see.

The problem of being able to predict human behaviour looms large for environmental policy-makers, for whom it represents a major area of uncertainty. Environmental problems, as we have seen, elicit calls for more 'facts', 'hard' data and numerical indices. The ability to make predictions, on the basis of these facts, provides the scientific community, and the policy community to which it is linked, with its overriding social authority. The failure to deliver concrete predictions ('knowledge') undermines the authority of science and those in government.

It is hardly surprising, then, that those who make environmental policy should be so concerned about public responses to environmental risks and choices. The authority lies with science, and government, rather than consumers and the public at large. Public concern therefore needs to be carefully monitored and controlled, rather than opened up as a legitimate area of concern of democratic government. Scientific knowledge about environmental risks and choices is closely linked to the political and social authority of those who make policy. Rather than devolve power and decision-making to the people who meet environmental problems in their everyday lives, governments tend to assume 'responsibility' for environmental problems, divesting local people of the ability to take decisions for themselves. The principal objective of environmental policy is then the maintenance of the credibility of science and government, rather than the establishment of clear choices from which people can make personal decisions that affect their own lives. Until such time as unequivocal evidence places environmental knowledge in jeopardy, it is a public good, which the public has little access to, and little chance of seriously challenging. Environmental policy is part of regulatory science, a dialogue carried out, under restricted conditions, between scientists, administrators and politicians.

In practice, the terrain covered by environmental policy is much narrower than the public interest in the environment. As a facet of regulatory science, environmental policy is largely occupied with the (unforeseen) consequences of the failure of interventions in science and technology. Most environmental policy concerns, such as the disposal of waste, pollution and health concerns, arise from changes in our use of technology being 'handed over' to regulatory science. They are depicted in the literature as part of the responsive or reactive mode, which characterises environmental policy interventions (Holling 1978). Environmental policy is thus an adaptation which seeks to reduce or reverse the full impact of decisions over the way we use resources and the environment. Environmental problems are significant for policy when they can be addressed within the existing institutional framework which is established in advance of environmental considerations. Institutions such as the World Bank, which has sought to give more attention to the environment, have been forced to recognise that this applies to every programme area with which they are involved (Schramm and Warford 1989). However, it is a much more radical step to *anticipate environmental problems, by giving attention to underlying social commitments* which societies usually regard as self-evidently good, or simply above political consideration.

Human commitments and environmental knowledge in the North

The public interest in the environment extends well beyond the confines of what is usually defined as environmental policy. It includes all our prior social commitments, which are usually depicted as economic or technological. Examples of prior commitments include: individual motorised mobility, increased energy use, improved standard of living, and military security. These are not policy goals for most industrial societies, but rather internalised assumptions which industrial societies and their members take for granted. This is the sense in which they are 'prior' commitments. Each of these assumptions, and many others, carry enormous implications for environmental policy, and for the way we routinely manage our environment. In particular, the prior social commitment to, for example, private motor transport limits the effectiveness of environmental policy to address problems associated with cumulative impacts. The cumulative social impact of the commitment to more private cars is a great deal of road congestion, air pollution and global warming. The cumulative social impact of increased levels of energy consumption is a significant problem in disposing of wastes, and (in some cases) heavily depleted non-renewable resources. However, many of the cumulative impacts of individual decisions are unknown, or at least vague. Some technologies, such as the telephone or word-processor, seem environmentally benign, while others, such as nuclear power, the internal combustion engine, toxic chemicals and tobacco, seem environmentally malign.

It is clear to us today that most of the adverse effects of the technologies we employ are only discovered retrospectively, while the development of these technologies is pushed forward at breathtaking speed. The 'precautionary principle' is very rarely invoked in the development of technologies, precisely because this process corresponds with underlying social commitments which are rarely acknowledged.

However, as the environmental problems we confront become more severe, and in some cases (such as global warming and the loss of species) threaten our very survival, we need to do more than limit the damage we do to the environment. We need, in fact, to explore our underlying commitments, and the technologies on which they depend, before we can address the new order of environmental problems that are posed today.

The current preoccupation with environmental policy, then, can be interpreted in one of two ways. Either it is evidence of our concern and the priority we have chosen to give to environmental factors. This broadly, is the position of most governments in the industrialised world. Alternatively, the current preoccupation with environmental policy is evidence of environmental neglect, pointing to our failure to examine the underlying assumptions on which our economic growth has been purchased. If the first hypothesis is correct, we are up to the challenge, without needing to examine further the consequences of our behaviour. Environmental policy will identify the problems and deliver the goods.

If the second hypothesis is true, however, we are faced with a much greater challenge. This is to regard the management of a consensus over environmental issues as part of the problem, rather than part of the solution. By seeking to maintain the social authority of science, and assure the public that governments are able to use science to redress environmental problems successfully, we are effectively disabling ourselves. What we require is not an instrumentalist epistemology, such as that provided by environmental economics, but one which enables us to explore, critically, the way in which our environmental choices are socially constructed. We would need to move outside our current 'mind set', which considers that everybody has the right to pursue a higher standard of living (and energy use/waste production/resource depletion/habitat destruction) and explore environmental choices at source. This will take us into the domain not simply of what is happening to the environment and how it can be avoided, but also a shared responsibility for ensuring that environmental problems are avoided in the first place. This will require a new relationship between the political and regulatory agencies responsible for taking care of the environment and the 'public'. It would no longer be adequate for environmental policy to be carried out on the strength of the social authority of 'expert opinion'. Each society, including our own, would seek to define many environmental risks and dangers as avoidable, through redefining social commitments, rather than as part of their description of the 'natural'. This is a tall order, of course, but it is precisely the challenge that faces us today.

Human commitments and environmental knowledge in the South

What happens if we take our discussion one stage further, and consider the effects of regulatory policy and the neglect of prior social commitments in the developing countries? Conflicts in the South over resources and the environment are now part of a global picture, and provide evidence of public concern with global issues. This global concern was given a boost by the publication of the report of the World Commission on Environment and Development (Brundtland 1987) and the meeting of the United Nations Conference on Environment and Development (UNCED) in Brazil in June 1992. It is difficult today to separate many environmental problems in the North from their effects, and frequently their causes, in the South.

The first aspect of global environmental concerns which merits close attention is the way these concerns are represented in the discourse that surrounds them. Again, the postmodernist perspective can be illuminating, for we in the North tend to view environmental problems in the South in terms of our own preoccupations; through the lens of our own environmental concerns. Issues like the loss of tropical forests are represented in terms of values which are resonant in the North, such as the retention of wilderness areas or the conservation of animal species. Such issues, which mirror anxieties within the industrialised societies themselves, are real enough, but reflect the insecurities of these societies, rather than of people in the South. Indeed, poor people in developing countries are frequently blamed for contributing to the problem through overpopulation.

Concern for the state of the planet provides a powerful moral purpose, and confers legitimacy on film stars and politicians alike. It is no accident that rock singers rush to defend threatened Amazonian peoples, and perhaps no discredit to them, but it should alert us to the way that the environmental crisis is represented in the media. The mass media are the window through which we see the world, and the people who perform in that window acquire global celebrity. The environment which is under threat in the South can only be brought to our sitting rooms via television, and the images which television presents are packaged consumer items, whether they are images of poverty and distress or advertising videos. We watch the struggle of the Cayapo indians against a backdrop of Westerns, Australian soaps and Californian crime movies, from which they cannot be readily distinguished.

It should come as no surprise, then, that our discourse about the environment is not 'theirs', not that of the people in the South. Our understanding of the source of global problems is not located in the concerns of the less developed countries, still less of those of the rural poor. It is located in our own concerns, within the anxieties we feel about the consequences of getting and spending in our own societies. Hence the difficulty news programmes face in meeting the short attention span of their audiences. When one famine or flood follows another, what has been aptly termed 'disaster fatigue' soon sets in. Similarly, a willingness on the part of some television programme makers to explain the connections between development and environmental problems, always a difficult task, meets with audience figures that are almost derisory. It is the mass audience's passion for natural history films, many of them with an 'environmental' message, that is the true counterpart to Green consumerism, enabling us to choose more sustainable products, without raising the chimera of global sustainability, or straying too far from our television sets.

The way that environmental issues are communicated internationally suggests that the points raised earlier in this paper have equal relevance to the South. Much of the armoury of environmental economics, to be effective, require priority to be given to environmental quality. As Pearce *et al.* (1989) note, there is considerable evidence that the quality of the environment is a growing concern in the industrialised countries. Environmental economics has succeeded in addressing these concerns. By exploring the environmental costs of economic growth, and the economic value of conserving resources, it has sought to incorporate what would otherwise be divergent schools of thought.

In the South environmental problems are just as pressing, but their occurrence is linked to scarcity, rather than plenty, to poverty rather than riches. Assumptions about the priority we can attach to the environment over the short term are less easily made when the livelihood of most of the population leaves little room for manoeuvre. In Mexico, for example, recent government decrees have prohibited the cutting down of tropical forest in the state of Yucatán. Nevertheless, much of the coast has been stripped of vegetation in preparation for still more extensive tourist development, aimed at foreigners. In the view of poor farmers in the area it devalues them to be given less attention than the trees and animals, and to witness, in addition, the avarice of entrepreneurs in the tourist industry. To be urged to act more sustainably in this context makes little sense, since environmental quality only represents additional welfare for one privileged section of the population. Environmental economists like Pearce have argued that tropical forests are destroyed because the marginal opportunity costs of preserving them are rarely calculated (Pearce *et al.* 1989, 45–6). However, poor people do make calculations of the opportunity costs attached to removing tropical forests, although these calculations take account of their labour, rather than the physical resource which they are destroying. It is a calculation forced on them by poverty.

Conclusion

This paper has argued that before endorsing recent developments in environmental economics it is worth asking how we arrive at the models we employ, and how much they tell us about human behaviour and perceptions. Environmental policy has been extended and refined by recent work in environmental economics, but none of it has examined the underlying social commitments which determine the way in which we use the environment. In the search to provide universal models, and in the quantitative form used within the physical sciences, economics risks ignoring cultural differences which help to explain behaviour. It ignores the role of policy in gaining social authority for science and policy-makers, regardless of the 'truth' of scientific evidence, or the risks and dangers to which the public is exposed, often through ignorance. In particular, it ignores the cumulative social impact of individual choices, which neo-classical economics pays little attention to.

In addition, before confidently embarking on cross-cultural exercises in environmental policy, we need to consider the links between environmental knowledge and power in developing countries. We need to reflect upon the part played by 'development' in devaluing non-professional knowledge and seeking to marginalise political struggles for control over environmental goods. These are urgent tasks, which are receiving belated attention (UNRISD 1990) but this is not the place to pursue these lines of enquiry. They constitute additional evidence that if the environment is looked upon as marginal to the social sciences we are all the poorer for it.

References

Bourdieu, Pierre (1984) *Distinction: a social critique of the judgement of taste* (Routledge, London).
Brundtland, H.G. (1987) *Our Common Future*, Report of the World Commission on Environment and Development (Oxford University Press, Oxford).
Foster, H. (ed.) (1984) *Postmodern culture* (Pluto, London).
Gleick, James (1987) *Chaos* (Cardinal, London).
Giddens, A. (1981) *A Contemporary Critique of Historical Materialism* (Macmillan, London).
Hardin, G. (1968) The tragedy of the commons. *Science* 162, 1243–8.
Harvey, David (1990) *The Condition of Postmodernity* (Blackwells, Oxford).
Holling, C.S. (ed.) (1978) *Adaptive Environmental Assessment and Management* (John Wiley, Chichester).
Lash, Scott and Urry, John (1987) *The End of Organized Capitalism* (Polity Press, Oxford).

Norgaard, R. (1991) *Development Betrayed* (Routledge, London).

Pearce, D., Markandya, A. and Barbier, E. (1989) *Blueprint for a Green Economy* (Earthscan, London).

Schramm, G. and Warford, J. (eds) (1989) *Environmental Management and Economic Development* (Johns Hopkins University Press, Baltimore, MD).

Thompson, Michael (1991) Policy making in the face of uncertainty. MS.

UNRISD (United Nations Research Institute for Social Development) (1990) *Sustainable Development through People's Participation in Resource Management*, (UNRISD, Geneva).

Chapter 5

Valuation of the Environment, Methods and Techniques: The Contingent Valuation Method

Ian J. Bateman and R. Kerry Turner

Introduction and overview

The valuation of environmental resources

Sustainable development (SD) has become a catch-all phrase for forms of economic development which highlight the need to retain an 'acceptable' level of environmental quality and to conserve nature's assets. But from the conventional economic perspective, the sustainability issue has at its core the phenomenon of market failure and its correction via 'proper' resource pricing. Thus, what is required is a strategy that ensures an intertemporally efficient allocation of environmental resources through price corrections based on individual preference value.

A large environmental economics literature has therefore grown up since the late 1960s, encompassing a range of monetary valuation methods and techniques designed to 'price' the spectrum of environmental goods and services provided by the biosphere. Because of the fact that many environmental goods and services are non-marketed commodities, the valuation methods utilised involve market-adjusted, surrogate and simulated-market approaches.

As far as conventional economic theory is concerned, the value of all environmental assets can be measured by the preferences of individuals for the conservation or utilisation of these commodities. Given their existing preferences and tastes, individuals will hold a number of values which, in turn, result in objects being given various assigned values. In order, in principle, to arrive at an aggregate measure of value (total economic value) economists begin by distinguishing user values from non-user values (Pearce and Turner 1990a).

Figure 5.1 illustrates the use and non-use values which a multiattribute environmental asset, such as a woodland, provides. By definition, use values derive from the actual use of the environment. Slightly more

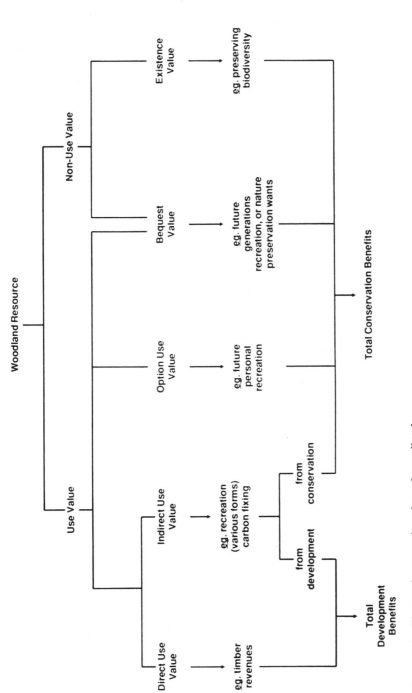

Figure 5.1 The total economic value of woodland

complex are values expressed through options to use the environment (option values) in the future. They are essentially expressions of preference (willingness to pay) for the conservation of environmental systems or components of systems against some probability that the individual will make use of them at a later date. Provided the uncertainty concerning future use is an uncertainty relating to the 'supply' of the environment, economic theory indicates that this option value is likely to be positive (Weisbrod 1964; Cicchetti and Freeman 1971; Krutilla and Fisher 1975; Kriström 1990). A related form of value is bequest value, a willingness to pay to preserve the environment for the benefit of one's descendants. It is not a use value for the current individual valuer, but a potential future use or non-use value for his or her descendants.

Non-use values are more problematic. They suggest non-instrumental values which are in the real nature of the thing but unassociated with actual use, or even the option to use the thing. Instead such values are taken to be entities that reflect people's preferences, but include concern for, sympathy with and respect for the rights or welfare of non-human beings. These values are still anthropocentric but may include a recognition of the value of the very existence of certain species or whole ecosystems. Total economic value is then made up of actual use value plus option value plus existence value.

During the 1980s more extensive use of monetary valuation methods was combined with technical improvements in techniques. The result is a large literature consisting of a wide diversity of valuation case studies, both in terms of environmental assets and valuation methods (Pearce and Markandya 1989; Turner *et al.* 1992b; Smith 1992a).

Approaches to valuation

Figure 5.2 illustrates one way in which the various approaches and methods of monetary valuation can be classified, in the context of environmental resources. Two basic approaches are distinguished, those which value a commodity via a demand curve (Marshallian or Hicksian) and those which do not and therefore fail to provide 'true' valuation information and welfare measures. These latter methods are, however, still useful heuristic tools in any cost–benefit appraisal of projects, policies or course of action (Turner *et al.* 1992a).

Recent debates, both within the economics profession and between economists and non-economists, about the 'usefulness' (i.e. the reliability and validity) of one of the valuation methods, the contingent valuation method (CVM), have brought to the surface several general and fundamental questions. Thus, the theory, methodology and application

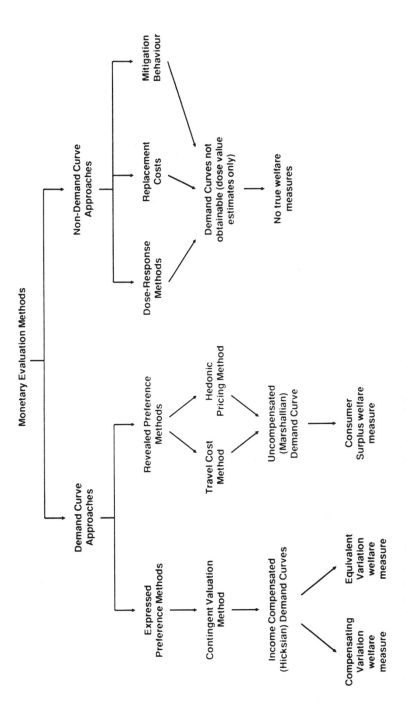

Figure 5.2 Methods for the monetary evaluation of the environment

of non-market valuation of environmental resources have all come under close scrutiny. Harris and Brown (1992), for example, have noted that who gains or loses from some environmental change and how motivations (feelings of responsibility and/or self-interest) influence the value judgements of those gainers and losers, all have important implications for CVM.

The contingent valuation method

The CVM requires that individuals express their preferences for some environmental resources, or change in resource status, by answering questions about hypothetical choices. The very nature of this methodology has therefore meant that CVM has been subject to criticism from both economic and psychological experimentalists, whose growing research focus has been the problem of preference elicitation. This criticism has in turn caused supporters of CVM to pay much more attention to a testing protocol in which questions of method reliability and validity are directly addressed.

The respondents to a CVM questionnaire will be asked a variety of questions about how much they would be willing to pay (WTP) to ensure a welfare gain from a change in the provision of a non-market environmental commodity; or how much they would be willing to accept (WTA) in compensation to endure a welfare loss from a reduced level of provision. A basic question for the implementation of the CVM is therefore whether WTP or WTA is the most appropriate indicator of value in a given situation.

For cost–benefit analysis based on the Hicks–Kaldor compensation test, WTP would seem to be the appropriate measure for gainers from some resource-allocation decision, and WTA the proper measure for losers from that same reallocation. But, as Harris and Brown (1992) have pointed out, it is often not easy conclusively to identify gainers and losers since this judgement is itself influenced by the valuer's own perspective.

Willig (1976) claimed that WTP and WTA measures should, in the absence of strong income effects, produce estimates of monetary value that are fairly close (within 5%). However, since 1976 strong evidence has been accumulated which shows that, for given environmental goods, WTA is significantly greater than WTP (over 40% divergence). In addition, WTA valuations seem to have greater variance than WTP ones, and are less accurate predictors of actual buying and selling decisions.

The format of the questions used to elicit valuations may be *continuous* (or 'open-ended'), i.e. asking respondents to state WTP or

WTA without any prompts concerning possible answers, or *discrete* (or 'dichotomous'), i.e. presenting the respondent with a single buying price or selling price which must be accepted or rejected. Many intermediate formats are also possible, e.g. bidding games. These differences in format can produce systematically different responses (Desvousges *et al.* 1987; Loomis 1990).

A number of explanations have been offered for the differences in valuations elicited by different formats:

(i)   There may be income effects, as predicted by Hicksian consumer theory. In a recent paper Hanemann (1991) has argued that such effects could account for some observed WTP/WTA differences for public goods. He has calculated that a WTA measure five times greater than WTP can be justified in cases where the elasticity of substitution is low and/or the WTP/income rate is high, i.e. for unique, irreplaceable environmental assets about which individuals care a great deal.

(ii)   A psychological phenomenon, loss aversion, may be important especially in the case of potential losers in a resource change when WTA questions are related to giving up things, rights or privilege (Eberle and Hayden 1991; Bishop and Heberlein 1979). Valuations may be made relative to *reference points*, losses being weighted more heavily than gains. Such effects, which could account for some WTP/WTA differences, have been found experimentally (see, for example, Knetsch and Sinden 1984). Similarly *anchoring* effects (or starting point bias) may cause differences between responses to discrete and continuous formats (Green and Tunstall 1991).

(iii) WTA questions may be less readily understandable than WTP ones, since most people have more experience of buying goods, paying taxes, etc. than of selling (Hoehn and Randall 1987; Coursey *et al.* 1986). Similarly, continuous questions may be less readily understandable than discrete ones, since most people have more experience of choosing whether or not to pay stated prices than of stating valuations.

(iv)   The continuous format may have a stronger tendency than the discrete format to suggest opportunities for free riding.

(v)   Respondents may act strategically, i.e. make guesses about how their answers will be used and then give the answers that they believe will serve their interests best.

Overall, it is likely that merely identifying gainers and losers in some resource change situation will be insufficient to determine whether WTP or WTA is the most appropriate indicator of value. We need to know

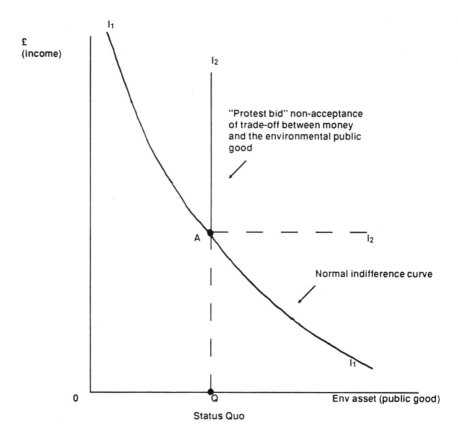

Figure 5.3 'Protest bidding' over the loss of implicit environmental property rights

more about the motives of the valuer (Harris and Brown 1992). Economics has much to learn from psychological research in this context. In fact, some of CVM's strongest critics are to be found outside the economics profession, in the ranks of philosophers, psychologists, political scientists and scientists.

Debate outside economics

Sagoff (1988) has argued that economics makes a 'category mistake' in its approach to environmental valuation. For him, it is not preferences but attitudes that determine people's environmental valuations. Thus people may not be willing to consider market-like transactions (assumed by CVM) involving public resources. CVM surveys pick this effect up in the form of refusals and 'protest bids' (see Figure 5.3). Some combination of individual preferences and public (collectively held) preferences will be held by any given individual who by necessity has to operate in daily life as both a consumer and a citizen. Thus the environment can be both a purchased commodity and a moral or ethical concern (Turner 1988a; 1988b).

According to Sagoff, environmental economics has no role to play in the determination of the goals of environmental policy. Environmental protection standards are determined by political, cultural and historical factors, not by preference-based values. If economics has a role it is restricted to revealing the costs (social opportunity costs) of the pre-emptive environmental standards. But if action is taken on the basis of the opportunity cost analysis then an implicit valuation has been made. Nevertheless, from this viewpoint, there is no role for direct monetary valuation (preference-based) of the benefits of environmental protection policy.

Other critics do not go as far as completely rejecting the validity of WTP/WTA measures of value, but instead argue that economic values are only partial values for many environmental resources. Thus Brennan (1992) seeks to distinguish so-called transformative value from the economic value of environmental assets.

Many environmental assets, according to Brennan, possess additional transformative properties such that 'exposure' to the assets cause a change in people's preferences. The impact of transformative values on people's preferences is said to be completely unpredictable and the degree of impact (when it occurs) will vary significantly from person to person. If natural things and systems possess transformative results, then, it is argued, they cannot be priced by economic analysis.

As far as we can see, this transformative value argument can be accommodated within the total economic value approach. The transformative value is equivalent to a use or existence value, except that it is latent, and for some individuals may never actually exist, i.e. their preferences are never changed (transformed) by contact with or knowledge of the specific thing that possesses the transformative property. At the policy level, if it is the case that natural things and systems possess transformative values then a conservation strategy based on the total economic value principle (in particular, option and bequest values) would be sufficient to guarantee their future existence. So individuals may value (exhibit a WTP for) natural things and systems in order to retain the option to use them, or be transformed by them, some time in the future. Bequest value similarly expresses an individual's wish to retain options for their descendants.

Some scientists have argued that the full contribution of component species and processes to the aggregate life-support service provided by ecosystems has not been captured in economic values (Ehrlich and Ehrlich 1992). There does seem to be a sense in which this scientific critique of the partial nature of economic valuation has some validity – not in relation to individual species and processes but in terms of the prior value of the aggregate ecosystem structure and its life-support capacity.

Since it is the case that the component parts of a system are contingent on the existence and functioning of the whole, then putting an aggregate value on ecosystems is rather more complicated a matter than has previously been supposed in the economics literature. Taking wetland ecosystems as an example, the total wetland is the source of *primary value* (PV) (Turner 1992). The existence of the wetland structure (all its components, their interrelationships and the interrelationships with the abiotic environment) is prior to the range of function/service values. The concept of a total economic value (TEV) has two limitations, rather than one as previously supposed. TEV may fail fully to encapsulate the total secondary value (TSV) provided by an ecosystem, because in practice some of the functions and processes are difficult to analyse (scientifically) as well as to value in monetary terms. But in addition, TEV fails to capture the PV of ecosystems; indeed, this 'existence' or 'glue' value (e) notion is very difficult to measure in direct value terms since it is a non-preference, but still instrumental, type of value.

We believe that this primary and secondary ecosystem value classification goes some way towards satisfying many scientists' concerns about the 'partial' nature of the conventional economic valuation approach (Ehrlich and Ehrlich 1992). It is also a classification that avoids the instrumental versus non-instrumental value in nature debate which we believe has become rather sterile.

More formally:

each ecosystem provides a source or stock of primary value = PV = e ('existence' or 'glue' value of the ecosystem);
the existence of a 'healthy' ecosystem (i.e. one which is stable and/or resilient) provides a range of functions and services (secondary values) = TSV;
$TV = TSV + e$ (a prior system 'existence' or 'glue' value);
so total ecosystem value = TV;
and $TV > TSV$;
$TSV = TEV$ (total economic value)
where $TEV = UV$ (use value) + NUV (non-use value);

$TV = (TEV, e)$;
and $TV = (TEV, 0) = 0$;
and $TV = (0, e) \geqslant 0$.

What is clear is that the components of TEV (use and non-use values) cannot simply be aggregated. There are often trade-offs between different types of use value and between direct and indirect use values. Smith (1992a) has also pointed out that the partitioning of use and non-use values may be problematic, if it is the case that use values may well depend on the level of services attributed to non-use values. The TEV approach has to be used with care and with a full awareness of its limitations.

Towards an understanding of the valuation process

Figure 5.4 summarises the various elements thought to comprise the full valuation process. The interaction between a person and an object (to be valued) involves perception of the objects and a process whereby relevant held values, beliefs and dispositions come to the fore. Perception and beliefs are interrelated and together result in an unobservable sense of value (unity), which may then be expressed as an assigned value and certain behaviour (Brown and Slovic 1988). Brown and Slovic conclude that the valuation context may affect how objects are perceived, the beliefs that become relevant, the utility experienced and the value assigned.

Information (existing and new) plays a key role in the valuation process. An individual's familiarity with the environmental commodity/context and the resulting perceptions are dependent on both the information stock and the provision of new information. The type and form of information supplied is particularly important in situations where direct perception is not possible and recourse to 'expert' knowledge is required.

Perceptions, information and beliefs all then feed into motivation. Harris and Brown (1992) identify what they call a responsibility motive in the environmental loss context. The motive is best represented as a spectrum of feelings extending from personal responsibility to a more general concern for the environment unrelated to use value. Randall (1987) has argued that all non-use values have their basis in the motive of altruism – interpersonal, intergenerational and Q-altruism (based on the knowledge that some asset Q itself benefits from being undisturbed). What this discussion of motivation does is to question the simplistic 'rational economic person' psychological assumptions that underpin conventional economic analysis. The motive of self-interest is only one

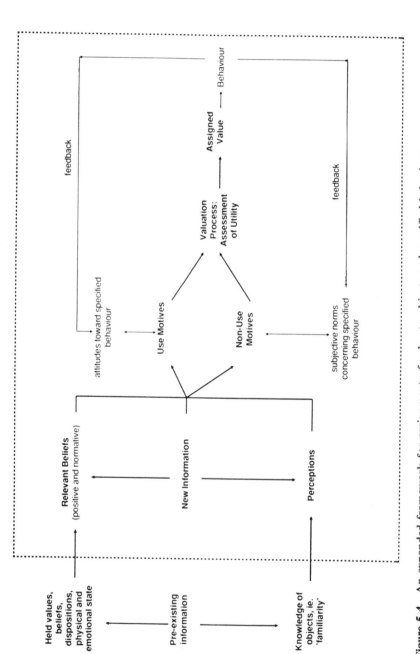

Figure 5.4 An expanded framework for assignment of value to objects and specified behaviour

Source: Adapted from Harris and Brown (1992) and Ajzen and Fishbein (1977)

of a number of human motivations and need not be the dominant one.

Maslovian psychology, for example, substitutes the concept of human needs for human wants and portrays needs in a hierarchical structure (Maslow 1970). Instead of an individual facing a flat plane of substitutional wants (as in conventional economics), Maslow conceives of the same individual attempting to satisfy levels of need. The satisfaction of higher-level needs leads to a process of 'self-actualisation'. Self-actualised individuals would be expected to possess a strong responsibility motivation and hold non-use values. Such individuals might well be prepared to pay to maintain some environmental asset regardless of the benefits they themselves receive from that asset.

Because of the divergence between WTP and WTA valuations, many practitioners have taken the pragmatic decision to regard stated WTP valuations as reliable measures of true WTP and therefore to use CVM only in cases in which WTP is the appropriate measure of benefit. But this raises the question as to what is the exact set of cases in which WTP is appropriate. Harris and Brown (1992) have argued that WTP is in fact the appropriate measure of welfare change for a majority of situations. They identify only self-interested losers from a resource change as the appropriate group to be surveyed with WTA format questions.

A mail survey of Idaho taxpayers undertaken (in 1988) by the same researchers indicated that 53% of the sample felt that their state should pay for the loss of non-game wildlife with tax dollars, implying that *all* taxpayers should pay to cover this loss (a WTP rather than WTA approach). Only 32% of survey respondents said that only those responsible for the loss should pay to prevent it (WTA approach). Thus altruism and moral responsibility will have an important role to play in influencing policy judgements.

We examine further this issue of WTP/WTA divergences and the implications for CVM in a following section, where we present some theoretical analysis which indicates that a degree of divergence (beyond the Willig limits) is to be expected and that there is an appropriate WTP measure for welfare loss.

Future role for monetary valuation methods

Overall, there are grounds for cautious optimism about the role that economic valuation methods can play in environmental policy and management. While there will be a limit to the coverage of meaningful monetary valuation in the environmental context, the establishment of this limit is still an open research question (see Figure 5.5). There is a need to move beyond the current situation in which asset valuation

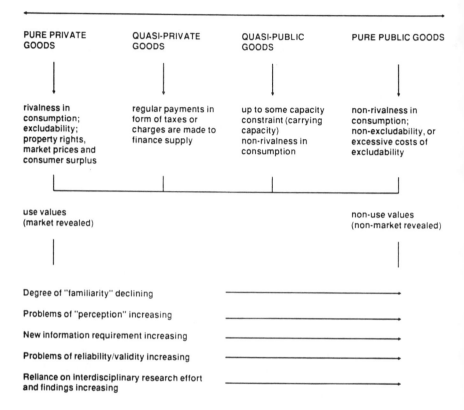

Figure 5.5   Environmental commodities spectrum

information exists in the form of a diverse set of individual environmental asset valuation studies. Policy-makers will require classes of environmental commodities and value measures which are robust enough to be capable of being consistently aggregated and disaggregated and possibly transposed across national boundaries.

The remaining sections of this paper are devoted to an analysis of CVM. The next section examines the basic economic theory underpinning CVM and the divergence between WTP and WTA measures of value. The section following that analyses a number of methodological issues: reliability and the nature of bias in CVM, as well as validity issues. The final section presents some conclusions and recommendations for a 'best practice' protocol in the application of CVM.

The contingent valuation method: Method and economic theory[1]

Hanley (1990) identifies six distinct phases involved in the practical application of CVM which we have interpreted as follows:

Stage 1: Preparation
(i)   Set up the hypothetical market: individuals may be presented with two basic variants:
       (a)   How much are you willing to pay (WTP) for a welfare gain?
       (b)   How much are you willing to accept (WTA) in compensation for a welfare loss?
(ii)   Define the elicitation method. The major alternatives are:
       (a)   Open ended: 'how much are you willing to pay?' This approach produces a continuous bid variable and may therefore be analysed using OLS/ML approaches.
       (b)   Take-it-or-leave-it (dichotomous choice): 'Are you willing to pay £X?', the amount X being systematically stepped across the sample to test individuals' responses to different bid levels. This approach produces a discrete bid variable and requires logit-type analysis.
       (c)   A recent variant upon the dichotomous approach is to supplement the initial question with an iterative second round (double-bound) question (see Hanemann *et al.* 1991). For example, if the respondent answers 'yes' to the £X bid then he is asked if he is WTP £2X (or £0.5X if he answered 'no' to the initial question).[2]
       (d)   Other elicitation methods include the use of payment cards and bidding games with suggested starting points.
(iii) Provide information regarding:
       (a)   the quantity/quality change in provision of the good
       (b)   who will pay for the good
       (c)   who will use the good
(iv)   Define the payment vehicle, for example:
       (a)   higher taxes
       (b)   entrance fees
       (c)   donation to a charitable trust

Stage 2: Survey
   Obtaining responses to the questionnaire. Interview can be either on-site (face to face; users only), house to house (face to face; users and non-users) or by mail or telephone (remote; users and non-users).

Stage 3: Calculation

Calculate the mean WTP (or WTA) from responses. This commonly involves the omission of protest votes[3] and/or the use of trimmed means. In a dichotomous choice format experiment the mean is obtained by calculating the expected value of the dependent variable (WTP or WTA).[4]

Stage 4: Estimation

A bid curve can be estimated to investigate the determinants of WTP bids. For a continuous question format OLS estimation techniques are often employed. Typically, in WTP scenarios, the bid curve will relate bids (WTP_i) to visits (Q_{ij}), income (Y_i), social factors such as education (S_i), and other explanatory variables (X_i). A parameter of the environmental quality of the site (E_i) may also be included.

$$WTP_i = f(Q_{ij}Y_i,\ S_i,\ X_i,\ E_i)$$

There is no theoretical correct form of this function. However, if a log-log function is chosen then the coefficients are elasticities. In such a case the bid curve allows us to estimate changes in mean WTP_i arising from changes in E_i. Indeed, if the other relationships are sufficiently stable then we can use this curve to evaluate changes to other strongly related environmental goods, for example impacts of water quality change upon wetland quality.

If a dichotomous payment format has been used then a logit approach is required, relating the probability of a 'yes' answer to each suggested sum to the explanatory variables listed above.[5]

Stage 5: Aggregation

This is required in order to move up from mean WTP to total value. This entails decisions about, for example, moving between household and individual data, and distinguishing the relevant population.

Stage 6: Appraisal

Was the CVM successful?

To answer the question posed in stage 6 we need to consider the technical acceptability of the evaluation estimates produced by CVM. However, before we do so it should be emphasised that in practice this is only one of the criteria upon which both CVM and all other evaluation methods are likely to be judged. Figure 5.6 illustrates four facets of method acceptability: technical (whether the evaluation estimates are valid and reliable, i.e. theoretical and methodological acceptability);

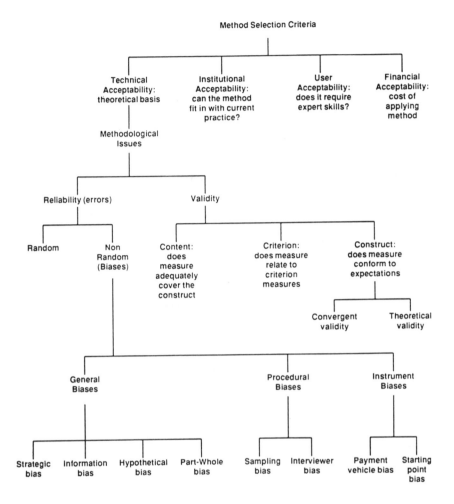

Figure 5.6   Criteria for the selection of a monetary evaluation method

institutional (whether decision-makers can incorporate the method into their framework of analysis); user (whether analysts sufficiently comprehend the technique so as to put it into practice); and financial (whether the cost of application is reasonable). This chapter concerns itself only with issues of technical acceptability. However, while analysts and academics may see this of prime interest, it should be noted that the other acceptability issues may in the event prove as important in the selection of an evaluation method.

Welfare change measures and the CVM: A theoretical overview

In estimating monetary values for environmental resources we are concerned with how changes in the provision of environmental public goods impact upon individuals' utility. Traditionally the welfare gain or loss from such changes of provision has been approximated by changes in consumer surplus;[6] the area underneath the ordinary (Marshallian) demand curves and above the price level.[7]

The Marshallian demand curve tracks the 'full price effect' which occurs when the provision of a good changes. Typically it has been used to show how much the quantity consumed of a normal good increases when its price falls. A practical problem therefore arises in estimating the Marshallian demand curve for an unpriced environmental public good. Without private property characteristics, such as rival consumption and excludability, a good cannot be traded in a market and the price/consumption information required to estimate the Marshallian demand curve will not be directly observable. One solution is to estimate the Marshallian demand curve via a surrogate market, for example using incurred travel costs as a proxy for the recreational value of an open-access leisure site. However, a more fundamental theoretical problem remains in that the presence of income effects means that consumer surplus itself can give an inaccurate measure of the welfare change resulting from a change in good provision.

In the case of environmental public goods the individual is usually faced with a quantity rather than a price constraint, the good often being unpriced. Furthermore, these goods often have much higher income elasticities[8] than those associated with many market goods (Bateman *et al.* 1992). The consequently large income effect arising from a change in quantity provision may undermine the consumer surplus measure of welfare change. In order to move from the ambiguity of consumer surplus to a theoretically more accurate measure of welfare change we therefore need to compensate for the income effect by holding real income constant, i.e. moving from using the ordinary Marshallian demand curve to the compensated (Hicksian) demand curve.

The Hicksian approach evaluates welfare change as the money income adjustment necessary to maintain a constant level of utility before and after the change of provision. Two such welfare change measures are feasible for such an approach. The 'compensating variation' (*CV*) is the money income adjustment (welfare change) necessary to keep an individual at his initial level of utility (U_0) throughout the change of provision, while the 'equivalent variation' (*EV*) is the money income adjustment (welfare change) necessary to maintain an individual at his final level of utility (U_1) throughout the provision change.

We therefore have two approaches to measuring welfare changes. Furthermore, these changes can be either positive (a welfare gain) or negative (a welfare loss) giving us four possible scenarios. For a proposed welfare gain (i.e. a change in provision which increases utility, for example more recreation or less pollution) the CV measure tells us how much money income the individual should be willing to give up (WTP) to ensure that the change occurs,[9] while the EV measure tells us how much extra money income would have to be given to an individual (WTA) for that person to attain the final improved utility level in the absence of the provision change occurring.[10] For a proposed welfare loss (i.e. a change in provision which decreases utility, for example less recreation or more pollution) the EV measure will now show how much an individual is WTP to prevent the welfare loss occurring,[11] while the CV measure now shows the individual's WTA compensation for allowing the welfare loss to occur.[12]

These variation measures (CV and EV) only strictly apply where the consumer is free to vary continuously (i.e. non-discretely) the quantity of the good consumed. Where the consumer is constrained to consume only discrete or fixed quantities (as for most environmental public goods) then we should consider compensating *surplus* (CpS) and equivalent *surplus* (ES) measures in place of CV and EV, respectively. Appendix 5.1 discusses in more detail the relationship between welfare measures for price- and quantity-constrained goods. Table 5.1 defines four economic welfare change measures.

The upper panel of Figure 5.7 shows a utility curve analysis of welfare gain and loss measures in the context of an unpriced, quantity-constrained environmental good X_1. Provision of X_1 is shown on the

Table 5.1 Welfare measures

	Proposed change	Measure	Continuous Consumption Function	Non-Continuous Consumption Function
Type 1	Welfare gain	WTP to ensure that change occurs	CV_{WTP}	CpS_{WTP}
Type 2	Welfare gain	WTA if gain does not occur	EV_{WTA}	ES_{WTA}
Type 3	Welfare loss	WTP to avoid loss occurring	EV_{WTP}	ES_{WTP}
Type 4	Welfare loss	WTA if loss does occur	CV_{WTA}	CpS_{WTA}

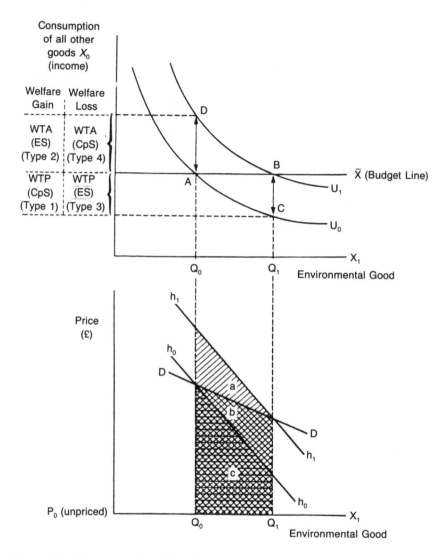

Figure 5.7 Compensated welfare change measures for an unpriced quantity-constrained good

horizontal axis, while the vertical axis shows income as a money-composite of all other consumption X_0. Because X_1 is unpriced, the budget line is shown as the horizontal line $\bar{X}$ with initial consumption of X_1 being quantity- rather than price-constrained at Q_0, corresponding to point A on initial utility curve U_0.[13] Suppose that a welfare gain is proposed, increasing provision of X_1 from Q_0 to Q_1. This is shown as a move from point A on U_0 along the budget line to point B on U_1. This corresponds to the full price effect shown by the Marshallian demand curve DD in the lower panel and the corresponding increase in consumer surplus shown by the shaded areas $b + c$. Despite X_1 being itself unpriced, its increased provision will still have an income effect by releasing some of that income previously spent upon priced goods (e.g. if Q is recreation then its increased provision relieves spending upon other priced recreation goods). Consumer surplus is therefore only an approximate measure of the true welfare change. We can compensate for the income effect and obtain a correct welfare change measure by asking how much the individual is WTP to ensure that the increase in provision does occur (a type 1 scenario in Table 5.1). The individual should be prepared to give up the amount of income BC which returns him to point C on his initial utility curve U_0 but with the increased provision Q_1. The corresponding compensated demand curve $h_0 h_0$ is shown in the lower panel and it is the shaded area c under this curve which correctly measures the welfare change for this scenario (CpS_{WTP}).

Now suppose that the same proposed welfare gain (Q_0 to Q_1) is not implemented (type 2 scenario). The authorities could still raise the individual's utility from U_0 to U_1 by increasing money income by the amount AD (the equivalent value of extra income which individuals are WTA to forgo the welfare gain change in provision). This moves the individual to point D on U_1 and maps out the compensated demand curve $h_1 h_1$ in the lower panel. The correct welfare measure for this scenario is therefore the equivalent surplus ES_{WTA} (the shaded area $a + b + c$ in the lower panel). Note then that for a welfare gain we have $CpS_{WTP} <$ consumer surplus $< ES_{WTA}$, in short WTP is less than WTA.

Now consider a proposed welfare loss, say a decrease in the provision of the same unpriced environmental good from Q_1 to Q_0. Here the individual will start at point B, and the initial utility curve will be U_1. Faced with a fall to point A on new utility level U_0, the individual will be WTP the amount BC to avoid the loss (ES_{WTP}; type 3 scenario).[14] However, if the welfare loss change in provision does occur, then the authorities can still compensate the individual by giving him extra income AD to return him to his initial utility level U_1, i.e. CpS_{WTA} (type 4 scenario).[15] Note that for the welfare loss we now have $ES_{WTP} <$ consumer surplus $< CpS_{WTA}$. Therefore, for either gains or losses, the WTA measure exceeds WTP; however, the derivation of these measures (i.e. CpS or ES) changes. Appendix 5.2 presents an expenditure function approach to welfare measures.

In summary, as we have seen, there are theoretical problems with the consumer surplus measure of welfare change. Yet, because of the impossibility of mapping utility functions, consumer surplus measures have often been calculated as best practical estimates of welfare change. The CVM approach, in eliciting explicit statements of how much income consumers are WTP to ensure that a welfare gain occurs (or prevent a welfare loss occurring) or how much income they are WTA to endure a welfare loss (or forgo a welfare gain), is, in theory, directly estimating the true Hicksian welfare measures of these changes (see Appendix 5.2 for formal proof). Although in later sections we address several important methodological criticisms of the empirical method, this theoretical ability to estimate true welfare measures represents a considerable potential advance over other approaches and deserves emphasis.

Theoretical and empirical asymmetry of welfare measures:
WTP versus WTA

At first glance we might have expected there to be no difference in the amount which consumers would be WTP for a specific welfare gain compared to the amount which they would be WTA in compensation for an equivalent loss; indeed, neo-classical utility theory might well lead us to expect such a result (Schoemaker 1982). However, as Figure 5.7 illustrates, there is a theoretical asymmetry between WTP and WTA measures such that WTP for the welfare gain (i.e. move from A to B, type 1 change; CpS_{WTP}) is exceeded by WTA compensation for the welfare loss (i.e. move from B to A, type 4 change; Cps_{WTA}).

In his seminal articles, Willig (1973; 1976) showed that, for priced normal goods in most plausible situations, the deviation between compensating and equivalent variation measures should be relatively small (thus promoting consumer surplus as a valid welfare measure). The Willig limits suggest that Hicksian WTP and WTA measures should generally lie within 2% either side of the Marshallian consumer surplus. These results using Hicksian analysis were formulated for price changes (Hicks 1943). We can show that this asymmetry is, in theory, slightly more pronounced for unpriced goods subject to quantity constraints (see Appendix 5.1).

Nevertheless, these limits in no way provide a theoretical explanation of the very wide WTP/WTA asymmetry found in empirical testing. Table 5.2 shows that in practice CVM studies have recorded very wide divergence between WTP and WTA. These large divergences have caused considerable concern about the validity of CVM. We therefore need to consider whether such a pronounced empirical asymmetry is indicative of a fundamentally flawed methodology or whether it has any theoretical plausibility.

Table 5.2 Empirical divergences between WTP and WTA

Study		WTA/WTP
Knetsch and Sinden (1984)		4.0
Coursey *et al.* (1983)	(i)	3.8
	(ii)	1.6
Brookshire *et al.* (1980)	(i)	1.6
	(ii)	2.6
	(iii)	6.5
Bishop and Heberlein (1979)		4.8
Banford *et al.* (1977)	(i)	2.8
	(ii)	4.2
Hammack and Brown (1974)		4.2

Source: adapted from Pearce and Markandya (1989).

Varian (1984) derives the following approximation of the Willig formulae:

$$\frac{CV - CS}{CS} \simeq \frac{CS \cdot \eta}{2Y^0}$$

where

CV = compensating variation (here WTP)
CS = Marshallian consumer surplus
η = income elasticity of demand
Y^0 = initial income (expenditure).

Willig (1973; 1976) shows that, for the priced good case, such errors are likely to be small.[16] However, this error will clearly increase with greater income elasticity.[17] More importantly in the environmental context, when we consider unpriced goods the income elasticity of demand term is not strictly relevant. Randall and Stoll (1980) show that income elasticity (η) should be replaced by the 'price flexibility of income' (ϵ) and reformulate the Willig limits as:

$$\frac{CS - WTP}{CS} \simeq \frac{\epsilon CS}{2Y} \qquad (5.1)$$

and

$$WTA - WTP \simeq \frac{\epsilon CS^2}{Y} \qquad (5.2)$$

where:

CS = Marshallian consumer surplus
WTP = willingness to pay (for a welfare gain this equals CpS_{WTP} [type 1])
WTA = willingness to accept compensation (for a welfare gain this equals ES_{WTA} [type 2])
Y = Mean respondents income
ϵ = Price flexibility of income

Randall and Stoll (1980) estimate ϵ in a manner analogous to an ordinary income elasticity by regressing WTP upon the quantity of the good, income and other significant explanatory variables. From this they estimate that, under reasonable assumptions, measures of WTP and WTA for quantity constrained goods should be within 5% of each other. CVM practitioners concluded from this that the wide empirical divergence of WTA above WTP was merely a methodological glitch which could effectively be ignored and that WTP sums were valid approximations to WTA (see, for example, Desvousges et al. 1983).

In a significant re-analysis of theory, Hanemann (1986; 1991) shows the Randall and Stoll (1980) derivation of the price flexibility of income (ϵ) to be inexact, demonstrating instead that:

$$\epsilon = \frac{\eta}{\sigma} \qquad (5.3)$$

where

η = income elasticity for the environmental good
σ = elasticity of substitution between this and all other goods

Using what they term 'the not-too-unreasonable values of, say, $\eta = 2$, and $\sigma = 0.1$', so that $\epsilon = 20$, Mitchell and Carson (1989) apply the above formulae to their earlier empirical work on the evaluation of water quality improvements (Mitchell and Carson 1981). In this work they found an average WTP of $250 (with average income = $18,000). We can therefore rewrite equation (5.3) as:

$$\epsilon CS^2 - 2Y.CS + 2Y\ (WTP) = 0$$

Substituting in values for ϵ, Y and WTP gives a consumer surplus (CS) of \$300 and substituting this into equation (5.2) gives a WTA of \$350. On the basis of these assumptions, WTA is shown to be some 40% larger than WTP in this example. Furthermore, while they state that higher values of η are unlikely, Mitchell and Carson (1989) state that 'much smaller values of σ for a number of public goods are quite plausible'. Using the same empirical data we can deduce that for WTA to be double WTP requires $\sigma = 0.0625$; while for WTA to be triple WTP requires $\sigma = 0.05$. Such substitution elasticities describe progressively superior goods (some environmental goods appear to fit this profile rather well).

In an important extension of his work in this area, Hanemann (1991) simulates WTP and WTA levels for a generalised CES utility model under a variety of assumptions.[18] Hanemann confirms the inverse relationship between the elasticity of substitution measure and the WTA/WTP ratio, i.e. for unique and irreplaceable environmental goods (Hanemann cites Yosemite National Park as an example) with very low substitution elasticities. In this context, we should expect WTA to be much greater than WTP. Hanemann also demonstrates that the same result still holds with a much higher elasticity of substitution ($\sigma \simeq 1$) where the ratio of WTP to income is high, i.e. where the proposed change matters a lot to the individual concerned. Under both these scenarios, Hanemann demonstrates that standard theory can explain levels of WTA more than five times the magnitude of WTP. Furthermore, the Hanemann formulae confirm that, where elasticity of substitution is not low and the WTP/income ratio is not excessively high (a scenario typical of many market-priced private goods), then WTP and WTA will not diverge very significantly.

These findings extend rather than refute the original Willig limits. Indeed, they show that the observed WTP/WTA asymmetry does have a theoretical basis and we should expect such asymmetry to occur where we are evaluating environmental goods which are in some significant way unique, irreplaceable or lacking substitutability. Such asymmetry, rather than being a methodological glitch, should actually be interpreted as theoretical backing for the internal consistency of the CVM.

However, we also recognise that other arguments have been put forward to explain the apparent WTP/WTA asymmetry. We highlight one such approach, prospect theory, below before formulating our conclusions.

Brookshire et al (1980) illustrate the standard Willig-type divergence between CV (WTP) and EV (WTA) as the smooth 'standard evaluation curve' illustrated in Figure 5.8. Here an individual is initially at the

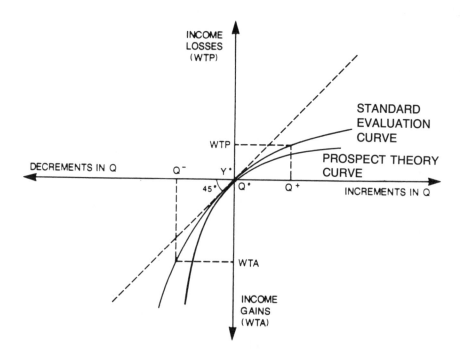

Figure 5.8 Total value of changes in the provision of an environmental good
Source: Adapted from Brookshire *et al.* (1980); Kalneman and Tversky (1979)

origin with income Y^0 and an initial non-zero allocation of an environmental good Q equal to Q^0. Under standard theory, for an increment in Q (from Q^0 to Q^+) the individual has a WTP equal to the distance Y^0WTP, while for an equal decrement in Q (from Q^0 to Q^-) the individual has a WTA equal to the distance Y^0WTA. The relevant factor here is the smooth nature of the standard evaluation curve and the consequent relatively small divergence between WTP and WTA. In their 'Prospect Theory', Kahneman and Tversky (1979) postulate that individuals will have a psychological affinity for the status quo such that, while they may be willing to pay for increments, they are very unwilling to contemplate a reduction in their initial allocation of the good in question.[19] In such a model the Prospect Theory evaluation curve is kinked at the initial allocation 'reference point' such that WTA is related not to WTP but to that reference point and exceeds WTP very significantly. Prospect theory provides a fundamental challenge to standard microeconomic theory regarding all goods and as such has found few friends. However it is a challenge which deserves serious consideration and is a focus for ongoing research.

WTP/WTA asymmetry: Summary and conclusions

Critics of the CVM have persistently cited the empirical asymmetry of WTA over WTP sums as evidence of the unreliability of the method. However, from the above discussion we can see that, far from being a pure methodological glitch, the marked differences between WTA and WTP sums observed in CVM studies do have a strong theoretical basis stemming directly from neo-classical economics.

Such an argument supports the use of either WTP or WTA question formats as appropriate and, as we observed earlier, Harris and Brown (1992) have set out the conditions under which WTP is the correct welfare measure. It is important at what is still a formative stage in the development of CVM that we do not abandon the theoretical high ground because of empirical problems posed, in particular, by past WTA results. Such results are perceived with unease by many economists, used as they are to value via payment rather than compensation. Nevertheless, it may well be that WTA responses are not methodological artefacts but rather pointers towards the narrowness of the basic psychological assumptions underlying conventional utilitarian value theory.

However, uncertainty regarding WTA scenarios does not imply a necessary weakness in WTP formats. Notice from Figure 5.7 that we have two potential WTP scenarios; WTP to ensure that a welfare gain occurs (type 1); and WTP to forgo a welfare loss (type 3), i.e. from the latter we have a WTP measure which is valid for certain types of welfare loss. Mitchell and Carson (1989) highlight this in their 'property rights approach' to welfare measures. Here they focus upon collective property rights and in particular those 'public goods which require annual payments or their equivalents to maintain a *given level* of the good' (emphasis added), for example, maintenance payments such as those ultimately made by taxpayers to ensure standards of water quality, air quality and many other non-excludable goods (see also Harris and Brown 1992). The collective property right ensures both that such goods are available to all members of society and that these access rights cannot be sold. It is this latter facet which, they argue, makes the WTA format inappropriate for evaluating such goods.

The correct welfare change measure can now be found, Mitchell and Carson (1989) argue, by examining whether the appropriate *given level* of public good provision, i.e. that amount which the public consider is the 'correct' amount which should be provided, is currently available or not. For a *proposed* welfare gain the *given level* may not be that currently available and the appropriate measure 'is the amount which the consumer is willing to pay for the improvement which leaves her just as well off before the change as after' (ibid.), i.e. CpS_{WTP}, a type 1 change.[20] However, when considering a proposed welfare loss the *given level* may be that currently available.[21] Now the appropriate welfare measure 'is the amount of income the consumer is willing to pay to forego the reduction in the quality level of the good and still be as well off as before' (ibid.), i.e. ES_{WTP}, a type 3 change.[22] In field applications, the appropriate CVM instrument should first inform respondents of their current payments for the current quality level and then ask for 'the maximum payment (which could be the present payment) that he/she is willing to make to preserve this quality level before he/she would prefer a quality reduction' (ibid.).

In summary, we argue that the observed asymmetry between WTP and WTA sums does have theoretical justification and can be seen as support for rather than a rejection of CVM. We recognise competition between theories justifying this asymmetry and currently advise against the present use of WTA formats in empirical tests. However, we do find strong theoretical backing for the empirical use of WTP scenarios to evaluate welfare gains and certain welfare losses. We now need to consider whether the theoretical justification of WTP formats can be matched with empirical accuracy in practical applications. Specifically, we need to consider some methodological issues surrounding the CVM.

CVM: Methodological issues

As indicated in Figure 5.6, those methodological issues most pertinent to the CVM can be roughly divided into validity, reliability and bias categories. Validity refers to the degree to which the CVM evaluation correctly indicates the 'true' value of the asset under investigation, bias being a common cause of low validity. Reliability refers to the consistency or repeatability of CVM estimates. Of course, reliability and validity need not be synonymous; for example, a particular CVM instrument may, in repeated trials, yield a consistent value estimate for a particular asset. However, if these trials are all subject to a bias then the results will not be valid. Bateman *et al.* (1991) consider reliability and validity in terms of a standard generalised linear model (GLM):[23]

$$y = ax + b + e$$

where

y = the measured value of the variable
x = the true value of the variable
a, b = constants
e = residual error

The reliability of the CVM instrument can then be measured by e, while a and b reflect validity; the instrument being absolutely valid if $a = 1$; $b = 0$ and e is a random variable. Where e is a non-random variable then a bias is likely to be present.

Reliability

In CVM surveys reliability is associated with the degree to which the variance of WTP responses can be attributed to random error, with reliability being inversely related to the degree of non-randomness. Notice that reliability says nothing about the validity of estimates.

Variance in WTP responses derives from three sources: true random error; sampling procedure; and the questionnaire/interview itself (instrument variance). True random error is essential to the statistical process, while induced sampling procedure error is a potential problem inherent in any statistical survey and can usually be acceptably minimised by ensuring that a statistically significant sample size is used. It is instrument variance which is of most concern here.

Following the recommendations of Rowe and Chestnut (1982), a good CVM instrument should be informative; clearly understood; credible; 'realistic by relying upon established patterns of behaviour and legal institutions'; and 'have uniform application to all respondents'. The further a particular CVM scenario moves from these norms – for example, the less familiar the respondent is with the environmental good or the construct of its valuation[24] – the more likely it is that such an instrument will increase the variance of responses. Realism and familiarity are therefore at a premium in constructing CVM scenarios.

Several commentators (Mitchell and Carson 1989; Kriström 1990; Hanley 1990) advocate the subsequent retesting of a particular CVM scenario as a test of the reliability of estimates from an initial test. According to Kriström (1990): 'If the same experiment is repeated a number of times with different samples and careful statistical analysis reveals no correlation between the variables collected then this is a warning flag' indicating low reliability. Few such replicability tests have been carried out, mainly due to the high resource cost involved. However, one test by Heberlein (1986) did find a substantial correlation between WTP amounts despite a one-year interval between test and retest.[25] Other studies by Loehman and De (1982) and Loomis (1989; 1990) generally support the reliability of CVM instruments.

Bias issues

CVM is an expressed-preference valuation method and as such is inherently susceptible to various types of bias. The conventional classification partitions bias into general, procedural and instrument types (following Smith and Desvousges 1986; and Kriström 1990).

General biases

Free-riding and strategic bias   Probably more than any other arguments, free-riding and strategic behaviour are the problems which economists have focused upon in criticising CVM. This is because neo-classical theory describes the 'rational' individual as essentially selfish. Developing this idea, economists therefore expect such an individual 'to pretend to have less interest in a given collective activity than he really has' (Samuelson 1954); and therefore to understate his WTP for a public good on the assumption that others will pay for its provision, which he will then enjoy, i.e. free-riding. Varian (1984) notes that such behaviour in CV studies will depend upon both the respondent's perceived payment obligation and his expectation about the provision of the good. If an individual feels that the payments of others will be sufficient to ensure provision of a good, or that others will not pay as much for the good, then he has an incentive to free-ride by lowering his WTP bid below his true valuation.[26] However, if an individual is particularly keen upon a good and calculates that the decision regarding provision depends upon the mean valuation of the sample, then he may behave strategically and overstate his true WTP in an effort to raise that mean and thereby ensure provision, i.e. strategic over-bidding. Mitchell and Carson (1989) identify a matrix of situations in which various types of strategic behaviour, both over- and underpayment, might theoretically occur (Milon 1989). In particular they highlight the high incentive to free-ride (understate WTP) where the provision of the good is seen as likely irrespective of the preference expressed.

There have been a number of empirical investigations of the free-rider theorem. Brookshire *et al.* (1976) and Schulze *et al.* (1981) argue that, assuming that true WTP bids are theoretically normally distributed, then free-riding should disturb this causing, in a WTP scenario, a distribution bias towards zero. Using such an approach, Brookshire *et al.* (1976) test for and reject the presence of strategic bias. However, Rowe *et al.* (1980) criticise the underlying assumption of such a test, stating that bimodal distributions can be posited upon the income characteristics of the respondent population. Certainly in a recent large sample experiment, Bateman *et al.* (1992) found highly skewed income and bid distribution. A more typical approach is adopted by Brubaker (1982), where respondents were asked to bid for a $50 shopping voucher under three scenarios: S_1, in which the n highest bidders were guaranteed a voucher; S_2, in which respondents were told that vouchers would be provided for all as long as the total WTP of all respondents exceeded a specific amount; and S_3, in which respondents were told that all those giving any positive WTP would receive a voucher. Brubaker assumed that S_1 would provide the true WTP, while S_2 had a weak incentive to free-ride compared to S_3 where a strong free-ride response was expected. The mean WTP results were (S_1 = 33.99) > (S_2 = 27.07) > (S_3 = £23.96$). These results appear to bear out the expectations of strategic behaviour. However, with further analysis, only the first two, S_1 and

Table 5.3 Stated WTP as a percentage of true WTP in the presence of a free-ride incentive

Study	Percentage of true WTP[1]
Schneider and Pommerehne (1981)[2]	96
Marwell and Ames (1981)[2]	84
Brubaker [S_2] (1982)[2]	80
Christiansen (1982)[2]	79
Bohm (1972)[2]	74
Brubaker [S_3] (1982)[3]	70
Schneider and Pommerehne (1981)[3]	61

1. The true WTP being measured in an auction where the winning bid(s) received the good.
2. In these experiments a group threshold WTP was required for provision, i.e. there was a relatively weak free-ride incentive.
3. In these experiments provision of the good was guaranteed irrespective of the (non-zero) WTP sum offered by the respondent, i.e. there was a relatively strong free-ride incentive.

Source: Adapted from Mitchell and Carson (1989).

S_2, are significantly different at the 5% level. This experiment tends to indicate that while free-riding does occur, it appears to be less prevalent than standard neo-classical theory would predict and does not invalidate CVM exercises. Table 5.3 compiles results from a number of these studies.

The findings indicate that where respondents were told that a certain threshold total WTP was required from the population before the good was provided (i.e. weak free-ride incentives) then stated WTP was between 74% and 96% of true WTP. The fact that stated WTP is still somewhat below 'true' WTP in such situations is not surprising and accords with the theoretical conclusions of Hoehn and Randall (1987). An analogy can be made to credit card type transactions where there is some distance, either temporally or spatially, between the agreement to purchase and the payment for goods by the purchaser. It is well accepted that such distance increases an individual's willingness to agree to a purchase price above that which would exist if immediate payment were necessary.

Notice also that those experiments where provision of the good was guaranteed irrespective of stated WTP (i.e. strong free-ride incentive) are subject to the highest deviation of stated WTP from true WTP, i.e. the highest free-ride effect.

Another important caveat to Table 5.3 is that all the WTP sums shown (including the 'true' WTP auction sum against which they are referenced) were obtained using an open-ended format where respondents could vary WTP by any amount. It has been pointed out (Loomis 1987; Kriström 1990) that this is likely to produce different responses to a discrete response approach. In this latter approach respondents are asked to state whether or not they are WTP a suggested sum for a given level of provision, these suggested sums being varied across the sample in a systematic manner. Kriström (1990) points out that this latter approach more accurately reflects a market decision, while Loomis (1987) argues that the insecurity of supply inherent in a negative discrete response will help to minimise free-riding behaviour. Hoehn and Randall (1987) extend this to argue that an individual's optimal strategy is truth-telling within a discrete valuation questionnaire. WTP responses under a continuous valuation scenario therefore correspond to a lower bound on true valuation (Kriström 1990).

Garrod and Willis (1990) follow the conclusions of Barnett and Yandle (1973) in suggesting that the theoretical presence of a free-riding incentive can only be removed by implementing a property-rights approach in which respondents' provision of a good is relative to their given WTP. However, as the authors point out, such approaches have limited applicability in the context of most environmental public goods.

We suggest that the move from considering private to environmental public goods is in itself a major explanatory factor regarding the apparent lack of free-riding problems with the latter. The non-use and altruistic values attached to environmental public goods act as a counter-incentive to free-riding tendencies (although the potential for overstating WTP may be greater).

Interestingly, there is considerable evidence to suggest that strategic over-bidding (overstatement of WTP in order to inflate mean WTP) may be less of a problem in practice than utilitarian value theory might initially indicate. Rowe *et al.* (1980) include a strategic bias test question, informing people of mean WTP and allowing respondents to revise their bids in the light of this information. Results from this experiment show a very low rate of bid revision, indicating that WTP bids are good reflections of true evaluations and are not subject to significant strategic behaviour.[27] Such a conclusion is reinforced by Schulze *et al.* (1981), whose survey of six CVM studies failed to reveal any conclusive evidence of strategic bias, a result confirmed in a more recent review by Milon (1989).

These results provide several important guidelines for optimal CVM design:

(i)   Provision of the good should be made conditional upon individuals' responses rather than automatic, thus removing the strong free-rider incentive.

(ii) Strategic bidding should not pose a serious problem for CVM studies. However, there is evidence of free-riding behaviour in open-ended format experiments.

(iii) Because of free-riding behaviour, stated WTP from an open-ended study is likely to be a downwardly biased estimate of true WTP.

(iv) Given this, open-ended format CVM studies should be interpreted as providing lower-bound evaluations of the good under investigation.

(v)   Dichotomous choice format CVM studies should not present respondents with free-riding opportunities, in order to avoid downward bias.

Hypothetical bias Despite the focus of attention upon free-riding and strategic behaviour, the most technique-specific problem facing CVM studies is whether the hypothetical nature of the market renders responses meaningless or not, i.e. whether respondents' declared intentions (WTP statements) can be taken as meaningful guides to behaviour (true value).

There is debate about the very nature of any such hypothetical bias.

Freeman (1986) sees the impact of an increasingly hypothetical scenario as being increased bid variance, while Mitchell and Carson (1989) extend this to reject the entire notion of hypothetical bias, referring instead to situations of low model reliability. However, many commentators (Schulze *et al.* 1981; Bishop *et al.* 1983; Randall *et al.* 1983) are convinced that the use of hypothetical rather than real markets can in certain circumstances produce its own distinct bias problems.

Research into the predictive ability of hypothetical markets has followed two paths; studies of the attitude-behaviour relationship; and experiments examining the substitution of real for hypothetical markets.

Market research, political polls and consumer surveys all operate on the premise that stated attitudes or intention are significantly reliable indicators of actual behaviour. One explanation of the linkages involved in this relationship is given by the Fishbein–Ajzen attitude–behaviour model illustrated in Figure 5.4.

The Fishbein–Ajzen model postulates that individuals have both positive and normative beliefs concerning behaviour. It can be argued that these are analogous to the private and public preferences, emphasised by Turner (1988a) which respectively correspond to the self-seeking and altruistic elements inherent in CVM responses. These beliefs in turn create attitudes and subjective norms concerning behaviour which combine as an intention to perform a specific behaviour. This intention will then influence the specific behaviour actually exhibited. This actual behaviour will then feed back into the modification of positive and normative beliefs, and so the cycle continues. Therefore, each stage in the cycle influences the next. However, this influence is not perfect – for example, intentions may not perfectly predetermine actions. Similarly, the feedback loops provide a dynamic adjustment system so that, for example, a recent visit to the countryside may well affect a respondent's WTP to preserve wildlife habitat. This, however, is a reflection of reality present in the consumption of all goods, marketed or not, and does not pose a special problem for the CVM technique.

In later work, Ajzen and Fishbein (1977) develop three hypotheses from their model indicating how the attitude–behaviour link can be maximised (i.e. minimising hypothetical bias). Firstly, attitude (CVM response) will best predict behaviour (true value) where the specified attitude (WTP scenario) closely corresponds to the specified behaviour (the precise good measured). So, asking WTP for a general environmental improvement will be a poor indicator of the willingness to pay higher taxes for improving the water quality of a specific river. Scenario misspecification, either as constructed by the interviewer or as perceived by the respondent, will obviously cause hypothetical bias. Secondly, 'the fewer the intervening stages between a component in the Fishbein–Ajzen

model and behaviour, the greater the predictive power of that component' (Mitchell and Carson 1989). Thus intentions are a much better
predictor of behaviour than attitudes, while attitudes are better predictors
than beliefs, because in both cases fewer influence relationships are
involved. In a study of unleaded petrol consumption, Heberlein and
Black (1976) found an attitude–behaviour correlation of just 0.12 but an
intention–behaviour correlation of 0.59. Thirdly, attitude will be a better
predictor of behaviour when the respondent is dealing with familiar
behaviour situations.[28] Hanley (1990) sees this as a source of error with
respect to environmental goods where, unlike marketed goods, there is
no opportunity to learn by experience of purchasing. However, he feels
that, in the main, this error will be associated with WTA scenarios,
where respondents will be very unfamiliar with the selling rather than
purchasing role, and less significant in WTP situations with which
respondents have both greater experience and empathy.

Surveys of experimental tests (Schuman and Johnson 1976; Hill 1981;
Mitchell and Carson 1989; Rowe and Chestnut 1983) support the conclusion that, while hypothetical bias does appear a significant problem in
WTA format studies, it can be reduced to an insignificant level in WTP
format studies. The most common approach to such testing is that put
forward by Bishop and Heberlein in a series of experiments in the late
1970s and 1980s. These tests consist of comparing respondents'
'hypothetical' WTA or WTP bids with 'true' bids obtained from
simulated markets in which real money compensation/payments were
used. The first of these (Bishop and Heberlein 1979) centred upon WTA
compensation by duck-hunters in order to give up their annual hunting
permits. While one sample of hunters were asked hypothetical WTA, a
separate sample were mailed a negotiable cheque to be honoured if
respondents mailed back their hunting permits. By varying the amounts
of these cheques across the second sample, a mean 'true' WTA was
established. Comparing this with the hypothetical WTA sum obtained
from the first sample, Bishop and Heberlein noted that while 90% of the
second (actual compensation) sample accepted a $50 cheque as compen-
sation for forgoing their hunting permits, only 40% of the first
(hypothetical compensation) sample had a WTA of $50 or less.
Comparison of relevant consumer surplus measures from the two
scenarios led the authors to conclude that the WTA sum obtained from
the hypothetical scenario was significantly higher than the true WTA, i.e.
that hypothetical bias was present in the WTA format.[29]

While Mitchell and Carson (1989), in a reworking of this data, claim
that the observed difference is not statistically significant (arising, they
claim, from truncation decisions in the original analysis), the deviation
in real and hypothetical WTA sums reported by Bishop and Heberlein is

not surprising. To reiterate the point raised by Hanley (1990), consumers are not experienced in selling and receiving payment for goods, particularly public goods, and there may well be a natural tendency to overstate expectations in such a costless environment.

A series of experiments in the mid-1980s (see Bishop *et al.* 1984; Bishop and Heberlein 1985; Heberlein and Bishop 1986) examined hypothetical and actual WTP for deer-hunting permits in Wisconsin. In the last of these experiments both actual and hypothetical WTP were determined via take-it-or-leave-it discrete question frameworks, with respondents being told that this was either an actual or a hypothetical transaction. The results showed actual and hypothetical WTP differences to be much smaller than for the WTA format and generally not significant at the 10% level.[30] In a further experiment, Kealy *et al.* (1990) asked respondents their

(i)   WTP for private goods (chocolate bars);
(ii) WTP for public goods (reduced acid rain);
(iii) WTP for public good (reduced acid rain) with an explicitly stated future actual payment obligation.

In each case the hypothetical market was then made real with actual payments being requested (in cases (i) and (ii) respondents were not forewarned that such a request would be made). Stated WTP as a percentage of actual WTP was shown to be 75%, 72% and 95%, respectively. This indicates, firstly, that hypothetical bias appears no worse for public than for private goods, and secondly, an indication that the stated responses may subsequently lead to actual requests for such payments causes a marked decrease in hypothetical bias.[31] The deviations between actual and stated WTP in cases (i) and (ii) are in line with the discussions of free-riding and results of Table 5.3.

These results indicate that the divergence between actual and hypothetical WTP is much less than that for WTA, i.e. WTP formats, unlike WTA formats, do not appear to suffer significantly from hypothetical bias. The familiarity of respondents with payment rather than compensation scenarios, suggested by Hanley (1990), would appear to be pertinent here.[32]

Recent investigations into hypothetical bias include that by Seip and Strand (1990), who asked 101 respondents for their WTP to join a national nature conservation group; of these 62 gave bids above the current group membership fee. However, when this latter group were subsequently sent membership invitations only 10% did actually join. Navrud (1991) criticises this study both for attempting to present the private membership good as the public good of environmental quality,

and for encouraging overstatement (i.e. bias) by not sufficiently informing respondents of the likelihood of actual payments being sought.

In conclusion, we suggest that hypothetical bias may be minimised by:

(i)   using WTP scenarios;
(ii)   making the hypothetical market as accurate, lifelike and believable
       as possible;
(iii) motivating respondents into putting sufficient effort into the
       bidding process; and
(iv)   investigating the impact of elicitation method (open-ended,
       dichotomous choice, etc.) upon stated WTP. As Brookshire *et al.*
       (1980) conclude, it is the credibility of the scenario presented which
       is the key to minimising hypothetical bias.

Part–whole (mental account) bias   Tversky and Kahneman (1981), in
considering decision rationality, argue that individuals see groups of
goods, rather than specific goods, as the basis for utility maximisation.
Extending this, Kahneman and Knetsch (1992a; 1992b) and Quiggin
(1991), contend that CVM may be fatally flawed by 'part–whole' (or
'mental account') bias, occurring where an individual's WTP responses
fail to distinguish between the specific good which is under analysis (the
'part') and the wider group of goods (the 'whole') into which that
specific good falls (see also Kneese 1984; and Hoevenagel 1990). If this
were the case, then, 'when respondents are asked to value some
environmental good they may in fact make that valuation on the basis
of a much wider range of environmental goods' (Willis and Garrod
1991). Mental account problems can arise if respondents ignore the calls
upon their income made by other environmental goods. In theory,
therefore, then a respondent could, if asked sufficient CVM questions,
pledge more than his entire income.

The potential for part–whole bias is well documented (see, for example, Walbert 1984; Thaler 1985; Hoevenagel 1990). However, the major
recent empirical support for such a criticism is provided by a Kahneman
and Knetsch (1992a) study. In this, respondents were asked their WTP
to maintain the quality of fishing in lakes in Ontario. The authors
reported no significant difference between mean WTP for a small
number of lakes (about 1% of the total lakes in Ontario) and mean WTP
for all lakes in Ontario. Kahneman and Knetsch concluded from this that
the WTP statements elicited in CVM studies referred to a 'purchase of
moral satisfaction' or a 'warm glow of giving' (see Andreoni 1990),
rather than a payment for a good.

An initial criticism of the Kahneman and Knetsch paper was provided
by Mitchell (1991), who pointed out that this particular study relied upon

both a poor instrument design (using telephone surveys, thus relying upon a weak medium of description and dialogue with a high potential for low respondent commitment to the survey); and poor information (a single-sentence description was used, arguably providing vague information and thus eliciting a vague valuation, potentially based upon knowledge of the 'whole' rather than the 'part').

These criticisms are expanded upon by Smith (1992b), who concludes that it is the question framing itself, rather than some underlying theoretical problem, which results in the reported part–whole phenomena. Indeed, Smith claims that the question framing used by Kahneman and Knetsch 'does not satisfy the criteria that Kahneman (see Kahneman and Tversky 1982) helped to develop in his earlier research on how people interpret valuation questions'.

While the criticisms of Mitchell and Smith may be sufficient to discount the particular results of Kahneman and Knetsch (1992a) indicating a fundamental theoretical flaw in CVM, they are insufficient to prove that the methodological problems of part–whole bias are an insignificant problem in empirical CVM studies. An empirical investigation is undertaken by Willis and Garrod (1991) who accept the potential for part–whole bias and explicitly set out to minimise it. The authors set the part–whole problem in the context of the theory of two-stage budgeting (Deaton and Muellbauer 1980; Tversky and Kahneman 1981) where total income is, in the first stage, allocated to various broad categories of expenditure, such as housing, food and recreation, and then, in the second stage, subdivided within categories among specific items, such as forest recreation and water recreation (see Figure 5.9). The individual therefore has several mental accounts, each referring to a category, using 'each account to evaluate some multi-attribute option with regard to the particular multi-attribute reference set that represents the category' (Kahneman and Tversky 1984). The authors, by this construct, therefore link their defence of CVM directly to Kahneman *et al.*'s previous work.

Willis and Garrod (1991) recognise that, in responding to questionnaires, individuals may omit to consider all available relevant material (see Slovic 1972). In particular, they may not consider the limits to, and other demands upon, their relevant (recreation) mental account. This omission lies at the root of part–whole bias, and Willis and Garrod address this problem explicitly by asking the respondent to calculate his or her 'total yearly budget for all environmental issues including those donations and subscriptions that he or she might already have made'.[33] This question directly addresses Tversky and Kahneman's (1981) fundamental point that decision making is highly context-sensitive.[34] The respondent is then reminded to consider all the other demands upon this limited budget before any WTP question is asked. In their empirical testing of

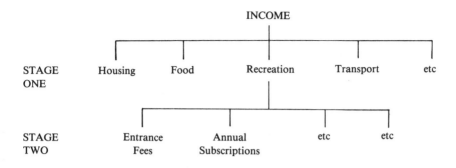

Figure 5.9   Two-stage budgets mental accounting

this approach, Willis and Garrod (1991) analyse WTP for the Yorkshire Dales national park. Both visitors and local households were interviewed, with all the visitors (group V) and half of the households (group A) being asked the recreation mental account question directly prior to the WTP question. The remaining households were used as a control group and not asked the mental account question prior to being asked for their WTP bid. A highly statistical significant difference was found between the mean WTP and the total recreation budgets of groups V and A (at the 1% level) indicating that, even if part–whole bias were present, it was not a serious problem. This result is also supported by the empirical work of Hoevenagel (1990; 1991); Magnussen (1991); Hoen and Winther (1991); and Strand and Taraldset (1991). In a recent large sample CVM survey Bateman *et al.* (1992) adopted the same approach as Willis and Garrod (1991), reporting no significant part–whole problems, with mean WTP being approximately 16% of annual recreation budget for users of the environmental resource under investigation.

An interesting mental account bias test is adopted by Navrud (1989a) in a study of WTP for the liming of Norwegian lakes as a counter to acidification and consequent fish loss. Here, along with a typical WTP question regarding the lakes, respondents were also asked their WTP for improvements in the quality/quantity of all the public goods affected by acid rain, a verbal description of the effect upon these goods being included in the questionnaire. Comparisons of mean WTP for the lakes with mean WTP for all public goods (including lakes) ranged between 59% and 91% over nine subsamples. Navrud, however, feels that this does not indicate a serious mental account bias, stating that 'the generally high percentage can be viewed as reasonable due to the fact that the current damage to public goods by acid rain in Norway is largest for the freshwater fish populations'.

Green, in Bateman *et al.* (1991), identifies certain variants of mental account type problems. He notes that where respondents are asked for evaluations of a category group of goods (e.g., recreation) then WTP may be less than the sum of WTP responses regarding the specific good contents of that category. This can be viewed as the corollary of the part–whole hypothesis in that, where the WTP question for a specific good is asked in the context of other goods then this amount may tend to be less than if it were asked in isolation of other goods.

However, the empirical evidence is mixed. While Rae (1982), Burness *et al.* (1983), Tolley and Randall (1983) and Strand and Taraldset (1991) confirm evidence of such a tendency, Brown and Green (1981), Schulze *et al.* (1983) and Rahmatian (1987) reject such a hypothesis. Willis and Garrod (1991) test a similar hypothesis by comparing mean WTP for group A (see above), asked a mental account question (i.e. spending on all environmental goods), with that for a control group not given such a question, concluding that the inclusion of this extra variable did not significantly affect mean WTP (at the 1% level). This satisfies Rae's (1982) criterion that CVM results may only be considered valid if the inclusion of further environmental goods in the questionnaire does not significantly alter WTP values. An extension of this contextual problem is the ordering effect observed by Tolly and Randall (1983) and Hoevenagel (1990), where it was noted that a good will elicit a higher WTP response if it is placed at the top of a list of goods to be evaluated, than if it is valued after other goods.

We can offer no firm conclusion regarding the contextual problem, noting that empirical evidence is currently mixed and requires supplementation before informed conclusions can be drawn. We feel that this problem can be addressed by running repeated tests, varying the position of goods within any evaluation list so that consensus results may be arrived at.

Regarding the overall part–whole issue, we see this as a potential problem which will be exacerbated by poor CVM instruments. However, with the specification of instruments which encourage considered responses to accurate unbiased information and testing as outlined above, we believe that the problem of part–whole bias can be reduced to acceptable levels.

Information bias   Does the quality of information presented to 'service' a hypothetical market affect the responses received? The answer is almost certainly 'yes'. Samples *et al.* (1985) compared responses from two experimental groups given varied levels of information regarding an endangered species (the humpback whale) with those responses received from a control group given constant information. It was found that increased information increased mean WTP by between 20% and 33%; however, statistical tests showed that while this test was significant at the 20% level, it was not significant at the 5% level.

In the study by Mitchell *et al.* (1988) two groups were given differing information regarding four Sites of Special Scientific Interest (SSSI). Again, additional information raised mean WTP, but this was shown not to be a statistically significant increase. A similar weak information bias result is found by Hanley and Munro (1991) in two CVM experiments regarding WTP for heathland and woodland preservation. They postulate a threshold effect of information build-up below which no bias is detectable but above which a weakly positive effect is found. A stronger result is provided by Bergstrom *et al.* (1985), whose study of bids to preserve prime farmland in the USA produced a 1% significance test that additional information had resulted in higher bids. However, such a finding is firmly challenged on both empirical and theoretical grounds by Boyle (1989). In an experiment regarding WTP for brown trout fisheries in Wisconsin, Boyle found no significant difference between mean WTP statements for three levels of information, although bid variance fell significantly as information increased. Boyle states that 'the argument that changes in accurate or true commodity description in the framing of CV questions will change value estimates is unwarranted as a blanket statement'.

A less extreme view is adopted by Randall *et al.* (1983), Carson (1989); Kriström (1990) and Hanley (1990), who argue that, since individuals do have preferences regarding environmental goods, their provision, distribution and funding, then information will always affect WTP but that this is no different from any other good, priced or not, i.e. this is an expected information input effect. The important factor, then, is to ensure that such information is seen to be true, constant across the sample, and not designed to induce bias towards a particular result, polemic and implicit value judgements being inadmissible. If this is possible the work of Samples *et al.* (1985) indicates that inherent information bias should not be an overriding problem.

Procedural bias

Aggregation bias A particular problem in the estimation of total economic value sums for spatially fixed environmental goods such as forests is that on-site surveys will ignore the non-use values held by non-visitors. We argue that such surveys can only claim to estimate user values and that supplementary random sample remote (off-site) surveys are necessary to estimate non-use values. Such studies (see, for example, Brookshire *et al.* 1983) have shown that when aggregated over the larger non-visitor populations, total non-use value may be significant and may even exceed total use values by a significant factor.[35]

The aggregation procedure itself can induce bias. The main issue is to define the relevant population at the pre-survey stage and then conduct standard diagnostics to validate the sample collected as being representative of the population.[36] However, the connection between the sample and the population is rarely perfect and certain adjustments may be justified; choice of adjustment procedure can, however, have a major impact upon aggregate estimates. In one experiment Loomis (1987) varied adjustment procedures to produce a 2.25 times difference in the range of aggregate benefit estimates.

A more fundamental question which arises is the choice of an appropriate welfare measure for aggregation. If the distribution of WTP bids is non-normal (perhaps Poisson or binomial) then the sample mean will have been biased by the major (usually upper) tail of the distribution. Similarly, reference to the median in such situations will not be valid for aggregation as it cannot be said to be representative of the sample. In a recent review, Duffield and Patterson (1991) therefore support the use of a truncated mean as the basic welfare measure for aggregation.[37]

Interviewer and respondent bias   We have argued that scenario misspecification (accidental or contrived) will induce hypothetical bias. Further, it can be argued that the very character of the interview or interviewer may influence responses. For example, if the interviewer in some way portrays the environmental good as morally desirable, or if the interviewer is highly educated (or attractive) then the respondent may feel inhibited about expressing a low WTP bid. Walsh *et al.* (1990) tests for the presence of interviewer bias by comparing WTP statements across several interviewers, some of whom were economics graduates. The authors found some evidence of interviewer bias in WTP bids such that those interviewers without a background in resource economics elicited WTP bids some 24% lower than those with such background knowledge.[38]

One approach to the problem of interviewer bias is to attempt to minimise it by using mail or telephone rather than face-to-face interviews. However, the former approaches entail a loss of the information opportunity afforded by the latter and may cause an increase in hypothetical bias if respondents feel less concerned to invest time in considering their bids. High non-response rates are also typical of mail surveys.[39] We would therefore generally caution against using such an approach. However, the 'total design method' advocated by Dillman (1978) has achieved considerable success in boosting response rates for various social surveys and the thorough application of its recommendations may be useful here. Furthermore, in a recent comparison of mail and in-person CVM surveys, Mannesto and Loomis (1991) found that assessing item non-response (rather than total non-response) and WTP functions showed 'more similarities between methods than differences'. They also found less resistance to sensitive (e.g., income) and complex, future-orientated questions using the mail approach and argue that it allowed for greater contemplation and reduced respondent stress. This strength must be traded off against the problem of non-response.

Another variant of this problem is the phenomenon of the 'good respondent'. Orne (1962) pointed out that the relationship between analyst and respondent is an interactive process, with the respondent seeking clues as to the purpose of the experiment. If this purpose is inadequately conveyed then the respondent may react in two ways, either not giving the questions due consideration or attempting to guess the 'correct' answers, i.e. he will try to be a 'good respondent' and give the answers which he feels that the analyst wants. The problem of low consideration can be assessed by recording and analysing the numbers of respondents who refuse to take part in the survey and the length of interview. The 'good respondent' problem may be exacerbated where the interviewer is held in high esteem by the respondent (Harris et al. 1989), resulting in responses which differ from true willingness to pay. Desvousges et al. (1983) found little evidence of such a bias but it should be noted that this study employed professional interviewers, a potential solution to such problems. Tunstall et al. (1988) further recommend that interviewers follow the wording of the questionnaire exactly and that respondents be presented with a choice of prepared responses so as to minimise over or understatement of true evaluations. Approaches designed to combat hypothetical bias (discussed above) may also be relevant here.

Many variants of procedural bias are endemic to social statistical surveys. For further discussion of these issues with respect to CVM, see Garrod and Willis (1991) and Bateman et al. (1991). Regarding wider survey issues, see Converse and Presser (1986).

Instrument-related biases

Payment vehicle bias   Rowe et al. (1980) found that WTP to preserve landscape quality was higher when an income tax increase was suggested than when an entrance fee was proposed, concluding that respondents viewed fee-paying as a debasement of the experience. Many other studies, for example, Desvousges *et al.* (1983), Brookshire and Coursey (1987) and, more recently, Navrud (1989b) have reported a similar effect. Tunstall *et al.* (1988) feel that efforts should be made to adopt a 'neutral' payment vehicle, i.e. one which does not affect WTP, and claim that using a charitable trust fund will promote such a result. However in a recent experiment Bateman *et al.* (1992) found that such a vehicle resulted both in high bid variance and a high bid refusal rate, with respondents stating (ironically) that they did not trust trust funds.

Choice of payment vehicle can therefore affect WTP. However, this should be viewed as a change in the good offered and researchers can effectively eliminate such problems by avoiding controversial payment methods and instead using that vehicle which is most likely to be used in real life to elicit payment for the good in question.

Starting point, anchoring and discrete bid level bias   Several studies have noted that the suggestion of an initial starting point in a bidding game can significantly influence the final bid – for example, the choice of a low (high) starting point leads to a low (high) mean WTP (see Desvousges *et al.* 1983; Boyle *et al.* 1985; Navrud 1989a; Green *et al.* 1990; Green and Tunstall 1991). While the use of starting points may reduce non-response and variance in open-ended questionnaires, commentators argue that the statistically observable bias this induces indicates that such 'bidding hints' lead respondents to take cognitive short-cuts to arrive at a decision rather than thinking seriously about their true WTP (Cummings *et al.* 1986; Mitchell and Carson 1989; Loomis 1990). It has also been noted that informing respondents as to the construction costs associated with a proposed environmental change can also affect resultant bids (Cronin and Herzeg 1982). One approach to this problem is to allow the respondent to choose a bid from a range shown on a payment card. Unfortunately such an approach of necessity produces 'anchoring' of bids within the range given on the card, with most respondents assuming that such a range contains the 'correct' valuation and outliers being effectively ignored (Kahneman and Tversky 1982; Roberts and Thompson 1983; Kahneman 1986; Harris *et al.* 1989). We suggest that both starting points and payment cards should not be used.

Dichotomous choice formats may be open to anchoring bias according to the levels of bid asked of respondents. If these levels do not reflect reality (e.g., if they do not go as high as respondents' true WTP would be and therefore truncate possible bids) then vital information will not be gathered. In setting optimal bid levels 'the range of take-it-or-leave-it bids used should be such that the lowest results in nearly 100% acceptance and the highest 100% rejection' (Bateman *et al.* 1991). Within this

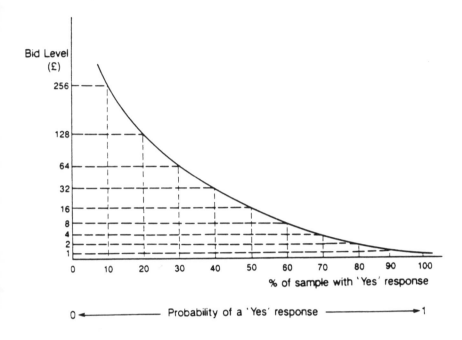

Figure 5.10 An example of optimal bid levels for a dichotomous choice format CVM study

range, bid levels should reflect the distribution of bids so that, optimally, each bid interval reflects the same proportion of the population. Such a system is illustrated in Figure 5.10, with optimum bid levels shown on the vertical axis.

Bishop and Heberlein (1979) set the upper boundary bid level according to a priori expectations and then assume a log-linear demand curve to define bid intervals. Boyle and Bishop (1988) suggest the prior implementation of an open-ended format CVM study in order to define the likely bounds for bids. Such an approach, they argue, may reduce the necessary sample size for a discrete question survey carried out in the absence of such information. The authors then use a random number approach to define bid levels within this range.

In a recent CVM study regarding WTP for the Norfolk Broads, Bateman *et al.* (1992) follow Boyle and Bishop (1988) and use a large-sample open-ended format WTP question to estimate the distribution and range of WTP bids. A bid function was then estimated in order to map out a probability of discrete bid acceptance curve. On the basis of

Table 5.4 Comparison of open ended (OE) and dichotomous choice (DC) format WTP responses

DC WTP bid level (£)	% of DC respondents accepting this amount as one they are WTP[1]	% of OE respondents stating WTP equal or larger than DC bid level[2]
1	95.00	92.06
5	95.15	86.38
10	90.27	80.42
20	89.43	67.20
50	69.12	47.49
100	57.85	29.76
200	41.59	10.85
500	20.83	3.04

[1] Based upon DC sample $n = 1770$.
[2] Based upon OE sample $n = 756$.

Source: Bateman *et al.* (1992)

this information, eight WTP bid levels were chosen for a dichotomous choice experiment with the expectation being (from the open-ended results) that the lowest take-it-or-leave-it WTP bid (£1) would be almost universally accepted while the highest bid (£500) would be almost universally rejected. The remaining bid levels were chosen to reflect the bid distribution indicated by the open-ended study. A large-sample dichotomous choice survey was then undertaken. A comparison was then made between the percentage of respondents in each dichotomous choice bid category stating that they would be WTP the bid level asked, and the percentage of respondents in the open-ended experiment who had stated a WTP greater than or equal to the dichotomous bid level. Table 5.4 illustrates the results of these comparisons across the full range of dichotomous bids. It can be clearly seen at all WTP levels that a dichotomous choice respondent is more likely to assent to the take-it-or-leave-it question 'are you willing to pay £X?' than an open-ended respondent is likely to state a WTP of £X or above. A chi-squared test confirmed the significance of these differences.

Several factors may well be influencing the results given in Table 5.4.

(i)   As already noted, open-ended format WTP studies are subject to free-rider bias and as such are likely to be biased downward from true WTP. This tendency is not a problem for dichotomous choice experiments (Kriström 1990).

(ii)   Dichotomous choice formats may be subject to a form of 'good respondent' or interviewer bias where the respondent feels that the sum he is being asked to pay is in some way a 'correct' amount. In such cases more acceptances to bid levels will be given than correspond to true WTP, upwardly biasing the mean WTP estimated by the dichotomous choice approach.

(iii) Dichotomous choice formats are likely to exhibit anchoring bias. Quite separately from the previous point, this may result in an upward rounding effect. Where a respondent's true WTP exceeds the dichotomous bid level this will not cause a problem. However, if a respondent has, for example, a large true WTP for, say, £150 and is asked whether he is willing to pay £200 then there may be an unwillingness to say 'no' to such a question if the respondent feels that this may be taken as a negative preference for the asset in question. As the difference between true WTP and the bid level is relatively small, the respondent may well decide to say 'yes' to the bid question in order to register a strong positive preference for the asset. Such an effect will result in a further upward biasing of dichotomous choice mean WTP.

The last two arguments will result in an upward biasing of dichotomous choice valuation estimates above true WTP. Taken in conjunction with the downward (from true WTP) biasing of open-ended format estimates due to free-riding, these results have important consequences for the practical application of CVM to project appraisal. We can conclude that open-ended approaches will provide a lower-bound WTP estimate below which true WTP is unlikely to lie, while dichotomous choice approaches provide an upper-bound WTP estimate above which true WTP is unlikely to lie.[40] Conducting both format experiments will provide a useful envelope of true WTP evaluation.

CVM bias: Conclusions
CVM surveys are naturally prone to a number of biases, some of which are specific while others are endemic to all survey questionnaires. We conclude that CVM surveys can be designed to reduce bias problems to an acceptable level so that a CVM evaluation can provide us with useful value estimation information. We now turn to consider the validity of the valuations produced by CVM studies.

Validity

Mitchell and Carson (1989) identify three categories of validity testing: content; criterion; and construct validity (including convergent and theoretical validity).

Content validity
Content validity is a concern over whether the measure estimated (WTP) can be said accurately and fully to correspond to the object under investigation (the construct). Pearce and Turner (1990b) point out both that the true construct (*CpS*; *ES*) will not be directly observable and that the pure public good nature of certain environmental goods (such as clean air) will make the necessarily subjective assessments of content validity extremely difficult to undertake in any structured or replicable manner. Analysts must decide for themselves whether a particular CVM questionnaire has asked 'the right questions in an appropriate manner' and whether the WTP measure is 'what respondents would actually pay for a public good if a market for it existed' (Mitchell and Carson 1989).

Cummings *et al.* (1986) propose five 'reference optimal conditions' for enhancing the validity of CVM studies.

(i)   Use familiar environmental goods.
(ii)   Respondents should have some valuation experience of the environmental change in question.
(iii)   The scenario should not have a high degree of uncertainty.
(iv)   WTA scenarios should not be used.
(v)   Use values are likely to be more accurate than non-use values.

In reviewing the literature, Garrod and Willis (1990) conclude that a general improvement in survey questionnaire design has meant that 'content validity has not been regarded as too great a problem in recent years'.

Criterion validity
One method used to assess the validity of CVM estimates is to compare these with the 'true' value (the criterion) of the good in question. For many environmental goods such a test is, of course, not feasible and is the reason why CVM experimentation is being undertaken. However, experiments such as those by Bishop and Heberlein and by Bohm, discussed previously, do provide us with such a test. These experiments compare the hypothetical $WTP_h(WTA_h)$ sums obtained by CVM type questions, with 'true' $WTP_s(WTA_s)$ as determined by simulated markets using real money payments (compensation). In their original study,

Bishop and Heberlein (1979) found that 'real' WTA_s fell outside the confidence interval of hypothetical WTA_h, while other critics have challenged the consequent conclusion that WTA_h is not a valid estimate of WTA_s (Carson and Mitchell 1983). When a WTP format was subsequently employed, Heberlein and Bishop (1986) concluded that, generally, hypothetical WTP_h was a valid estimate of real WTP_s. Hanley (1990) concludes that while such results reinforce an empirical rejection of the WTA format, they also provide significant justification for the use of WTP format CVM studies. While we are broadly in agreement with the latter point, we note the quasi-private nature of the hunting permits used in Bishop and Heberlein's experiments and question the validity of extending their findings to pure public goods such as clean air for which, we have previously suggested, substantially different elasticities of substitution may be relevant and alternative explanations and motivations (e.g., prospect theory) may be of increased importance.

Construct validity

One approach to validity testing is to examine whether the measures produced by CVM relate to other measures as predicted by theory. Two variants of this construct validity approach can be identified: theoretical validity, testing whether the CVM measure conforms to theoretical expectations; and convergent validity, testing whether the CVM measure is correctly correlated with other measures of the good in question.

Tests of theoretical validity have mainly centred upon examination of bid curve functions to see if they conform to theoretical expectations, for example, whether elasticities are correctly signed and have feasible sizes. In an early test of theoretical validity, Knetsch and Davis (1966) estimated bid curves for forest recreation, concluding that 'the economic consistency and rationality of the responses appeared to be high'.

A similar approach was adopted by Whittington *et al.* (1990) in an examination of WTP for water services in Haiti. Tests of the significance of explanatory variables found them to conform to standard expectations as defined by economic theory. A further variant of this approach is to examine the explanatory power of bid functions. However, the high variance associated with CVM and other social survey techniques tends to produce low R^2 statistics. Hanley (1990) recommends that a minimum R^2 value of 0.2 should be used, while Mitchell and Carson (1989) suggest an R^2 value of 0.15 as a minimum. However, psychologists are at pains to point out that the very nature of social survey techniques make R^2 statistics of limited use. Indeed, if it is the case that respondents are reluctant to bid zero amounts then the significance of the constant (and therefore R^2) will be over-inflated for low-value goods. A much stronger test is to examine specific relationships to see if they conform to theoretical expectations. So, for example, we should expect a significant positive and marginally diminishing relationship between income and WTP, with a similar relationship between visits to a site and WTP for it. The significance of coefficients can then be judged via simple t-statistic tests. Studies which do not establish significant relationships where theory indicates they should exist must therefore be treated with suspicion.[41]

A further theoretical test can be performed where measures of consumer surplus are available. These surplus measures can be compared to the WTP estimates obtained and, by manipulating the Willig equations discussed earlier, implied elasticity values can be calculated. Results can then also be compared with theoretical expectations and empirical findings. Finally, Smith *et al.* (1991) propose a variant of convergent testing using CVM to measure the demand for actual marketed commodities or programmes. CVM results can then be compared with real-world outcomes.

Convergent validity may at first seem reminiscent of criterion testing. However, in this context none of the comparative measures can claim to be 'truer' than any other. The most common approach is to compare CVM measures with those from revealed preference techniques such as travel costs (TCM) or hedonic pricing (HPM) methods. Cummings *et al.* (1986) detail four comparisons of CVM with TCM and two with property-based HP studies, reporting that the value estimates produced by the different approaches were within 60% of each other, with some being much closer. Mitchell and Carson (1989) report on a further nine comparisons and concur with this conclusion – see also Brookshire *et al.* (1982) and Smith *et al.* (1986).

A significant problem with such convergent validity testing is that the methods compared are usually measuring different constructs. For example, while CVM should in theory be providing estimates of aggregate use plus non-use values, the TCM only estimates use value. An important further distinction for site-specific environmental goods (e.g., forest recreation compared with the benefits of clean air) is whether the CVM study is carried out with just an 'on-site' sample or whether an off-site 'remote' population is also utilised (the first should theoretically hold mainly use values while the second should mainly exhibit non-use values). Clearly in the latter case, comparison with TCM results is questionable particularly where non-use values are thought to be significant.[42] A further theoretical problem is that, while TCM and HPM estimates derive from *ex-post* situations, CVM provides *ex-ante* measures, positing a potential information inconsistency in the comparison of these measures.

CVM validity: Conclusion

All of the validity tests reviewed can be criticised. The content validity test is in many respects fundamental; however, its operation cannot (as yet) be formalised and subjective judgement is the underlying operand. Criterion validity, as expressed through the comparison of hypothetical with real markets, provides perhaps the most substantive test of validity and, indeed, such tests indicate that WTP format questions can provide valid estimates of true value. Unfortunately, the empirical applicability of this test to pure public goods is restricted, and results from tests upon private or quasi-public goods can only be extended by inference to give validity to CVM estimates of pure public good values. The convergence testing form of construct validity has been subject to considerable practical application. However, we question the degree of comparability between measures obtained by the CVM, TCM and HPM methods on the grounds that they are measuring different underlying theoretical constructs. Construct analysis through theoretical validity testing, does, we feel, provide a defensible test of the theoretical appropriateness of the results obtained.

Conclusions and recommendations

Conclusions

CVM is a widely applicable and widely applied monetary evaluation method. It has the potential for application to a wider range of environmental goods than any of the other main monetary valuation techniques. We believe that CVM possesses a strong theoretical basis, with the unique advantage that it estimates income-compensated welfare measures. Furthermore, we have demonstrated that this theoretical basis is consistent with many of the empirical results obtained in practice (notably the observed asymmetry of WTP/WTA measures) which, rather than being symptomatic of a flaw in the technique, appear to have considerable theoretic justification.

Because of its nature as an expressed-preference survey technique, CVM is susceptible to bias and, indeed, while it is easy to instil bias into responses, the task of minimising such bias to an acceptable level is, we recognise, one which requires considerable skill. Certainly CVM cannot be currently classed as a non-expert technique. However, our analysis of bias issues is cautiously optimistic. In particular, we feel that the widely held notion that free-rider problems or strategic bias form a fatal Achilles' heel is poorly founded in fact.

While reliability testing is infrequently applied in practice, we do not see any major problems here. In contrast, however, validity testing still appears to be considerably under-researched. In particular, we are concerned that the emphasis upon convergent validity testing has paid too little attention to the basic fact that these comparative methods are estimating different values. Nevertheless, we see considerable reasons for optimism particularly in the application of criterion testing for quasi-private goods. For pure public goods there does seem scope for the use of bid curve analysis to provide theoretic validity testing by examining coefficient estimates for expected signs and magnitudes.

Young and Allen (1986), in criticising the CVM approach, felt that the method could 'only be justified if the expected deviation of the estimated valuation is more acceptable than complete ignorance'. Many of the items valued via the CVM have indeed traditionally been ignored by decision-makers. While cautious, our analysis of the CVM attempts to highlight the major pitfalls of empirical application and seeks to show how, by their avoidance, the technique can provide evaluations of environmental goods which are sufficiently valid and accurate to enhance significantly the economic analysis and appraisal of projects which in some way impact upon these goods.

Recommendations

To conclude this chapter, we provide the following list of recommendations for optimal CVM survey design. Our major recommendations are as follows:

- Only apply CVM to goods with which respondents have at least some familiarity (direct or indirect).
- The scenario should be realistic and believable, clearly understood, and not have a high degree of uncertainty.
- WTA scenarios should be avoided.
- The payment vehicle should be realistic and appropriate.
- Estimates of use value are likely to be more accurate than those for non-use value.
- Use both open-ended and dichotomous choice formats to provide lower and upper valuation boundary estimates.
- The survey should question intentions to behave rather than attitudes towards behaviour.
- The scenario should make provision of the good dependent upon behaviour.
- Adequate (rather than excessive) unbiased information should be

provided, with the impact of that information assessed via a control group who receive no information.
- Specific questions should be included to minimise part–whole (mental account) problems.
- Sample size must be statistically significant.
- Avoid starting points, bidding games and payment cards.
- Avoid any direct or implied-value cues either via information, questionnaire or interviewer.
- Choose carefully between face-to-face and remote (mail, etc.) approaches and ensure that the correct population is being sampled.
- CVM will work best where respondents have some experience of valuing the good in question.
- In dichotomous choice formats the upper bid level should be such that almost 100% rejection is achieved, while the lower bid level should achieve almost 100% acceptance.
- Analysts should consider carefully the removal of outliers and the use of trimmed means.
- Telling respondents that payments by others will be compulsory may reduce non-response but may increase free-riding and strategic bidding.
- Great care is required in the aggregation process.
- Theoretical validity testing (bid function estimation) should be carried out. Further validity tests should be included where feasible.
- Where possible, assess reliability by retesting at a later date.
- Report in full all results, including all sample statistics, details of information given, and a reprint of the full questionnaire.

Notes

1. Readers are referred to Mitchell and Carson (1989) as an excellent source for further reading.
2. Hanemann *et al.* (1991) evaluate such a double-bound dichotomous choice system, concluding that it yields lower variability estimates of mean WTP. Ongoing work at CSERGE is examining the feasibility of estimating a triple-bound dichotomous choice system.
3. Respondents who refuse to state a WTP or WTA for an asset (or state extreme amounts) are commonly termed 'protest voters'. They should not be confused with the respondents who state a considered zero valuation for the good in question. A high proportion of protest votes may well signify a fundamental weakness in a study (see Sagoff 1988; Eberle and Hayden 1991; and discussions of strategic bias).
4. See, for example, Kriström (1990).

5. Alternatively a probit approach may be used: see Cameron and James (1987) and Cameron (1988).

6. References to 'consumer surplus throughout this (and subsequent) chapters refer to the Marshallian consumer surplus measure.

7. Price may be zero or positive dependent upon the property rights of the good. In the case of environmental goods we are usually faced with unpriced, quantity-constrained, public goods. For an introductory text see Johansson (1991), and for further reading see Just *et al.* (1982) and Johansson (1987).

8. As will be shown below, because of their uniqueness, environmental goods often also have a very low elasticity of substitution (Shogren *et al.* 1993).

9. That is, the loss of money income which, after the increase in provision, returns the individual to his initial lower utility level.

10. That is, the increase in money income which raises the individual to the same final utility level as if the forgone welfare gain in provision had occurred.

11. That is, the maximum amount of money income which the individual is prepared to give up to prevent the welfare loss occurring, leaving him as well off as if it had occurred (at the final, lower utility level).

12. That is, the increase in money income which returns the individual to his initial (higher) utility level, given that the welfare loss change in provision does occur.

13. Note that equilibrium is not achieved at a point tangential to a utility curve. Although the individual would prefer to be at such a point (i.e. more along X from point A to a tangential point with a higher utility curve), consumption of X_1 is exogenously constrained at Q_0.

14. This is an equivalent surplus measure as the welfare change is measured from the new utility curve, here U_0.

15. Similarly, this is a compensating surplus measure as the initial utility level, U_1, is our welfare measure reference.

16. A result confirmed by Just *et al.* (1982) who also show that the Willig approach may be generalised to the multiple price change case (1982, 375–86).

17. This error will also increase for aggregate populations where there are large variations in income and/or income elasticity of demand between consumers.

18. See Deaton and Muellbauer (1980) for details of this and other utility systems.

19. The roots of such an idea may be traced back to Adam Smith (1790), who states: 'We may suffer more, it has already been observed, when we fall from a better to a worse situation, than we ever enjoy when we rise from a worse to a better.'

20. Mitchell and Carson (1989) argue that, for welfare gains, individuals will see the post-gain level of provision as the 'correct' allocation.

21. Mitchell and Carson (1989) argue that, for welfare losses, individuals will see the pre-loss level of provision as the 'correct' allocation. Therefore, in both gain or loss situations, the largest provision level is seen as the reference quantity.

22. Note that Mitchell and Carson (1989, p.41) identify this as CS_{WTP}, arguing that the initial rather than the proposed utility level is the appropriate reference. However, in our typology developed in Figure 5.2 a payment to avoid a welfare loss corresponds to the type 4 ES_{WTP} measure.

23. See the test theory work of Lumsden (1976) and GLM theory of McCullagh and Nelder (1983). This model does not imply that any (for instance, linear) particular functional form is likely to be superior.

24. WTA scenarios are inherently less familiar to consumers than are WTP formats.

25. Mitchell and Carson (1989) also highlight similar findings from non-CVM survey retests.

26. Similarly, for a WTA format, the lack of a budget constraint may lead the respondent to overstate WTA.

27. In a test of strategic behaviour, Rowe et al. (1980) report that only one respondent exhibited marked free-rider tendencies; that respondent was an economics professor. See also Marwell and Ames (1981).

28. Ajzen and Peterson (1988) extend these attitude–behaviour criteria by emphasising that behaviour must be under the volitional control of the respondent, that lags between the measurement of intention and prediction of behaviour will be problematic and that levels of generality in the measures of intention and behaviour should be identical, i.e. the extrapolation of intention towards one environmental asset as a predictor of intentions toward a wider asset set is highly dubious.

29. A similar conclusion is reached by Rowe and Chestnut (1983), who conclude that poor definition of the implied property right in WTA studies is a major cause of hypothetical bias.

30. Statistical confidence tests are also reported by Rowe et al. (1980). Bid function analysis can also be useful; for example, if WTP bids were completely hypothetical there should be no relationship between them and respondents' income.

31. See also the work by Dickie et al. (1987) and Brookshire and Coursey (1987).

32. Note that the above results appertain to open-ended rather than dichotomous choice formats; subsequent discussions of anchoring bias suggest that stronger forces distort actual and state WTP responses in the latter format.

33. The mental account question reads as follows: 'Bearing in mind how many good causes there are looking for our support today, could you estimate how much money your household spends on a voluntary basis each year on the enjoyment and preservation of the countryside – this might include any donations to countryside causes, membership of conservation or countryside organisations and charges for entry to reserves/parks etc' (Willis and Garrod 1991).

34. Such an approach may also provide some insurance against strategic behaviour.

35. An important point to note here is the criticism that when non-users, unfamiliar with an environmental asset, are asked for their WTP to preserve

that asset, they may state some small sum as a token of charitable concern – the moral satisfaction argument of Kahneman and Knetsch (1992a). If such sums are accepted as true evaluations, aggregation over the entire non-visitor population (which is likely to be large) may produce considerable sums far in excess of user value. However, until the validity of such amounts can be established, such aggregations should be treated with caution. Ongoing research at CSERGE is currently addressing this issue.

36. See Mitchell and Carson (1989) and Hanley (1990). This will be a particular problem for mail surveys where response rates are low, as it is likely that responses will be biased towards those with a particular interest in the good and therefore unrepresentative of the general population. Response rates significantly below 40% are common in such surveys.

37. In discrete take-it-or-leave-it models the mean will be given as the expected value of WTP. Truncation is then effected by calculating the expected value between set limits rather than up to infinity.

38. Significant at the 15% level.

39. It is important that face-to-face studies also report observed non-response rates (i.e. the number of randomly selected respondents who refuse to be interviewed) so that an accurate comparison with mail non-response can be carried out. In a recent experiment using both mail and on-site (face-to-face) surveys, Bateman et al. (1992) report a mail survey response rate of 30% and a face-to-face response rate of 78%.

40. Note that, following our discussion of the asymmetric valuation of gains and losses, it does not necessarily follow that WTA will not lie above the dichotomous choice, upper bound estimate of WTP (although it may do so).

41. Interestingly the absence of such relationships might be used to test our earlier assertion that non-user WTP valuations for poorly perceived public goods may exhibit small sum 'charitable' responses rather than genuine valuations.

42. Several studies have reported highly significant non-use values. For example, Walsh et al. (1990) report existence values equal to approximately 25% of total value. Other studies have exceeded this value (see Mitchell and Carson 1989).

References

Ajzen, I. and Fishbein, M. (1977). Attitude–behaviour relations: A theoretical analysis and review of empirical research. *Psychological Bulletin* 84(5), 888–918.

Ajzen, I. and Peterson, G.L. (1988). Contingent value measurement: The price of everything and the value of nothing? In Peterson, G.L., Driver, B.L. and Gregory, R. (eds), *Amenity Resource Valuation: Integrating Economics with Other Disciplines*. Venture Publishing, 65–76.

Andreoni, J. (1990). Impure altruism and donations to public goods: A theory of warm-glow giving. *Economic Journal* 100.

Banford, N., Knetsch, J. and Mauser, G. (1977). Compensating and equivalent measures of consumers surplus: further survey results. Department of Economics, Simon Fraser University, Vancouver.

Barnett, A.H. and Yandle, B. (1973). Allocating environmental resources. *Public Finance* 28, 11–19.

Bateman, I.J. (1991). A critical analysis of COBA and proposals for an extended cost benefit approach to transport decisions. In Hanna, J. (ed.), *What Are Roads Worth?: Fair Assessment for Transport Expenditure*. New Economics Foundation/Transport 2000, London.

Bateman, I.J., Green, C., Tunstall, S. and Turner, R.K. (1991). The Contingent Valuation Method. Report to the Transport and Road Research Laboratory, Department of Transport, Bracknell.

Bateman, I.J., Willis, K.G., Garrod, G., Doktor, P. and Turner, R.K. (1992). *A Contingent Valuation Study of the Norfolk Broads*. Report to the National Rivers Authority.

Bergstrom, J.C., Dillman, B.L. and Stoll, J.R. (1985). Public environmental amenity benefits of private land: The case of prime agricultural land. *Southern Journal of Agricultural Economics* 17(1).

Bishop, R. and Heberlein, T. (1979). Measuring values of extra market goods: Are indirect measures biased? *American Journal of Agricultural Economics* 61, 926–30.

Bishop, R.C. and Heberlein, T.A. (1985). Progress Report of the 1984 Sandhill study. Preliminary report to the Wisconsin Department of Natural Resources. Reprinted in Mitchell, R. and Carson, R., *Using Surveys to Value Public Goods*. Resources for the Future, Washington, DC.

Bishop, R.C., Heberlein, T.A. and Kealy, M.J. (1983). Hypothetical bias in contingent valuation: Results from a simulated market. *Natural Resources Journal* 23, 619–33.

Bishop, R.C., Heberlein, T.A., Welsh, M.P. and Baumgartner, R.A. (1984). Does contingent valuation work? A report on the Sandhill study. Paper presented at the Joint Meeting of the Association of Environmental and Resource Economists and the American Economics Association, Cornell University, New York, August.

Bohm, P. (1972). Estimating demand for public goods: An experiment. *European Economic Review* 3, 733–6.

Boyle, K.J. (1989). Commodity specification and the framing of contingent valuation questions. *Land Economics* 65, 57–63.

Boyle, K.J. and Bishop, R.C. (1988). Welfare measurements using contingent valuation: A comparison of techniques. *Journal of the American Agricultural Association*, February, pp.20–8.

Boyle, K.J., Bishop, R.C. and Welsh, M.P. (1985). Starting point bias in contingent valuation surveys. *Land Economics* 61, 188–94.

Brennan, A. (1992). Moral pluralism and the environment. *Environmental Values* 1, 15–53.

Brookshire, D.S. and Coursey, D.L. (1987). Measuring the value of a public good: an empirical comparison of elicitation procedures. *American Economic Review* 77, 554–66.

Brookshire, D.S., Ives, B.C. and Schulze, W.C. (1976). The valuation of aesthetic preferences. *Journal of Environmental Economics and Management* 3, 325–46.

Brookshire, D.S., Randall, A. and Stoll, J.R. (1980). Valuing increments and decrements in natural resource service flows. *American Journal of Agricultural Economics* 62, 478–88.

Brookshire, D.S., Thayer, M.A., Schulze, W.D. and d'Arge, R.C. (1982). Valuing public goods: a comparison of survey and hedonic approaches. *American Economic Review* 71, 165–77.

Brookshire, D.S. *et al.* (1983) Estimating option prices and existence values for wildlife resources, *Land Economics* 59, 1–15.

Brown, R.A. and Green, C.H. (1981). Threats to health and safety: Perceived risk and willingness to pay. *Social Science and Medicine* C, 15, 67–75.

Brown, T.C. and Slovic, P. (1988). Effects of context on economic measures of values. In Peterson, G.L., Driver, B.L. and Gregory, R. (eds), *Amenity Resource Valuation: Integrating Economics with Other Disciplines*. Venture State College, PA, pp.23–30.

Brubaker, E. (1982). Sixty-eight percent free revelation and thirty-two percent free ride? Demand disclosures under varying conditions of exclusion. In Smith, V.L. (ed.), *Research in Experimental Economics*, Vol.2. JAI Press, Greenwich, CT.

Burness, H.S., Cummings, R.G., Mehr, A.F. and Walbert, M.S. (1983). Valuing policies which reduce environmental risk. *Natural Resources Journal* 23, 675–82.

Cameron, T. (1988). A new paradigm for valuing non-market goods using refendum data. *Journal of Environmental Economics and Management* 15, 355–79.

Cameron, T. and James, M. (1987). Efficient estimation methods for use with closed ended contingent valuation survey data. *Review of Economics and Statistics* 69, 269–76.

Carson, R.T. (1989). Constructed markets. Mimeo. Reprinted in Kriström, B., *Valuing Environmental Benefits Using the Contingent Valuation Method – An Econometric Analysis*. Umeå Economic Studies No.219, University of Umeå, Sweden.

Carson, R.T. and Mitchell, R.C. (1983). *A Reestimation of Bishop and Heberlein's Simulated Market – Hypothetical Markets – Travel Cost Results under Alternative Assumptions*. Discussion Paper D-107. Resources for the Future, Washington, DC.

Christiansen, G.B. (1982). Evidence for determining the optimal mechanism for providing collective goods. *American Economist* 26, 57–61.

Cicchetti, C.J. and Freeman, A.M. III (1971). Optional demand and consumer's surplus: Further comment. *Quarterly Journal of Economics* 85, 528–39.

Converse, J.M. and Presser, S. (1986). *Survey Questions: Handcrafting the Standardised Questionnaire*. Sage, Beverly Hills, CA.

Coursey, D.L., Schulze, D.W. and Hovis, J. (1983). A comparison of alternative valuation mechanisms for non-market commodities. University of Wyoming.

Coursey, D.L., Hovis, J. and Schulze, W.D. (1986). The disparity between willingness to accept and willingness to pay measures of value. *Quarterly Journal of Economics* 102, 679–90.

Cronin, F.J. and Herzeg, K. (1982). Valuing nonmarket goods through contingent markets. In Cummings, R.G., Brookshire, D.S. and Schulze, W.D. (eds), *Valuing Environmental Goods*. Rowman and Allenheld, Totowa, NJ, 1986.

Cummings, R.G., Brookshire, D.S. and Schulze, W.D. (eds) (1986). *Valuing Environmental Goods: A State of the Arts Assessment of the Contingent Method*. Rowman and Allenheld, Totowa, NJ.

Deaton, A. and Muellbauer, J. (1980). *Economics and Consumer Behaviour*. Cambridge University Press, Cambridge.

Desvousges, W.H., Smith, V.K. and McGivney, M.P. (1983). *A Comparison of Alternative Approaches for Estimating Recreation and Related Benefits of Water Quality Improvements*, EPA Report 230-05-83-001. US Environmental Protection Agency, Office of Policy Analysis, Washington, DC.

Desvousges, W.H. *et al.* (1987). Option price estimates for water quality improvement. *Journal of Environmental Economics and Management* 14, 248–67.

Dickie, M., Fisher, A. and Gerking, S. (1987). Market transactions and hypothetical demand data: A comparative study. *Journal of the American Statistical Association* 82, 69–75.

Dillman, D.A. (1978). *Mail and Telephone Surveys: The Total Design Method*. Wiley, New York.

Duffield, J.W. and Patterson, D.A. (1991). Inference and optimal design for a welfare measure in dichotomous choice contingent valuation. *Land Economics* 67, 225–39.

Eberle, W.D. and Hayden, F.G. (1991). Critique of contingent valuation and travel cost method for valuing natural resources and ecosystems. *Journal of Economic Issues* 25, 649–85.

Ehrlich, P.R. and Ehrlich, A.G. (1992). The value of biodiversity. *Ambio* 21, 219–26.

Freeman, A.M. (1986). On assessing the state of the arts of the contingent valuation method of valuing environmental changes. In Cummings, R.G., Brookshire, D.S. and Schulze, W.D. (eds), *Valuing Environmental Goods*. Rowman and Allenheld, Totowa, NJ.

Garrod, G.D. and Willis, K.G. (1990). *Contingent Valuation Techniques: A Review of their Unbiasedness, Efficiency and Consistency*, Countryside Change Initiative: Working Paper No.10. Countryside Change Unit, University of Newcastle upon Tyne.

Garrod, G. and Willis, K.G. (1991). *The Hedonic Price Method and the Valuation of Countryside Characteristics*, Countryside Change Unit Working Paper 14. Countryside Change Unit, University of Newcastle upon Tyne.

Green, C.H. and Tunstall, S.M. (1991). The evaluation of river quality improvements by the contingent valuation method. *Applied Economics* 23, 1135–46.

Green, C.H., Tunstall, S.M., N'Jai, A. and Rodgers, A. (1990). The economic evaluation of environmental goods. *Project Appraisal* 5(2), 70–82.

Hammack, J. and Brown, G.M. Jr (1974). *Waterfowl and Wetlands: Toward Bioeconomic Analysis*. Johns Hopkins University Press (for Resources for the Future), Baltimore, MD.

Hanemann, W.M. (1986). Willingness to pay and willingness to accept: How much can they differ? Draft manuscript, Department of Agriculture and Resource Economics, University of California, Berkeley. Reprinted in Mitchell, R. and Carson, R. *Using Surveys to Value Public Goods*. Resources for the Future, Washington, D.C.

Hanemann, W.M. (1991). Willingness to pay and willingness to accept: How much can they differ? *American Economic Review* 81, 635–47.

Hanemann, W.M., Loomis, J. and Kanninen, B. (1991). Statistical efficiency of double-bounded dichotomous choice contingent valuation. *American Journal of Agricultural Economics* 73, 1255–63.

Hanley, N.D. (1990). *Valuation of Environmental Effects: Final Report – Stage One*. Industry Department of Scotland and the Scottish Development Agency, Edinburgh.

Hanley, N.D. and Munro, A. (1991). *Design Bias in Contingent Valuation Studies: The Impact of Information*, Discussion Paper in Economics No.13. Department of Economics, University of Stirling.

Harris, C.C. and Brown, G. (1992). Gain, loss and personal responsibility: The role of motivation in resource valuation decision-making. *Ecological Economics* 5, 73–92.

Harris, C.C., Driver, B.L. and McLaughlin, M.J. (1989). Improving the contingent valuation method: A psychological approach. *Journal of Environmental Economics and Management* 17, 213–29.

Heberlein, T.A. (1986). Measuring resource values: the reliability and validity of dichotomous contingent valuation measures. Paper presented at the American Sociological Association Meeting, New York, August.

Heberlein, T. and Bishop, R. (1986). Assessing the validity of contingent valuation: Three experiments. *Science of the Total Environment* 56, 99–107.

Heberlein, T.A. and Black, J.S. (1976). Attitudinal specificity and the prediction of behaviour in a field setting. *Journal of Personality and Social Psychology* 33, 434–79.

Hicks, J.R. (1943). The four consumer surpluses. *Review of Economic Studies* 11, 31–41.

Hill, R.J. (1981). Attitudes and behaviour. In Rosenberg, M. and Turner, R.H. (eds), *Social Psychology: Sociological Perspectives*. Basic Books, New York.

Hoehn, J.P. and Randall, A. (1987). A satisfactory benefit cost indicator from contingent valuation. *Journal of Environmental Economics and Management* 14, 226–47.

Hoen, H.F. and Winther, G. (1991). *Attitudes to and Willingness to Pay for Multiple-Use Forestry and Preservation of Coniferous Forests in Norway*. Department of Forestry, Agricultural University of Norway.

Hoevenagel, R. (1990). The validity of the contingent valuation method: Some aspects on the basis of three Dutch studies. Paper presented at the first annual meeting of the European Association of Environmental and Resource Economists (EAERE). Venice, 17–20 April 1990.

Hoevenagel, R. (1991). A contingent valuation (CV) experiment to test 'part-whole' bias. Paper presented at the second annual meeting of EAERE, Stockholm, 10–14 June.

Johansson, P.-O. (1987). *The Economic Theory and Measurement of Environmental Benefits*. Cambridge University Press, Cambridge.

Johansson, P.-O. (1991). *An Introduction to Modern Welfare Economics*. Cambridge University Press, Cambridge.

Just, R.E., Hueth, D.L. and Schmitz, A. (1982). *Applied Welfare Economics and Public Policy*. Prentice Hall, Englewood Cliffs, NJ.

Kahneman, D. (1986). Comments. In Cummings, R.G., Brookshire, D.S. and Schulze, W.D. (eds), *Valuing Environmental Goods*. Rowman and Allenheld, Totowa, NJ.

Kahneman, D. and Knetsch, J.L. (1992a). Valuing public goods: The purchase of moral satisfaction. *Journal of Environmental Economics and Management* 22, 57–70.

Kahneman, D. and Knetsch, J.L. (1992b). Contingent valuation and the value of public goods: Reply. *Journal of Environmental Economics and Management* 22, 90–4.

Kahneman, D. and Tversky, A. (1979). Prospect theory: An analysis of decisions under risk. *Econometrica* 47, 263–91.

Kahneman, D. and Tversky, A. (1982). The psychology of preferences. *Scientific American*, January, pp.2136–41.

Kahneman, D. and Tversky, A. (1984). Choices, values and frames. *American Psychologist* 39, 341–50.

Kealy, M.J., Montgomery, M. and Dovido, J.F. (1990). Reliability and predictive validity of contingent values: Does the nature of the good matter? *Journal of Environmental Economics and Management* 19, 244–63.

Kneese, A.V. (1984). *Measuring the Benefits of Clean Air and Water*. Resources for the Future, Washington, DC.

Knetsch, J.L. and Davis, R.K. (1966). Comparisons of methods for recreation evaluation. In Kneese, Allen V. and Smith, Stephen C. (eds), *Water Research*. Johns Hopkins University Press (for Resources for the Future), Baltimore, MD.

Knetsch, J. and Sinden, J. (1984). Willingness to pay and compensation demanded: Experimental evidence of an unexpected disparity in measures of value. *Quarterly Journal of Economics* 99, 507–21.

Kriström, B. (1990). *Valuing Environmental Benefits Using the Contingent Valuation Method — An Econometric Analysis*, Umeå Economic Studies No.219. University of Umeå, Sweden.

Krutilla, J.V. and Fisher, A.C. (1975). *The Economics of Natural Environments: Studies in the Valuation of Commodity and Amenity Resources*. Johns Hopkins University Press (for Resources for the Future), Baltimore, MD.

Loehman, E. and De, V.H. (1982). Application of stochastic choice modelling to policy analysis of public goods: A case study of air quality improvements. *Review of Economics and Statistics* 64, 474–80.

Loomis, J.B. (1987). Balancing public trust resources of Mono Lake and Los Angeles' water right: An economic approach. *Water Resources Research* 23, 1449–56.

Loomis, J.B. (1989). Test-retest reliability of the contingent valuation method: A comparison of general population and visitor response. *American Journal of Agricultural Economics* 71, 76–84.

Loomis, J.B. (1990). Comparative reliability of the dichotomous choice and open-ended contingent valuation techniques. *Journal of Environmental Economics and Management* 18, 78–85.

Lumsden, J. (1976). Test theory. *Annual Review of Psychology* 27, 251–80.

Magnussen, K. (1991). Valuation of reduced water pollution using the contingent valuation method – methodology and empirical results. Paper presented at the second annual meeting of EAERE, Stockholm, 10–14 June 1991.

Mannesto, G. and Loomis, J.B. (1991). Evaluation of mail and in-person contingent value surveys: Results of a study of recreational boaters. *Journal of Environmental Economics and Management* 32, 177–90.

Marwell, G. and Ames, R.E. (1981). Economists free ride, does anyone else? Experiments on the provision of public goods. *Journal of Public Economics* 15, 295–310.

Maslow, A. (1970). *Motivation and Personality*. Harper and Row, New York.

McCullagh, P. and Nelder, J.A. (1983). *Generalized Linear Models*. Chapman and Hall, London.

Milon, J.W. (1989). Contingent valuation experiments for strategic behaviour. *Journal of Environmental Economics and Management* 17, 293–308.

Mitchell, L.A.R., Willis, K.G. and Benson, J.F. (1988). *A Review and Empirical Tests of Bias in Contingent Valuations of Wildlife Resources*, Working Paper No.5. Department of Town and Country Planning, University of Newcastle upon Tyne.

Mitchell, R. and Carson, R. (1981). An experiment in determining willingness to pay for national water quality improvements. Draft report to the US Environmental Protection Agency, Washington, DC.

Mitchell, R. and Carson, R. (1989). *Using Surveys to Value Public Goods: The Contingent Valuation Method*. Resources for the Future, Washington, DC.

Mitchell, R.C. (1991). Comments regarding Kahneman and Knetsch; made at the second Annual Conference of the European Association of Environmental and Resource Economists, Stockholm, 11–14 June.

Navrud, S. (1989a). Estimating social benefits of environmental improvements from reduced acid depositions: A contingent valuation survey. In Folmer, H. and van Ierland, E., *Valuation Methods and Policy Making in Environmental Economics*. Elsevier, Amsterdam.

Navrud, S. (1989b). *The Use of Benefits Estimates in Environmental Decision Making: Case Study on Norway*. OECD, Paris.

Navrud, S. (1991). Willingness to pay for preservation of species: An experiment with actual payments. Paper presented at the second Annual Conference of the European Association of Environmental and Resource Economists, Stockholm, 11–14 June.

Orne, M.T. (1962). On the social psychology of the psychological experiment. *American Psychologist* 17, 776–89.

Pearce, D.W. and Markandya, A. (1989). *The Benefits of Environmental Policies*. OECD, Paris.

Pearce, D.W. and Turner, R.K. (1990a). *Economics of Natural Resources and the Environment*. Harvester Wheatsheaf, Hemel Hempstead.

Pearce, D.W. and Turner, R.K. (1990b). *The Use of Benefit Estimates in*

Environmental Decision-Making. Report to the Environmental Directorate. OECD, Paris.

Quiggin, J. (1991). Total valuation for Kakadu National Park. Department of Agriculture and Resource Economics, University of Maryland.

Rae, D.A. (1982). Benefits of visual air quality in Cincinnati. Report to Electric Power Research Institute, Charles River Associates, Boston.

Rahmatian, M. (1987). Component value analysis. Air quality in the Grand Canyon National Park. *Journal of Environmental Management* 24, 217–23.

Randall, A. (1987) The total value dilemma. In Peterson, G.L. and Sorg, C.F. (eds), *Toward the Measurement of Total Economic Value*, U.S.D.A. For Serv. Gen. Tech. Report RM–148, Rocky Mountain Forest and Range Experiment Station, Fort Collins, Co., pp. 3–13.

Randall, A. and Stoll, J.R. (1980). Consumer's surplus in commodity space. *American Economic Review* 70, 449–55.

Randall, A., Hoehn, J.P. and Brookshire, D.S. (1983). Contingent valuation surveys for evaluating environmental assets. *Natural Resources Journal* 23, 635–48.

Roberts, K.J. and Thompson, M.E. (1983). An empirical application of the contingent valuation method to value marine resources. In Cummings, R.G., Brookshire, D.S. and Schulze, W.D. (eds), *Valuing Environmental Goods.* Rowman and Allenheld, Totowa, NJ, 1986.

Rowe, R.D. and Chestnut, L.G. (1982). *The Value of Visibility: Economic Theory and Applications for Air Pollution Control.* Abt Books, Cambridge, MA.

Rowe, R.D. and Chestnut, L.G. (1983). Valuing environmental commodities: revisited. *Land Economics* 59, 404–10.

Rowe, R., d'Arge, R. and Brookshire, D. (1980). An experiment on the economic value of visibility. *Journal of Environmental Economics and Management* 7, 1–19.

Sagoff, M. (1988). Some problems with environmental economics. *Environmental Ethics* 10, 55–74.

Samples, K.C., Dixon, J.A. and Gowen, M.M. (1985). Information disclosure and endangered species valuation. Paper presented at the annual meeting of the American Agricultural Economics Association, Iowa State University, Ames, Iowa, 4–7 August.

Samuelson, P. (1954). The pure theory of public expenditure. *Review of Economics and Statistics* 36, 387–9.

Schneider, F. and Pommerehne, W.W. (1981). Free riding and collective action: An experiment in public microeconomics. *Quarterly Journal of Economics* 97, 689–702.

Schoemaker, P.J.H. (1982). The expected utility model: its variants, purposes, evidence and limitations. *Journal of Economic Literature* 20, 529–63.

Schulze, W.D., d'Arge, R.C. and Brookshire, D.S. (1981). Valuing environmental commodities: some recent experiments. *Land Economics* 57, 151–69.

Schulze, W.D., Brookshire, D.S., Walther, E.G., MacFarland, K.K., Thayer, M.A., Whitworth, R.L., Ben-David, S., Malm, W. and Molenar, J. (1983). The economic benefits of preserving visibility in the national parklands of the South-West. *Natural Resources Journal* 23, 149–73.

Schuman, H. and Johnson, M.P. (1976). Attitudes and behaviour. In Inkeles, A. (ed.), *Annual Review of Sociology*, Vol.2, pp.161–207, Annual Reviews, Inc.

Seip, K. and Strand, J. (1990). Willingness to pay for environmental goods in Norway: A contingent valuation study with real payments. Paper presented at the first annual conference of the European Association of Environmental and Resource Economists, 17–20 April.

Shogren, J.F., Shin, S.Y., Hayes, D.J. and Kliebenstein, J.B. (1993). Experimental evidence on the divergence between measures of value. *Journal of Environmental Economics and Management* forthcoming.

Slovic, P. (1972). *From Shakespeare to Simon: Speculations – and Some Evidence – about Man's Ability to Process Information*, O.R.I. Research Monograph Volume 12. Eugene, Oregon.

Smith, A. (1790). The theory of moral sentiments. In Kriström, B., *Valuing Environmental Benefits Using the Contingent Valuation Method – An Econometric Analysis*, Umeå Economic Studies No.219, University of Umeå, Sweden, 1990.

Smith, V.K. (1992a). Non market valuation of environmental resources: An interpretative essay. Draft copy of unpublished manuscript.

Smith, V.K. (1992b). Arbitrary values, good causes and premature verdicts. *Journal of Environmental Economics and Management* 22, 71–89.

Smith, V.K. and Desvousges, W.H. (1986). *Measuring Water Quality Benefits*. Kluwer-Nijhoff, Boston.

Smith, V.K., Desvousges, W.H. and Fisher, A. (1986). A comparison of direct and indirect methods for estimating environmental benefits. *American Journal of Agricultural Economics* 68, 280–9.

Smith, V.K., Liu, J.L., Altaf, A., Jamal, H. and Whittington, D. (1991). How reliable are contingent valuation surveys for policies in developing countries. Draft report. Department of Economics, North Carolina State University.

Strand, J. and Taraldset, A. (1991). The valuation of environmental goods in Norway: A contingent valuation study with multiple bias testing. Memorandum no.2/91. Department of Economics, University of Oslo.

Thaler, R. (1985). Mental accounting and consumers' choice. *Marketing Science* 4, 199–214.

Tolley, G.S. and Randall, A., with G. Blomquist, R. Fabian, G. Fishelson, A. Frankel, J. Hoehn, R. Krumm and E. Mensah (1983). *Establishing and Valuing the Effects of Improved Visibility in the Eastern United States*. Interim Report to the US Environmental Protection Agency, quoted in Garrod and Willis (1990).

Tunstall, S., Green, C.H. and Lord, J. (1988). The evaluation of environmental goods by the contingent valuation method. Report by the Flood Hazard Research Centre, Middlesex Polytechnic.

Turner, R.K. (ed.) (1988a). *Sustainable Environmental Management: Principles and Practice*. Belhaven, London.

Turner, R.K. (1988b). Wetland conservation: Economics and ethics. In Collard, D., Pearce, D.W. and Ulph, D. (eds), *Economics, Growth and Sustainable Development*. Macmillan, London.

Turner, R.K. (1992). *Speculations on Weak and Strong Sustainability*, GEC Working Paper 92–26. CSERGE, University of East Anglia, Norwich and University College London.

Turner, R.K., Bateman, I.J. and Brooke, J.S. (1992a). Valuing the benefits of coastal defence: a case study of the Aldeburgh sea-defence scheme. In Coker, A. and Richards, C. (eds), *Valuing the Environment*. Belhaven Press, London, pp.77–100.

Turner, R.K., Bateman, I.J. and Pearce, D.W. (1992b). United Kingdom. In Navrud, S. (ed.), *Valuing the Environment: The European Experience*. Scandinavian University Press, Oslo.

Tversky, A. and Kahneman, D. (1981). The framing of decisions and the psychology of choice. *Science* 211, 453–8.

Varian, H.R. (1984). *Microeconomic Analysis*. Norton, New York.

Walbert, M.S. (1984). Valuing policies which reduce environmental risk: An assessment of the contingent valuation method. PhD thesis, University of New Mexico. Reprinted in Mitchell, R. and Carson, R., *Using Surveys to Value Public Goods*. Resources for the Future, Washington, DC.

Walsh, R.G., Bjonback, R.D., Aiken, R.A. and Rosenthal, D.H. (1990). Estimating the public benefits of protecting forest quality. *Journal of Environmental Management* 30, 175–89.

Weisbrod, B.A. (1964). Collective-consumption services of individual-consumption goods. *Quarterly Journal of Economics* 78, 471–7.

Whittington, D., Briscoe, J., Mu, X. and Barron, W. (1990). Estimating the willingness to pay for water services in developing countries: A case study of the use of contingent valuation surveys in southern Haiti. *Economic Development and Cultural Change* 38, 293–311.

Willig, R.D. (1973). *Consumer's Surplus: A Rigorous Cookbook*, Technical Report No.98. Institute for Mathematical Studies in the Social Sciences, Stanford University.

Willig, R.D. (1976). Consumer's surplus without apology. *American Economic Review* 66, 587–97.

Willis, K.G. and Garrod, G.D. (1991). *Landscape Values: A Contingent Valuation Approach and Case Study of the Yorkshire Dales National Park*, Countryside Change Working Paper 21. Countryside Change Unit, University of Newcastle upon Tyne.

Young, T. and Allen, P.G. (1986). Methods for valuing countryside amenity: An overview. *Journal of Agricultural Economics* 37, 349–64.

Appendix 5.1: Welfare measures for priced and unpriced goods

Compensating variation, in the context of a proposed price fall welfare gain, is defined as 'the loss of income which would just offset the fall in price, leaving the consumer no better off than before' (Hicks 1943), i.e. the consumer's WTP to have the welfare-improving change implemented. However, many environmental public goods do not have prices. In such situations, if supply were infinite then consumer surplus would equal the entire area under the Marshallian demand curve. However, in reality supply is not infinite and such environmental goods are quantity-constrained by institutional or other factors.

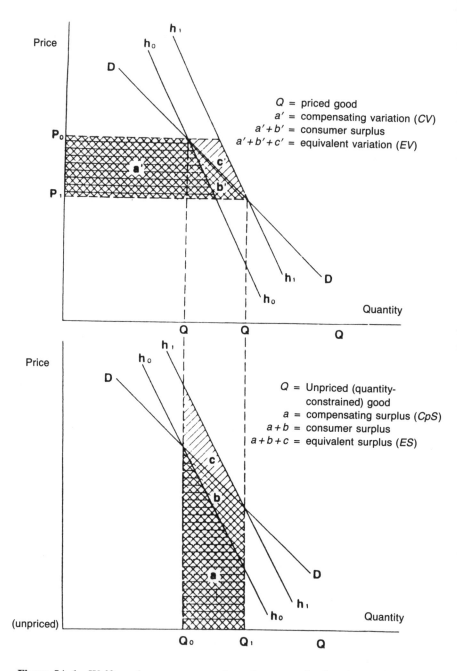

Figure 5A.1   Welfare change measures for price-constrained and quantity-constrained goods

Figure 5A.1 illustrates a welfare change to a single good Q. In the upper panel Q is depicted as a priced private good while in the lower panel Q is an unpriced public good. Such potential changes of property rights within the same good are demonstrated by cases such as woodland recreation drives which can be either marketed or unpriced (however, the diagram is intended to focus attention upon the general private/public good split). Suppose that in both cases the authorities propose to increase consumption from Q_0 to Q_1. In the upper panel we might envisage this occurring via a price subsidy, lowering prices from P_0 to P_1, while in the lower panel this may arise simply by increasing available provision (e.g., by constructing a new forest drive).

Assuming that, in the upper panel, price is a continuous variable, then the appropriate welfare measures will be compensating variation (CV), equivalent variation (EV), and consumer surplus (price constrained; pCS). The lower panel illustrates discrete quantity changes for an unpriced public good for which the appropriate welfare measures are compensating surplus (CpS), equivalent surplus (ES) and consumer surplus (quantity constrained; qCS).

From Figure 5A.1 we can readily demonstrate that, although we are looking at an identical change in provision, the relative welfare measures generated for priced and unpriced goods will not be equal. 'What the compensating variation measures is the change in income required to offset the fall in price, not the change in income required to offset the rise in quantity acquired (compensating surplus). It becomes apparent that these are not the same thing' (Hicks 1943).

While it is not central to our discussion, consider first the different measures of consumer surplus which are obtained by either integrating horizontally (upper panel; $pCS = a' + b'$) or vertically (lower panel; $qCS = a + b$). This arises due to an accounting artefact; in analysis either approach will provide equal measures of net social gain when net revenue changes are also included (under a unitary elasticity demand curve both measures of consumer surplus would be equal). Turning to our main focus of interest, notice now the difference in the size of the compensating triangles; b' in the upper panel and b in the lower. Similarly, notice a difference in the size of the equivalent triangles; c' (upper panel) and c (lower panel). It can be seen that the relative differences involved depend crucially upon the relative slopes of the Hicksian and Marshallian demand curves, i.e. upon the substitution (S) and income (I) effects.

For any normal good ($I > 0$), then, our price constrained welfare measures will be:

$$[CV = a'] < [pCS = a' + b'] < [EV = a' + b' + c']$$

and for our unpriced, quantity constrained case;

$$[Cps = a] < [qCS = a + b] < [ES = a + b + c]$$

In a seminal article, Hicks (1943) uses an intuitive geometric approach to demonstrate the arithmetic difference between these quantity- and price-constrained welfare measures. Figure 5A.2 simplifies Hicks's original figure and follows his analysis. In effect, it superimposes both panels of the previous figure on common axes. In the priced-good case, CV is the amount of income which the consumer must give up in return for a price fall from H to h (i.e. to move from point P to p) to stay at a constant initial utility level. This will be equal to the area $HPCh$. To evaluate arithmetically the difference between CV and CpS, Hicks treats the quantity-constrained good as if it is priced by extending the compensated demand curve h_0h_0 beyond PC down to the point M.

Along h_0h_0 utility is still constant at its initial level (U_0): the consumer still has the same quantity at M as at p. However, he will have

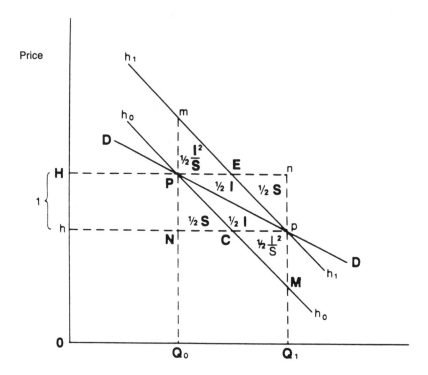

Figure 5A.2   Hicks's analysis

had to give up further income equal to the triangle CMp. Therefore, the CpS will equal $HPCh + CMp$, as this corresponds to the quantity-constrained case, this amount will be the WTP to ensure an increased provision of the environmental good. Similarly while the EV will be $HEph$, the ES will be $HEph + EmP$, i.e. the WTA compensation to forgo the increased provision of the environmental good.

Therefore, for a welfare gain change to the provision of a normal good

$$CpS < CV < CS < EV < ES$$

We can now define the arithmetic relation of these amounts. By defining the price change Hh as being a single unit change we can assume that all the lines in Figure 5A.2 are straight and parallel. Furthermore, for our marginal price change, as NC is the substitution effect (S), so area PNC must equal $\frac{1}{2}S$, as will area pnE. Similarly, as Cp is the income effect $(I$, here positive for a normal good), so area PCp must equal $\frac{1}{2}I$, as will area pEP. PCN and MCp are similar triangles, thus

$$\frac{\text{area of } PCN}{\text{area of } MCp} = \frac{NC^2}{Cp^2} \qquad \text{(similar triangles)}$$

Also:

$$\frac{NC}{Cp} = \frac{\text{area of } PCN}{\text{area of } PCp} \qquad \text{(common height; PN)}$$

$$= \frac{S}{I} \qquad \text{(ratio of substitution to income effect)}$$

Therefore

$$\frac{\text{area of } PCN}{\text{area of } MCp} = \frac{S^2}{I^2}$$

Now if we substitute in that area $PCN = \frac{1}{2}S$ we can therefore state that;

$$\text{area of } MCp = \frac{I^2}{2S} = \text{area of } EmP$$

$\downarrow$ (difference between CpS and CV) $\qquad$ $\downarrow$ (difference between ES and EV)

If we define area $HPNh = r$ and $Hnph = R$, noting that $R = r + I + S$, we can now define all four welfare measures with respect to substitution and income effects;

$$CV = r + \tfrac{1}{2}S \qquad\qquad EV = r + \tfrac{1}{2}S + I$$

$$CpS = r + \tfrac{1}{2}S - \frac{I^2}{2S} \qquad\qquad ES = r + \tfrac{1}{2}S + I + \frac{I^2}{2S}$$

For a welfare gain the compensating measures will be negative (WTP income) whereas the equivalent measures will be positive (WTA compensation); r will also be negative here as consumers spend more of their income. We can now vary I relative to S to analyse these welfare measures. Figure 5A.3 shows the situation for normal goods.

Note that throughout this range the relative order of welfare measures is maintained. However, for quantity-constrained goods with high income effects ($I > S$) the quantity-constrained welfare measures (CpS, ES) expand rapidly. Such conditions may well apply to environmental public goods.

Another interesting property of the ES measure (WTA compensation) arises when we drive the quantity consumed towards zero (see lower panel of Figure 5A.1). The ES measure is defined with respect to the compensated demand curve $h_1 h_1$. At very low consumption levels this will tend towards the vertical, i.e. ES will tend towards (and in the case of absolute necessities will become) infinity. Therefore, if we are considering reductions in the provision of goods which are considered necessities (which will include many environmental public goods) then we should expect very high or infinite bids with respect to WTA sums. This provides further theoretical support for the observed WTP/WTA asymmetry, specifically here when the WTA question proposes a reduction in provision to low levels of subsequent consumption.

As an extension to these notes we can postulate two conditions under which Hicks's 'four consumer surplus' measures (in our notation, CV, EV, CpS, and ES) may lead to high WTP or WTA sums. To do so we make an analogy with the diamonds and water phenomenon first noted by Adam Smith. Firstly, if individuals have high incomes and high material well-being then they may exhibit a high preference for environmental goods, i.e. a high income elasticity like diamonds. In such conditions the environment is treated as a luxury good and consequently high WTP sums are to be expected. Following Hanemann (1991), these will lead to even higher asymmetric WTA sums. Secondly, and conversely, now suppose that individuals have low allocations of a particular environmental good and perceive that good to be a basic

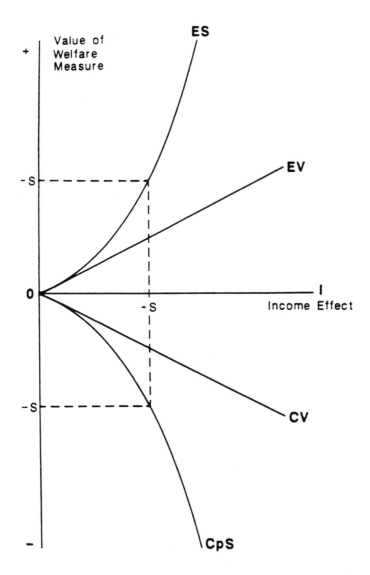

Figure 5A.3 Welfare change measures for price-constrained and quantity-constrained normal goods as *I* varies relative to *S*

Source: Adapted from Hicks (1943)

necessity. Here the analogy with water is apt as, at low provision, individuals will demand very high or infinite amounts of compensation (WTA) rather than endure a loss of the particular environmental good, i.e. low elasticity of substitution. We therefore have two very different scenarios, both of which can result in very high valuations of environmental goods.

Appendix 5.2: Using expenditure functions to define welfare measures

We can express CV and EV measures as functions of expenditure. Traditionally any utility equilibrium is expressed as maximising utility given income. We can use duality theory (for a full exposition of duality theory, see Deaton and Muellbauer 1980) to represent this as an expenditure problem thus:

$$\text{minimise } E = E(p, q, \bar{u}) = Y$$

where

E = expenditure
p = price vector for private goods
q = quantity of the environmental good (or bad) considered
$\bar{u}$ = some fixed level of utility (u)
Y = money income

Consider a proposed increase in q from q_0 to q_1; then the compensating variation measure of this welfare change will be the adjustment to income (Y) necessary to keep the consumer at his initial utility level (u_0) throughout. Therefore,

$$CV = [E(p, q_0, u_0) = Y_0] - [E(p, q_1, u_0) = Y_1] = Y_0 - Y_1$$

Now if q_1 is preferred to q_0 (q is a desired environmental good and the proposed increase in provision is therefore a welfare gain) then the individual will be WTP for the change (CV_{WTP}, i.e. Y_1 will be lower than Y_0. Similarly, if q_0 is preferred to q_1 (q is an environmental 'bad' and the proposed change is therefore a welfare loss) then the individual will have a WTA compensation to tolerate the change, i.e. Y_1 will have to be higher than Y_0 to maintain u constant at u_0.

Therefore, for a proposed change in provision of q from q_0 to q_1, we have;

CV = WTP if q_1 preferred to q_0 (i.e. a welfare gain; CV_{WTP} is +ve)
  = WTA if q_0 preferred to q_1 (i.e. a welfare gain; CV_{WTA} is −ve)

Again considering a proposed increase in q from q_0 to q_1, the equivalent variation measure of this welfare change will be the adjustment to income (Y) necessary to keep the consumer at his subsequent (post change) utility level (u_1) throughout. Therefore,

$$EV = [E(p, q_0, u_1) = Y^e_0] - [E(p, q_1, u_1) = Y^e_1] = Y^e_0 - Y^e_1$$

Now if q_1 is preferred to q_0 (i.e. the proposed increase in provision is a welfare gain) then the EV measures the amount of compensation which the individual would be WTA to forgo the change in provision occurring. Here $Y^e_0 < Y^e_1$, i.e. EV_{WTA} is negative. Similarly if q_0 is preferred to q_1 (i.e. the proposed increase is a welfare loss) then EV measures the amount which the individual would be WTP to prevent the change occurring. Here $Y^e_0 > Y^e_1$, i.e. EV_{WTP} is positive.

Therefore, for a proposed change in the provision of q from q_0 to q_1 we have;

EV = WTA if q_1 preferred to q_0 (i.e. a welfare gain; EV_{WTA} is −ve)
  = WTP if q_0 preferred to q_1 (i.e. a welfare loss; EV_{WTP} is +ve)

The surplus measures CpS and ES behave in a similar manner to CV and EV, although non-continuity causes certain problems.

Chapter 6

Valuation of the Environment, Methods and Techniques: Revealed Preference Methods

Ian Bateman[1]

Overview

Individuals' preference for and evaluations of environmental goods can in some circumstances be revealed by their purchases of certain marketed goods associated with the consumption of those environmental goods. Both the travel cost method and hedonic pricing method discussed in this chapter adopt revealed preference approaches to environmental evaluation. The TCM is typically applied to the estimation of the recreational value of a recreation site by analysing the travel expenditures (petrol, etc.) of visitors to that site, while the HPM often uses variation in house prices to estimate the value of local environmental quality. Both techniques only capture use values and thereby omit any non-use value elements of the environmental goods under investigation. As such these techniques may underestimate the total economic value of such goods. However, use values will often be of prime importance (and acceptability) to decision-makers and both of these evaluation techniques have been widely applied.

The travel cost method

Introduction

The original idea behind the travel cost method (TCM) can be traced back to a letter from Hotelling – first reported in Prewitt (1949) – to the Director of the US National Park Service in which he suggested that the costs incurred by visitors could be used to develop a measure of the recreation value of the sites visited. However, it was Clawson (1959) and Clawson and Knetsch (1966) who first developed empirical models along these lines.

The TCM is a survey technique. A questionnaire is prepared and administered to a sample of visitors at a site in order to ascertain their place of residence; necessary demographic and attitudinal information; frequency of visit to this and other sites; and trip information such as purposefulness, length, associated costs, etc. From these data, visit costs can be calculated and related, with other relevant factors, to visit frequency so that a demand relationship may be established. In the simplest case this demand function can then be used to estimate the recreation value of the whole site, while in more advanced studies attempts can be made to develop demand equations for the differing attributes of recreation sites and values evaluated for these individual attributes.

Theoretical issues

The demand function estimated by TCM is an uncompensated ordinary demand curve incorporating income effects, and the welfare measure obtained from it will be that of Marshallian consumer surplus.

The method
In essence the TCM evaluates the recreational use value for a specific recreation site by relating demand for that site (measured as site visits) to its price (measured as the costs of a visit). A simple TCM model can be defined by a 'trip-generation function' (tgf) such as

$$V = f\,(C,X) \tag{6.1}$$

where

V = visits to a site
C = visit costs
X = other socioeconomic variables which significantly explain V.

The literature can be divided into two basic variants of this model according to the particular definition of the dependent variable V. The individual travel cost method (ITCM) simply defines the dependent variable as the number of site visits made by each visitor over a specific period, say one year. The zonal travel cost method (ZTCM), on the other hand, portions the entire area from which visitors originate into a set of visitor zones and then defines the dependent variable as the visitor rate (i.e. the number of visits made from a particular zone in a period divided by the population of that zone).

The ZTCM approach redefines the tgf as

$$V_{hj} / N_h = f(C_h, X_h) \qquad (6.2)$$

where

V_{hj} = visits from zone h to site j
N_h = population of zone h
C_h = visit costs from zone h
X_h = socioeconomic explanatory variables in zone h.

The visitor rate, V_{hj}/N_h, is often calculated as visits per 1000 population in zone h.

The underlying theory of the TCM is presented with reference to the zonal variant, and discussion of the differences between this and the individual variant is presented subsequently before consideration of more general issues.

The zonal travel cost method
Discussion of the ZTCM is illustrated by reference to a constructed example (detailed in Table 6.1) which estimates the recreation value of a hypothetical site. The method proceeds as follows:

(i)   Data on the number of visits made by households in a period (say annually) and their origin are collected through on-site surveys.
(ii)   The area encompassing all visitor origins is subdivided into zones of increasing travel cost (column 1 of Table 6.1); and the total population (number of households) in each zone noted (column 2).
(iii)   The number of household visits per zone (column 3) is calculated by allocating sampled household visits to their relevant zone of origin.
(iv)   The household average visit rate in each zone (column 4) is calculated by dividing the number of household visits in each zone (column 3) by the zonal population (number of households, column 2). Note that this will often not be a whole number and commonly less than one.
(v)   The zonal average cost of a visit (column 5) is calculated with reference to the distance from the trip origin to the site.
(vi)   A demand curve is then fitted relating the zonal average price of a trip (travel cost) to the zonal average number of visits per household. This curve estimates demand for the 'whole recreation experience' rather than just the time spent on-site. In our hypothetical example, this demand is explained purely by visit cost

Table 6.1 Consumer surplus estimates for the whole recreation experience using the ZTCM

Zone no. (h)	Zonal population (no. of households)[1] (N_h)	No. of household visits to site p.a.[2] (V_{nj})	Average no. of visits per house-hold p.a.[3] (V_{hj}/N_h)	Average travel cost per house-hold visit[4] (C_h)	Consumer surplus per household p.a. (£)	Consumer surplus per household visit (£)	Total consumer surplus p.a. (£)
Column number 1	2	3	4	5	6	7	8
1	10,000	12,500	1.25	0.16	2.60	2.08	26,040
2	30,000	30,000	1.00	1.00	1.67	1.67	50,100
3	10,000	7,500	0.75	1.83	0.94	1.25	9,400
4	5,000	2,500	0.50	2.66	0.42	0.84	2,100
5	10,000	2,500	0.25	3.50	0.10	0.40	1,000

Total annual consumer surplus of the whole recreation experience = £88,640

Notes: All figures rounded to 2 decimal places. Trip generating function: $V_{nj}/N_h = 1.3 - 0.3C_h$.
1. from Census records.
2. from survey; annual totals derived by extrapolating from sample data according to available information regarding tourism rates.
3. column 4 = column 3/column 2.
4. either calculated with reference to zonal distance or by survey (see subsequent discussion of travel costs).

and the curve has the (unlikely) linear form given in equation (6.3).

$$V_{hj}/N_j = 1.3 - 0.3C_h \qquad (6.3)$$

where

V_{hj}/N_j = visit rate (average number of visits per household) from each zone

C_h = visit costs from each zone.

Figure 6.1 illustrates this particular whole recreation experience demand curve. The estimation of this curve involves the implicit assumption that households in all distance zones react in a similar manner to visit costs. They would all make the same number of trips if faced with the same costs, i.e. they are assumed to have identical tastes regarding the site.

(vii)   In each zone the household consumer surplus for all visits to the site (column 6) is calculated by integrating the demand curve (equation (6.3)) between the price (cost) of visits actually made from each zone and that price at which the visitor rate would fall to zero (i.e. the vertical intercept of the demand curve at point P in Figure 6.1).[2] Households in zone 3, for example, would have a consumer surplus equal to area ABP for all their trips to the site, i.e.:

$$\text{Consumer surplus for zone 3} = \int_{C_h=B}^{P} (1.3 - 0.3C_h).dC_h \qquad (6.4)$$

(viii) In order that annual total consumer surplus for the whole recreation experience can be estimated in each zone, total household consumer surplus must first be divided by the zonal average number of visits made by each household to obtain the zonal average consumer surplus per household visit (column 7). This can then be multiplied by the zonal average number of visits per annum (column 3) to obtain annual zonal consumer surplus (column 8).

(ix)   Calculating annual zonal consumer surplus (column 8) across all zones gives our estimate of total consumer surplus per annum for the whole recreational experience of visiting the site.

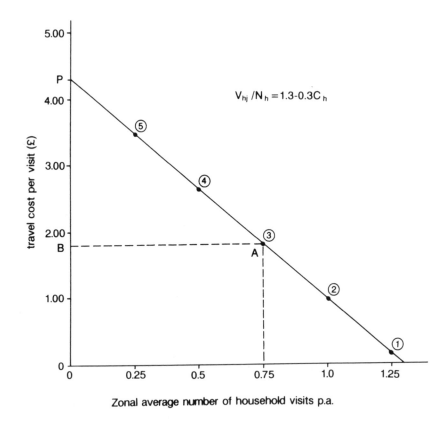

Figure 6.1 Demand curve for the whole recreation experience

One immediate problem with the above approach is that it yields value estimates for the whole recreational experience of a trip to a site rather than an evaluation of the site alone. Freeman (1979c) points out that the information gathered in a TCM survey only in fact defines one point on the demand curve for the on-site recreational experience. Many goods incur a travel cost for their consumption, but their price is set by the market. However, the market price of recreation is zero therefore the sum of all visits across all zones represents the demand for on-site recreation with a zero admission price. This point is shown as point A in Figure 6.2.

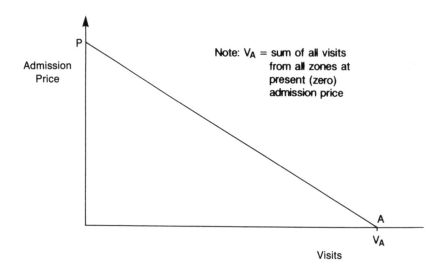

Figure 6.2   Demand curve for the on-site recreation experience

In estimating consumer surplus for the on-site recreation experience, most texts (Sinden and Worrell 1979; Hufschmidt *et al.* 1983) follow Clawson and Knetsch (1966) and estimate consumer surplus by first assuming that people would react to increases in admission price in the same way as they would react to increases in their travel costs, i.e. the demand curve function stays as estimated for the whole experience but each zone's travel cost is increased by an incremental admission cost and visits from each zone recalculated according to the estimated demand curve. Summing visits across all zones at each admission cost maps out the on-site experience demand curve. Integrating under this curve between the initial zero admission price and that admission price at which visits in all zones fall to zero estimates total consumer surplus for the on-site recreational experience. Table 6.2 extends our previous hypothetical example to illustrate this approach; by assuming that visitors react to admission fees in the same way as travel costs, we can use equation (6.3) to estimate the number of visitors at various admission fee levels as shown in Table 6.2.

We can now plot admission fees against the total number of visitors from all zones at each fee level to obtain a demand curve for the on-site recreational experience as shown in Figure 6.3. Consumer surplus estimates are obtained as usual by integrating under this curve between a zero admission price and that price at which the total number of

Table 6.2 Total annual visits to a site at various admission fee amounts.

Trip generating function: $V_{hj}/N_h = 1.3 - 0.3C_h$

where V_{hj}/N_h = visit rate

C_h = total visit cost = travel cost (from Table 6.1) + admission fee

Zone	Zonal population	Admission fee = £0.00			Admission fee = £1.00			Admission fee = £2.00		
		Total visit cost (£)	Visit rate	Number of visits	Total visit cost (£)	Visit rate	Number of visits	Total visit cost (£)	Visit rate	Number of visits
1	10,000	0.16	1.25	12,500	1.16	0.95	9,520	2.16	0.65	6,520
2	30,000	1.00	1.00	30,000	2.00	0.70	21,000	3.00	0.40	12,000
3	10,000	1.83	0.75	7,500	2.83	0.45	4,510	3.83	0.15	1,510
4	5,000	2.66	0.50	2,500	3.66	0.20	1,010	4.66	0.00	0
5	10,000	3.50	0.25	2,500	4.50	0.00	0	5.50	0.00	0
		Total visits (fee=£0) =		55,000	Total visits (fee=£1) =		36,040	Total visits (fee=£2) =		20,030

Zone	Zonal population	Admission fee = £3.00			Admission fee = £4.00		
		Total visit cost (£)	Visit rate	Number of visits	Total visit cost (£)	Visit rate	Number of visits
1	10,000	3.16	0.35	3,520	4.16	0.05	520
2	30,000	4.00	0.10	3,000	5.00	0.00	0
3	10,000	4.83	0.00	0	5.83	0.00	0
4	5,000	5.66	0.00	0	6.66	0.00	0
5	10,000	6.50	0.00	0	7.50	0.00	0
		Total visits (fee=£3) =		6,520	Total visits (fee=£4) =		520

Note: An admission fee of £4.33 or more will result in no visits being made from any zone.

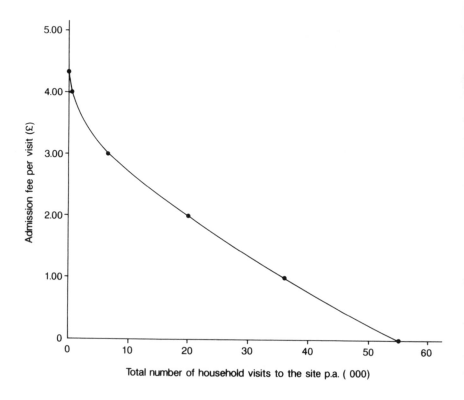

Figure 6.3   Demand curve for the on-site recreation experience

annual household visits falls to zero. Applying this approach, our example gives an estimated annual consumer surplus for the on-site recreation experience of £90,500.

The on-site demand curve estimates the maximum amount which people would be willing to pay for the recreation use value of a site once they have paid the cost of travel to the site. The relative magnitudes of the on-site and whole experience consumer surplus sums will depend upon the shapes of the relevant demand curves. The more concave the whole-experience demand curve the larger the relative size of the on-site value. In the above example the whole-experience value is some 2% smaller than the (additional) on-site value, while in both the hypothetical example used by Sinden and Worrell (1979) and the real-number experiments of Clawson and Knetsch (1966) this discrepancy was of the order of 5%. However, this larger discrepancy is to be expected because both of these studies calculated the whole-experience values by assuming that the consumer surplus of marginal users (which in effect meant the entire most distant zone) was zero. This further reduced total whole-experience consumer surplus compared to on-site values for which such an assumption was not employed (on-site valuations utilised the entire area under the relevant demand curve).

The weak link in the Clawson–Knetsch approach to on-site valuation is the need to assume that individuals will react in the same way to admission fees as they do to travel costs. The problems of vehicle bias, discussed with regard to the contingent valuation method (see Chapter 5), are pertinent here. If individuals have different willingness to pay for an environmental good because of the method of payment which is used, then it is likely that the above TCM assumption may well be violated.

In practice many TCM studied have rejected the Clawson–Knetsch approach to on-site valuation, preferring modification of the whole-experience demand curve.

A common approach is to ask visitors to evaluate how much of the utility of the whole recreation experience is due to the on-site experience. Typically visitors are asked to allocate percentage points to the on-site and off-site experience. This information can then be either used to reduce travel costs (i.e. evaluate how much of incurred costs can justifiably be said to have been purely related to the on-site experience) or directly entered into the trip-generating function as a separate continuous explanatory variable (for an example, see Willis and Garrod 1991). In either case the whole-experience demand function will be altered. The resultant curve will not be the same as the on-site demand curve as defined by Clawson and Knetsch above. However, its validity may well be more defensible in that it does not rely upon the previous assumption of travel cost effects perfectly duplicating admission price effects.

Extending the model

The simple model discussed above relies upon the assumption that visits are a function of their price, i.e. total visit costs. Total visit costs can be defined as the sum of money expenditure on travel (e.g., petrol costs), the opportunity cost of travel time and the opportunity cost of on-site time. More exactly we can define[3]

$$C_{hj} = PTC.D_{hj} + PTT_{hj}.TT_{hj} + PST_j.ST_j \qquad (6.5)$$

where

C_{hj} = Total visit cost from zone h to site j (visit price)

PTC = Money expenditure on travel (petrol, etc.) per mile or kilometre

D_{hj} = Distance from zone h to site j

PTT_{hj} = Opportunity cost per hour of travel time from zone h to site j (unlike PTC) as PTT may vary according to zone of origin of journey and site of destination.

TT_{hj} = Length of travel time from zone h to site j (hours)

PST_j = Opportunity cost per hour of on-site time at site j. This will probably vary by site but not by zone.

ST_j = Length of per visit on-site time at site j. This may or may not vary by zone of origin (here not); see later discussion of the value of time.

We now need to consider the other explanatory variables X in equation (6.1). These will include factors such as income levels, spending on other goods, the qualities of this and substitute sites, etc. Consideration of these factors leads us to specify a tgf such as

$$V_{hj}/N_h = f\,(C_{hj},\ P_{Vj},\ Q_j,\ SC_{hn},\ P_{Vn},\ Q_n,\ Y_h,\ P_x) \qquad (6.6)$$

where

V_{hj}/N_h = Visitor rate from zone h to site j

C_{hj} = Total visit cost from zone h to site j (see equation (6.5))

P_{Vj} = Entrance fee (may be zero) at site j

Q_j = Quality index at site j

SC_{hn} = Vector of total visit costs from zone h to n substitute sites (i.e. j is site number $n+1$)

P_{Vn} = Vector of entrance fees (may be zero) at n substitute sites

Q_n = Vector of quality indices of n substitute sites

P_x = Vector of private goods prices

Now the demand for visits to site j is a function (along with other variables) of the attributes of site j and all substitute sites. Further explanatory variables are plausible; for example, Bojö (1985) includes a dummy variable for the mode of transport used (car or train) which, in an empirical test, he finds significant.

Once adequate data are collected, the tgf may be estimated. In practice, because of data limitations, a reduced form of equation (6.6) is usually estimated (e.g. P_x is usually omitted). Furthermore, as discussed in the previous section, it is usually the modified whole-experience demand curve (equating to the on-site demand curve), rather than the theoretically pure Clawson–Knetsch on-site demand curve, which is calculated (i.e. whole-experience costs are allocated between off-site and on-site activities as discussed previously). This demand curve can then be mapped out by examining the partial derivative $\partial V_{hj}/\partial C_{hj}$. Consumer surplus in each zone h is then found as in the previous example, while total consumer surplus is found by summing across all zones.

The individual travel cost method (ITCM)
The fundamental difference between the ZTCM and ITCM is that the latter defines the dependent variable as V_{ij}, the number of visits made per period (annum) by individual i to site j. We can therefore rewrite the simple tgf of equation (6.1) as its ITCM equivalent:

$$V_{ij} = f\ (C_{ij},\ X_i) \tag{6.7}$$

where

V_{ij} = number of visits made per year by individual i to site j
C_{ij} = visit cost faced by individual i to visit site j
X_i = all other factors determining individual i's visits.

The demand curve produced by this model relates individuals' annual visits to the costs of those visits (i.e. there is no requirement to convert from zonal visitor rate to actual visits as in the ZTCM). As discussed previously, the above tgf relates to the whole recreational experience. On-site recreational experience demand curves can again be obtained as outlined by Clawson and Knetsch (1966) although, as before, many practical studies adopt a modified whole-experience approach by including as a separate variable a measure of how much of the visit's utility can be attributed to the on-site experience.[4]

The move from a zonal to an individual basis allows the specification of a number of individual-specific explanatory variables; for example, we could respecify our ITCM tgf as

$$V_{ij} = f\ (C_{ij},\ E_{ij},\ S_i,\ A_i,\ Y_i,\ H_i,\ N_i,\ M_i) \tag{6.8}$$

where

V_{ij} = number of visits made per year by individual i to site j
C_{ij} = individual i's total visit cost of visiting site j
E_{ij} = individual i's estimate of the proportion of the day's enjoyment which was contributed by the visit to site j
S_i = dummy variable: individual i's assessment of the availability of substitute sites
A_i = age of individual i
Y_i = income of individual i's household

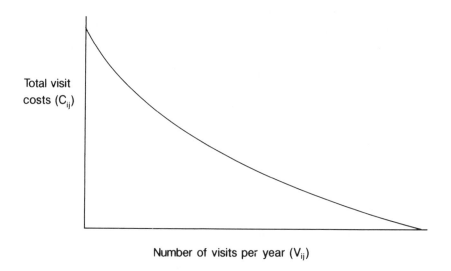

Number of visits per year (V_{ij})

Figure 6.4   An ITCM demand curve for a recreational site

H_i = size of individual i's household
N_i = size of individual i's party
M_i = dummy variable: whether individual i is a member of an outdoor or environmental organisation.

The tgf given in equation (6.8) will have a variety of possible exact specifications. The total visit cost variable (C_{ij}) will often be some simplification of equation (6.6). Furthermore, the dummy variables M_i and S_i may be specified either as single indicator variables or as a series of switches or as continuous variables, e.g. the number of substitute sites, or distances to these sites. The E_{ij} variable allows for the modification of the whole experience to the on-site experience and would usually be defined as a continuous variable between 0 and 1.[5] The move from zonal averages to an individual basis also allows realistic use of specific variables such as A_i, Y_i, H_i, N_i and M_i for which zonal averages would have little meaning in the tgf.

The demand curve for the site will be defined by the $\partial V_{ij}/\partial C_{ij}$ relationship as illustrated in Figure 6.4. Integrating under this curve gives us our ITCM estimate of consumer surplus per individual. Our estimate of consumer surplus for the site is then obtained by multiplying by the number of individuals visiting the site annually,[6] i.e.

$$\text{Total consumer surplus} = N_j. \int f(C_{ij}, X_i). \, dC_{ij} \qquad (6.9)$$

where N_j is the number of individual visits to site j per year and (C_{ij}, X_i) are defined as per equation (6.7).

Methodological issues

The central assumption
The underlying assumption that visit costs can in some way be taken as an indication of recreational value requires qualification. In a perceptive early study Gibson (1978) discusses cases where this assumption is invalid. Where individuals have changed their place of residence so as to be close to a site, the price of a trip becomes endogenous and the central assumption violated. In such a case the estimated demand curve will lie below the true demand curve and consumer surplus will be underestimated. Figure 6.5 illustrates such a case.

Very few empirical studies have taken account of this potentially highly important criticism. However, in a recent study, Parson (1991) argues that the endogeneity may be eliminated using an instrumental

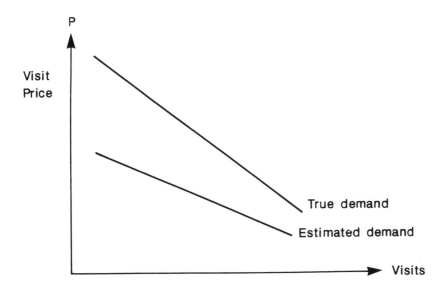

Figure 6.5 TCM demand estimates where individuals move house to be near a site

variables approach (place of work, job characteristics, etc.). A simple variant of this would be to include a survey question regarding the importance of proximity to the recreation site in deciding place of residency. A dummy variable could then be used to split up the responses, with significance tests determining the importance of this factor.

A second challenge to the central TCM assumption arises where the on-site time is not the only or even major objective of the trip. Cheshire and Stabler (1976) define three categories of visitor: the 'pure visitor' who is strongly site-orientated; 'transit visitors' who make multi-purpose trips; and 'meanderers' who gain utility primarily from the journey itself. While pure visitors pose no theoretical problem, transit visitors pose the problem of how journey costs are to be allocated among the sites visited. This problem also applies to meanderers, where the on-site time is by definition only a side issue in the trip decision and where travel time in particular may not represent a true opportunity cost, i.e. the utility of travel time may range from negative to positive across these visitor categories. These latter issues are discussed subsequently in the context of time costs.

Calculating visit costs

Following equation (6.5) we can decompose total visit costs into travel costs and time costs, the latter being subdivided into travel time and on-site time costs.

Travel costs   Here we are referring to the money expenditure necessary to reach a site ($PTC.D_{hj}$ in equation (6.5)). In calculating travel costs, Bojö (1985) simply multiplied household size by the economy class rail fair. However, such a simple approach is less applicable to car travel, where three cost calculation options exist:

(i)   petrol costs only (marginal costs);
(ii)   full car costs, i.e. petrol, insurance, maintenance costs, etc.;
(iii) perceived costs as estimated by respondents.

Clearly using option (ii) will raise visit costs above those of (i) and ultimately increase consumer surplus estimates. Hanley and Common (1987) apply both options to the same forest recreation data, finding that option (ii) gave a consumer surplus estimate more than twice as large as option (i).

Willis and Benson (1988) obtained a similar result in a study of visitors to wildlife areas in Yorkshire. Results for one of the sites studies are given in Table 6.3, showing that the move from defining travel costs as

Table 6.3 Impact upon estimated consumer surplus (CS) of alternative travel cost specifications

Case study: Wildlife visitors to Skipwith Common, Yorkshire
Method: ZTCM
Functional Form: Double log throughout

Travel cost specification	Travel cost coefficient	Model R^2	CS/visitor £	Visitors p.a.	Total CS estimate (£)
Petrol only	− 2.6667 (6.73)	0.83	0.59	15,235	9,001
Petrol plus standing charges	− 2.6050 (6.49)	0.83	1.02	15,235	15,574

Notes: CS/visitor rounded to nearest penny.
t-values given in brackets

Source: abstracted from Willis and Benson (1988).

petrol only to petrol plus standing charges made no significant difference to the explanatory power of the model (same functional form retained) and only a minor impact upon the cost coefficient (highly significant in both cases), that is to say, both assumptions had equal statistical validity. However, this translated into a major increase in consumer surplus per visitor (over 70% bigger for the full cost assumption) and thereby to total site consumer surplus.

The correct cost measure is that which visitors perceive as relevant to the visit. It may well be that visitors are poor at perceiving daily insurance and maintenance cost equivalents, or that they see these as sunk costs which do not enter the tgf, i.e. they only consider the marginal cost of a visit, equating this with marginal utility. Many analysts therefore apply option (i) in calculating these cost elements (see Bojö 1985; Sellar *et al.* 1985).

The use of perceived cost statements (option (iii) above) was pioneered in the UK by Christensen (1983).[7] A recent study (Bateman *et al.* 1993) examined the statistical performance of all three options, concluding that option (ii) performed significantly worse than the others, with option (i) marginally outperforming option (iii). This result suggests that respondents' actual marginal (petrol) cost provides a superior predictor of visits.

Time costs As indicated in equation (6.5), time enters the total travel cost function in two ways:

(i) PTT_{hj} = Opportunity cost per hour of travel time from zone h to site j (travel time cost);
(ii) PST_j = Opportunity cost per hour of on-site time at site j (on-site time cost).

The marginal utility of a visit will be influenced by both travel and on-site time and omission of either can be shown to lead to potential bias and indeterminacy in the welfare estimates. Consider first the utility maximisation problem:[8]

$$\text{maximise: } U = U(X, V_J, Z_{ij}) \qquad (6.10)$$

where

X = consumption of composite good
V_j = number of visits to site j
Z_{ij} = total distance travelled by individual i to site j (miles), this is equal to $V_j.D_{ij}$, where D_{ij} is the round-trip distance from individual i's home to site j,

subject to an income constraint:

$$M - PxX - \sum_j PV_j.V_j - \sum_j PTC.Z_{ij} = 0 \qquad (6.11)$$

where

M = income
Px = price of composite good
X = quantity of composite good
PV_j = entrance fee at site j
PTC = money expenditure on travel (petrol, etc.) per mile

and subject to the time constraint:

$$T - \sum_j ST_j . V_j - \sum_j t_{ij}Z_{ij} = 0 \qquad (6.12)$$

where

T = total recreation time (fixed)
ST_j = length of per visit on-site time at site j
t_{ij} = travel time per mile for individual i to site j.

To maximise utility the individual equates the marginal utility of visits with their total cost (money cost and time cost). Freeman (1979c) considers a simple additive utility function (although more complex forms yield consistent results)[9] which we shall use for illustrative purposes, substituting $V_j D_{ij}$ for Z_j in equation (6.10) to give the utility form:

$$U = U_1(X) + U_2(V_j) + U_3(V_j D_{ij}) \qquad (6.13)$$

We can now define the marginal utility of a visit as the partial differential of equation (6.13) with respect to V_j, namely:

$$\frac{\partial U_2}{\partial V_j} = \lambda PV_j + \lambda PTC \cdot D_{ij} + \mu ST_j + \underbrace{\mu t_{ij} D_{ij} - D_{ij} \frac{\partial U_3}{\partial V_j D_{ij}}} \qquad (6.14)$$

marginal utility of a visit to site j	opportunity cost of the entrance fee	opportunity cost of money travel costs (petrol, etc.)	opportunity cost of time spent on-site	net opportunity cost of time spent in travel[10]

where

λ = marginal utility of income
μ = marginal utility of time.

Equation (6.14) demonstrates the potential importance of on-site and travel time as determinants of visits. Furthermore, it shows that the relevant opportunity costs per hour need not be the same for these two items. However, determination of these opportunity costs raises considerable problems.

Empirical estimation of the value of travel and on-site time   Travel time values are particularly difficult to analyse in that, as noted previously, we have no definite a priori notion about whether travel time utility is positive or negative. If travel time has positive utility (i.e. individuals enjoy the travel as part of their recreational experience, as in the case of meanderers as previously defined) then using some general travel time cost figure to price this will overestimate the consumer surplus of a visit. Bojö (1985) does not include a travel time cost (i.e. he implicitly gives such time an opportunity cost of zero) on the grounds that 80% of survey respondents expressed a positive utility for travel time to the site under analysis. This approach assumes that ignoring residual travel time costs only leads to a minor underestimation of the true consumer surplus. If travel time has a negative utility (i.e. individuals actually dislike travelling to recreation sites) then the use of a generalised time cost may now underestimate total travel costs and consequently consumer surplus of a visit. Johansson (1987) point out that, in such cases

the estimated curve will be located inside and be less steep that the 'true' one, except possibly for those living very close to the recreation site, since the underestimation of costs increases in relation to distance from the visitor's zone of origin.

One practical approach to this problem is to apply a utility weighting to a standard travel time cost, the weighting being derived by direct questioning of respondents. By asking respondents to rate their enjoyment of the travel time alone an inverse index can be set up such that a respondent who hates travelling (pure visitor) is given an index value of 1 while a respondent who prefers travelling to visiting (meanderer) is given an index value of 0, with continuous gradations between these extremes. The resultant index score can be used to weight any per-hour travel time cost. However, this still leaves the problem of determining the unit cost of travel time.

One approach to the pricing of travel time is to examine its relationship with individuals' wage rates. The seminal work in this area is that of Cesario (1976) and Cesario and Knetsch (1970; 1976). This approach examined commuters' choice of transport to and from work (and relevant costs) to estimate an implicit value of travel time. Cesario (1976) concluded 'that, on the basis of evidence collected to date, the value of time with respect to nonwork travel is between one quarter and one half of the (individual's) wage rate', and subsequently used a value of one-third the wage rate to price travel time. An alternative approach is that of Nelson (1977), who calculated a marginal implicit price of proximity to the central business district with housing data for Washington, DC, from which he derived a value of time which, when related to wage rates, falls within the Cesario range. However, as he recognised at the time, Cesario's analysis only considers commuter time and there is no necessary reason why the marginal utility obtained should be applicable to recreation travel time.[11]

Common (1973) and McConnell and Strand (1981) used an iterative process whereby successive time values are substituted into the tgf, the final choice being determined where the explanatory power (R^2) of the model is maximised. Desvousges et al. (1983) applied the value of time results of Cesario (1976), McConnell and Strand (1981) and a full wage rate assumption to the following simple model of individual visitation patterns at 23 water recreation sites in the USA:

$$\text{Ln}(V_{ij}) = a_0 + a_1\,MC_j + a_2\,TC_j + a_3Y \qquad (6.15)$$

where

V_j = visits to site j
MC_j = mileage cost per visit to site j
TC_j = time cost per visit to site j
Y = household income.

Now if the Cesario (1976) estimation that the opportunity cost of travel time was approximately one-third of the wage rate is correct then we would expect that $a_2/a_1 = \frac{1}{3}$. Similarly, if the full wage was a better approximation of the value of travel time we would expect $a_1 = a_2$.

Testing equation (6.15) at the 10% significance level, Desvousges et al. (1983) rejected the McConnell and Strand (1981) approach, while both the Cesario (1976) and full-wage assumptions performed equally well, both being rejected in roughly 7 of the 23 cases. On the basis of these results Smith and Desvousges (1986) concluded that 'for practical purposes, there is no clearcut alternative to our using the full wage rate as a measure of the opportunity cost. Even though it may overstate the opportunity costs . . . none of the simple adaptations are superior'.

A further approach to the valuation of time, theoretically applicable to both in-travel and on-site time, is that of Bockstael et al. (1984), who examined labour supply functions, mapping the various relationships between earned income and leisure time which occur during the working week: normal work time; overtime working; second jobs; and recreation time. Unfortunately, because of the rigidities of normal working practice, individuals are usually unable to reallocate hours according to their personal preferences and this line of research has not received any significant empirical application.

Turning to consider the unit value of on-site time, if the length of time spent on-site were a constant for all visits to a particular site, then such costs could effectively be ignored as they would imply only an increase in absolute visit costs but not in marginal relationships. Figure 6.6 shows that, in such a situation, ignoring on-site costs results in the estimation of the lower demand curve D_{NST}, visit price P_{NST} and consumer surplus CS_{NST}. Adding in constant on-site time costs causes a vertical shift of the estimated demand curve to D_{ST} with increased visit price P_{ST} but identical magnitude consumer surplus CS_{ST}, i.e. omission of constant

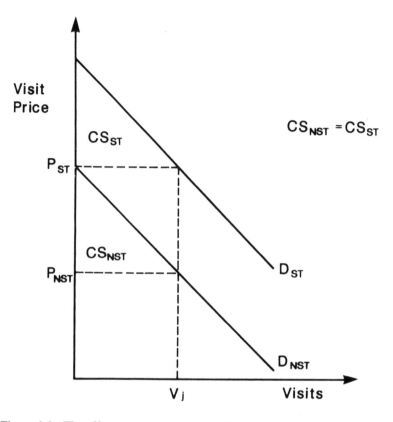

Figure 6.6 The effect upon consumer surplus estimates of adding in constant on-site time costs

on-site time costs does not produce a biased estimate of consumer surplus.

Bojö (1985) finds no evidence to refute an assumption of constant on-site time costs and therefore omits these from his analysis. However, there is no reason why this should necessarily be the case. Say, for example, that on-site time varied inversely with distance from the site, i.e. those living nearer the site spend longer on-site than those coming from far away (low on-site time costs).[12] Figure 6.7 illustrates such a situation. Demand curve D_{NST} is that estimated when all on-site time costs are omitted. Observed visits V_j then relate to a visit price P_{NST} giving consumer surplus CS_{NST}. Including increasing on-site time costs has little impact upon estimates of non-local visit price; however, those for visitors living near the site increase considerably, so that the new demand curve

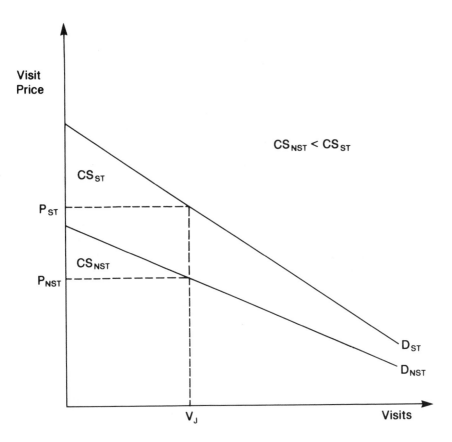

Figure 6.7   The effect upon consumer surplus estimates of adding in increasing on-site time costs

is D_{ST} with visit price P_{ST} and consumer surplus CS_{ST}. Notice that in this situation $CS_{NST} < CS_{ST}$, i.e. here the omission of on-site time costs leads to an underestimate of true consumer surplus.

We could plausibly reverse such a distance/on-site time cost assumption. If on-site time varied directly with distance from the site then the result of Figure 6.7 would be reversed, so that $CS_{NST} > CS_{ST}$.

In summary, we have noted that the opportunity cost of both travel time and on-site time are important arguments in the tgf and considerable problems can arise if these costs are ignored. We suggest that tests be carried out as per Bojö (1985) to obtain some estimate of the magnitude of the consumer surplus estimation error likely if time costs are ignored. If this error is not thought to be acceptably small then time costs must be included. However, we recommend that in such situations a sensitivity analysis be carried out using a range of time value estimates based on wage rates. As a working approximation, we suggest that values of 0.25, 0.5, 0.75 and full wage rate be used.

Site attributes (environmental quality and multicollinearity)
The trip generating function described in equation (6.6) highlights several independent variables as explanatory of visits, one of which is the site environmental quality variable, Q_j. In a simple single-stage analysis the entire aggregate function is estimated and statistically tested, often using ordinary least squares (OLS). While this is a common approach it is, strictly speaking, only valid for quantifiably unidimensional sites, that is, sites which possess only a single environmental quality attribute which can be measured in a quantitative manner. The reason for this is that where sites possess multiple attributes, these attributes should enter the tgf as separate variables. However, these attributes may themselves be highly correlated, i.e. a potential multicollinearity or 'suppressor variables' (Conger 1974) problem exists making single-stage OLS estimators invalid.

In reality, recreation sites very often provide multi-attribute services. For example, Vaughan and Russell (1982) include the explanatory variable Q_{kj}, the level of quality characteristic k at site j, where k may be 1 or more. If $k > 1$ then there may be multiple environmental quality factors significantly influencing visit rate. These factors may well be collinear – for example, wildlife parks which are large may also have many access routes, but both of these factors may be positively related to visits.

To illustrate the suppressor variables problem, consider a simplified version of the Smith and Desvousges (1986) lake-recreation study in which it is thought that the true ITCM trip generating function is the estimating equation:[13]

$$V_{ij} = b_0 + b_c\, C_{ij} + b_y\, Y_{ij} + b_s S_j + b_a A_j + e \qquad (6.16)$$

where

V_{ij} = visits by individual i to site j
C_{ij} = individual i's total costs (time and travel) of visiting site j
Y_{ij} = income of individual i visiting site j
S_j = shore length at site j
A_j = access points at site j
e = error term.

A suppressor variable problem occurs when at least two of the explanatory variables are highly collinear. Suppose that, as is quite possible, shore length (S_j) and access points (A_j) are strongly positively correlated (say, a Pearson correlation factor in excess of 0.5) with insignificant multicollinearity elsewhere in the model. Now suppose that we omit A_j and estimate the model to get an estimator for b_s, the coefficient on S_j (which we shall denote b_s^0, see model 1, Figure 6.8). By substituting A_j for S_j and re-estimating the model we can similarly estimate b_a, the coefficient on A_j (denoted b_a^0, see model 2, Figure 6.8). While they are admittedly imperfect, these estimators b_s^0 and b_a^0 are statistically valid measures of the relationships $\partial V_{ij}/\partial S_j$ and $\partial V_{ij}/\partial A_j$, respectively.

However, if we now re-estimate the full model in equation (6.16), including both S_j and A_j, then, in the presence of the high collinearity between these latter variables, our newly estimated coefficients b_s^1 and b_a^1 can be very different from b_s^0 and b_a^0.

Model 3 in Figure 6.8 illustrates one such potential outcome of such an experiment. Here we have the newly estimated coefficient on S_j (i.e. b_s^1) being roughly similar to that obtained in model 1 (i.e. b_s^0). However model 3 includes the explanatory variable A_j as well as S_j (and C_{ij} and Y_{ij}). The coefficient upon A_j is unstable in that the variation in the dependent which it explains is also explained by S_j and in this example the resulting coefficient upon A_j (b_a^1) is highly biased, being very different to the estimate of model 2 (b_a^0).

In reality a number of possible outcomes may arise from a suppressor variable problem. Coefficients may alter radically, even changing signs. Furthermore the significance of parameters becomes disturbed and may even spuriously increase.

Despite the potentially serious nature of the problem of suppressor variables in multi-attribute sites, no single definitive solution has yet been found. Clearly, a first step is to test for the presence of such a problem as, in its absence, single-state OLS estimation remains valid. However, if such a problem is confirmed one proposed course is to replace all site attribute variables with a single index of site attractiveness, thus removing collinearity (Talheim 1978; Ravenscraft and Dwyer 1978). However, such an index cannot be adequately set up without full knowledge of the functional relationship between demand and site attributes. As this relationship is dictated by individual preference for different attributes, the creation of a truly representative index is infeasible. Ideally, we would wish to respecify the individual's utility function in terms of the attributes of sites. Morey (1981; 1984; 1985) adopts various functional forms for the utility function which include sites' attributes and levels of use. By assuming budget-constrained utility maximisation we can obtain estimating equations

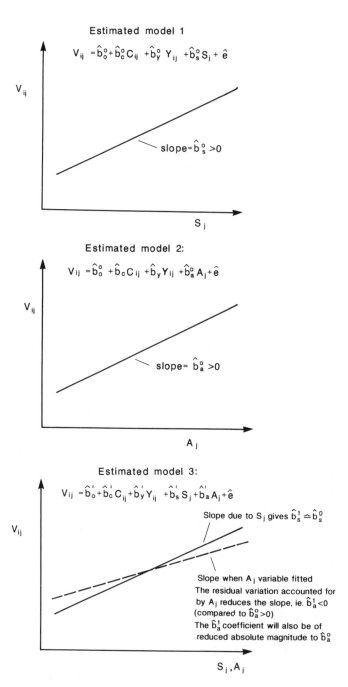

Figure 6.8 The suppressor variable problem

from which parameters (including those for site attributes) can be estimated. However, the need to specify the form of the utility function constitutes a weak link in this approach.

Another approach to the problem of suppressor variables is to use a two-stage generalised least-squares approach[14] (the generalised travel cost method). Such an approach is adopted in the Smith and Desvousges (1986) lake resources example given above. Here the authors postulate a 'true' trip generating function as per equation (6.16). However, in the absence of information, exact specification of this function with respect to the site characteristics shore (S_j) and access (A_j) is infeasible and the likelihood of suppressor variable effects is significant (although no tests are reported). Smith and Desvousges therefore use a two-stage approach, the first stage of which consists of omitting all potentially collinear site attribute variables and estimating the simple equation given in equation (6.17)[15]

$$V_{ij} = a_0 + a_1 C_{ij} + a_2 Y_{ij} + e_j \qquad (6.17)$$

with variables defined as per equation (6.16).

Estimation of equation (6.17) gives the estimated parameters a_0, a_1 and a_2 and generally accounts for significant variance in V_{ij}.[16] However, some of the remaining variability can be explained by the relationship between, for example, C_{ij} and S_j. Similarly, statistically (if not theoretically) certain variability can be accounted for in the relationships between C_{ij} and A_j; Y_{ij} and S_j; Y_{ij} and A_j. In other words, supposing that equation (6.17) gave $R^2 = 20\%$, then some of the remaining 80% variation can be explained by the relationship between the explanatory variables in equation (6.17) and the additional explanatory variables specified in the 'true' relationship given in equation (6.16). Therefore, the second stage of the technique involves the estimation of generalised demand functions for site attributes by regressing the parameter estimates of intercept, C_{ij} and Y_{ij} (i.e. $\hat{a}_0$, $\hat{a}_1$ and $\hat{a}_2$ respectively) on S_j and A_j thus:

$$\hat{a}_0 = f(S_j, A_j)$$
$$\hat{a}_1 = f(S_j, A_j)$$
$$\hat{a}_2 = f(S_j, A_j) \qquad (6.18)$$

In effect, we have now expressed the variation due to site attributes (S_j and A_j) as a function of the variation due to individual characteristics (C_{ij} and Y_i), i.e.

$$V_{ij} = f\,[C_{ij}, Y_{ij}, \{S_j = f(C_{ij}, Y_{ij})\}, \{A_j = f(C_{ij}, Y_{ij})\}] \qquad (6.19)$$

Because of the form of equation (6.19), if we now wish to calculate the relationship (partial derivative) between V_{ij} and a particular site attribute, say S_j, then we will need some information regarding the values of both C_{ij} and Y_{ij}. Smith and Desvousges (1986) address this problem by substituting in equation (6.19) values for mean costs ($C_{ij} = \bar{C}_{ij}$) and income ($Y_{ij} = \bar{Y}_{ij}$) at each site. Thus for example, the partial derivative $\partial V_{ij}/\partial S_j = f(\bar{C}_{ij}, \bar{Y}_{ij})$ then forms our estimate of the coefficient for the variable S_j.

By adopting such an approach Smith and Desvousges (1986) produce estimates of the impact of individual site attributes upon visits, i.e. they estimate attribute demand functions. This provides an important extra facility to the TCM in that such demand estimates allow the analyst to investigate which attributes contribute most to overall site demand and thus to welfare. In effect, such functions allow us to specify which attribute, or combination of attributes, individuals most enjoy at a recreation site and thus facilitate the optimum planning of site development and creation. However, it should be noted that, while this approach is valid, statisticians have no single agreed definitive approach to the treatment of suppressed variables.[17]

Finally it should also be noted that, alongside dilemma of multicollinearity, the ZTCM is inherently vulnerable to problems of heteroscedasticity (see discussion in Johansson, 1987; and Bateman, 1993). This arises because zonal observations are drawn from zones of varying population size and thereby have varying precision. A common approach to such problems is to transform the data by logarithms. However Christensen and Price (1982) show that a weighted least squares (WLS) approach is more appropriate and Price et al. (1986) accordingly weight demand curve observations by zonal population/visit rate.

Substitute sites

Equation (6.6) showed that substitute sites should impact upon visit demand in three ways: the visit price of the substitute sites; their entrance fees; and environmental quality at substitute sites. In practice such variables are rarely included in estimated forms (see, for example, Smith and Desvousges 1986), the major practical difficulty being the high data costs involved. In effect, a TCM survey would have to be performed at all significant substitute sites in order to provide the full data requirement.

The presence of substitute sites deflates recorded demand as shown in Figure 6.9. The further people live from a site the higher the probability that there are substitute sites closer to them than the site in question, i.e. the observed trip demand curve is depressed below the true demand curve at higher travel costs.

By concentrating upon site loss (i.e. a given site's 'contribution value' to

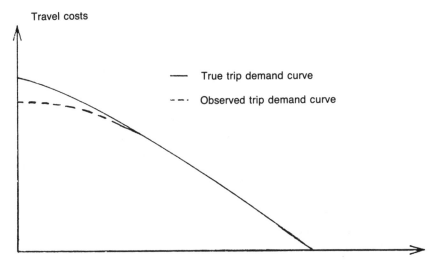

Figure 6.9   Impact of substitute sites

Source: Christensen (1983)

the total recreation value of all sites) rather than conventional TCM site value, Price (1978) and Connolly and Price (1991) show that, in fact, the presence of substitute sites leads the TCM systematically either to over- or underestimate true consumer surplus dependent upon the spatial relationship between sites and population centres. Assuming that population is randomly distributed, Price *et al.* (1986) argue that if recreation sites are clustered then the loss of one site will, on average, make little difference to the general proximity of population to sites, i.e. the conventional TCM site value will overestimate the value of site loss. Conversely, if sites are systematically spaced (particularly relevant for man-made recreation areas) then the loss of one site will induce a major site-proximity change for the nearby population and the TCM value will underestimate the true value of site loss. Only where sites (as well as population) are randomly distributed will these over- and underestimates on average cancel out and the TCM value accurately represent the true value of site loss.[18]

A number of solutions to the substitute sites problem have been put forward. Price (1979) addresses the problem 'by the simple expedient of basing visit rates, not on visits per year per 1000 population, but on visits per year per 1000 population for whom this is the nearest facility of its type'. However, this is at best a partial lower-boundary approach, ignoring distant visitors who presumably value their visits highly.

Burt and Brewer (1971) use their judgement to identify presumed substitute sites and enter the distances from respondents' homes to these sites as explanatory variables in the tgf. Such an approach is admittedly subjective; however, a more fundamental criticism is that it implicitly assumes homogeneity of sites, an improbable assumption. Greig (1977) imposes a predetermined, utility-based model linking visits to site characteristics. Such an approach may also be criticised both for lack of adequate prior information regarding the appropriate utility relationship and the need to define site characteristics. A hybrid of the Greig–Burt and Brewer approach could theoretically be constructed if data were available on actual visits to substitute sites. Given such data, we could run a Burt and Brewer substitute-distance model and compare predicted visitor rates under the homogeneity assumption with recorded actual visit rates. Differences between actual and predicted figures could then be used to provide information regarding the utility characteristics of the sites.

Connelly and Price (1991) suggest a fundamental change to the Clawson procedure by asking visitors hypothetical questions regarding their expected visit pattern if the site in question were to be closed. These responses could then be fed into the TCM model as proxy variables regarding substitute sites.

An interesting attempt to formulate a substitute availability index is given in Bojö (1985). The following index is constructed:

$$S_j = \sum_{k=1}^{n} \frac{P_j W_k}{P_k} \tag{6.20}$$

where

S_j = substitute availability from site j
P_j = travel cost to site j
P_k = travel cost to n substitute sites k
W_k = measure of the degree of substitutability between sites j and k.

Bojö measured W_k by questioning respondents as to their preferences for substitute sites. Unfortunately, his field experiment found that the majority of respondents all named one and the same site as their preferred substitute and it became impossible to operationalise the index.

The lack of adequate consideration of substitute sites remains a weakness in many TCM models.[19]

Congestion

A site becomes congested when the number of visitors at a site rises to the point where the supply of the characteristics of that site becomes restricted (i.e. the presence of marginal users diminishes the utility of other users). In extreme cases congestion will invalidate a TCM study as the observed visits correspond not to the standard demand-constrained system but to the intersection of an undefined demand curve with an unknown supply curve, i.e. the system becomes underidentified.

While Vaux and Williams (1977) feel that this problem is not of 'overriding importance', in an early experiment Stankey (1972) records that 82% of his sample felt that 'solitude' – not seeing many other people except those in your own party' was desirable. Johansson (1987) points out that site visitor numbers (X_v) may be a separate argument in the individual's utility function. Furthermore, this argument may be complex in that, where X_v is very low (or zero), utility may be impaired as people feel lonely or intimidated at the site (this will obviously not be so for all individuals). As X_v increases to a small number, utility may rise with the possibility of social interaction. However, as X_v becomes large utility may again decline as congestion sets in. The visit decision may therefore be dependent upon individuals' expectations of X_v. Such differences between expected and actual X_v might also prove significant in CVM-type studies.

The presence of congestion (or excess demand) means that the observed demand curve is an underestimate of true demand. The classic treatment of this problem is presented by Fisher and Krutilla (1972) as summarised in Figure 6.10. Here a zero admission price with no formal visitor restrictions (other than congestion) results in V_1 visits to a site with an observed demand curve (estimated using TCM procedures) D_1. Suppose that the site is recognised as being congested and a quantity restriction is placed upon visitors, reducing total visits to V_2. The consequent reduction in congestion would improve the utility of the site for the remaining visitors such that the observed demand curve expands out to D_2. The benefit of restricting visits can be seen to be the area P_1ZBP_2, while the cost of this restriction is ZV_2V_1, the net benefit being the difference between these sums (positive in Figure 6.10). This movement implies that the true demand curve of the non-congested site is given by the curve AB, i.e. TCM models underestimate consumer surplus in the face of congestion.

Christensen (1983) shows that we can repeat the Fisher–Krutilla analysis so as to map out the entire non-congested demand curve. However, because of the difficulty of imposing quantitative restrictions upon visitor numbers, it is difficult to envisage how a study might obtain the information necessary to estimate the divergence between observed

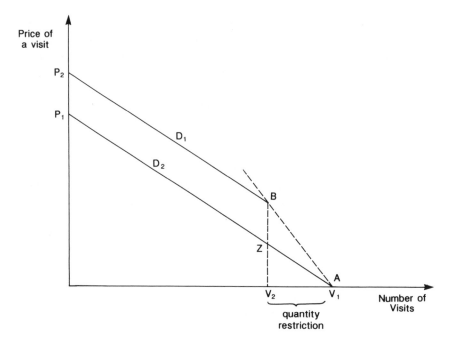

Figure 6.10   Fisher–Krutilla analysis of congested sites

and non-congested demand. Often the most practical approach is simply to test for the presence of site congestion and qualify TCM results in the light of this test.

Smith and Desvousges (1986) attempt to account for potential site congestion by eliciting the opinions of recreation site managers as to the level of site congestion. On the basis of received responses they concluded that congestion was not a significant factor at the sites studied and omitted it from further consideration. However, the use of site managers' rather than visitors' responses is questionable. Freeman (1979c) lists several references to the use of non-visitor samples drawn from the regional population of travel cost zones to examine how many present non-users would use the site if environmental quality were to be improved (Burt and Brewer 1971; Brown and Nawas 1973; Gum and Martin 1975), and such an approach could be extended to the analysis of congestion.

Functional form
Analysts are faced with a variety of functional forms under which the tgf

Table 6.4 Estimated trip generating functions for a forest recreation site

Functional form	Equation	Consumer surplus per capita	R^2
Quadratic	$\dfrac{V_i}{N_i} = 0.478 - 0.329\ TC_i + 0.05\ TC_i^2$ $\qquad\quad\ (4.06)\quad\ (3.47)\qquad\ (3.11)$	£0.32	0.34
Semi-log Independent	$\dfrac{V_i}{N_i} = 0.1523 - 0.146 \ln TC_i$ $\qquad\quad\ (3.91)\qquad (3.05)$	£0.56	0.24
Semi-log Dependent	$\ln\left(\dfrac{V_i}{N_i}\right) = -2.6 - 0.6\ TC_i$ $\qquad\qquad\quad (6.06)\quad\ (3.41)$	£1.70	0.37
Log-log	$\ln\left(\dfrac{V_i}{N_i}\right) = -2.76 - 1.7 \ln TC_i$ $\qquad\qquad\quad (8.39)\qquad (4.18)$	£15.13	0.37

Notes:
V_i/N_i = visits per capita from zone i (only one site considered)
TC_i = round trip travel costs (petrol plus any entry fee) from zone i
Figures in parenthesis are t-values

Source: Hanley (1989)

can be specified (linear, quadratic, semi-log and log-log). None of these is theoretically superior to the others. However, specification of a linear form exhibits a first derivative which will be a constant and is therefore theoretically problematic. Log forms may be useful for elasticity estimates and have the advantage of avoiding negative values for the dependent variable.[20]

An altered functional form (even if it has similar explanatory power) can have a highly significant impact upon the demand curve and resultant consumer surplus estimates. In a ZTCM study of recreational fishing in Grafham Reservoir (UK), Smith and Kavanagh (1969) found that both semi-log (dependent variables) and double-log functions fitted the data very well (with R^2 values of 0.91 and 0.97, respectively).[21] However, when the resultant demand curves were examined it was found that, at a zero admission price, while the semi-log form predicted 54,000 annual visits the double-log form predicted over 1,052,000 annual visits, with obvious consequences for consumer surplus estimates. Subsequent reestimation made little difference to this divergence. Table 6.4 details a similar result found by Hanley (1989) in his ZTCM study of forest recreation.

All the functional forms reported in Table 6.4 produce significant and correctly signed travel cost coefficients. In theory, the most appropriate functional form may be evaluated by examining relative degrees of explanation. However, R^2 tests are strictly non-comparable where the dependent variable changes (e.g., between linear and log forms) so that only semi-log (dependent variable) and log-log can be compared in this manner. In the above study both the quadratic and semi-log independent forms were subject to strong heteroscedasticity.[22] Transforming the dependent variable by natural logs solved this problem[23] but ruled out both of these forms. The semi-log dependent and double-log forms produced identical (and comparable) R^2 statistics but very different consumer surplus estimates. Hanley (1989) rejects the double-log form on the grounds that the consumer surplus estimates produced were very high compared to those reported in comparable UK studies using the TCM approach. A more consistent test would be a comparison of visitor rates predicted by the model with actual observed visitor rates using either a large sample Wilcoxon signed-rank test[24] or a Mann–Whitney U test[25] as appropriate.

Truncation bias and estimation procedure

TCM surveys can only sample those individuals who actually visit a site, i.e. non-visitors (zero visits often corresponding to higher visit price) are ignored. The truncation of non-visitors may bias our estimate of consumer surplus. This is illustrated in Figure 6.11.

Our estimated demand curve is D_{EST}. However, if the non-visitors (shown as points on the vertical axis) are also considered then the demand curve moves to D_{TRUE}. The size of this increase in consumer surplus is indeterminate, but some degree of truncation bias appears likely.

Pearce and Markandya (1989) point out that a further truncation will be introduced where least-squares estimation techniques are employed. The normal error distribution inherent in this technique allows the estimation of continuous and negative visitor rates rather than its discrete non-negative reality. This problem will not be solved by simply resorting to log dependent variable functional forms. OLS estimation is, strictly speaking, inappropriate for TCM models and should be replaced by procedures such as multinomial logit, or maximum likelihood (ML) approaches.

Empirical studies come to differing conclusions regarding the extent of variance between OLS (truncated) and ML (non-truncated) estimates of

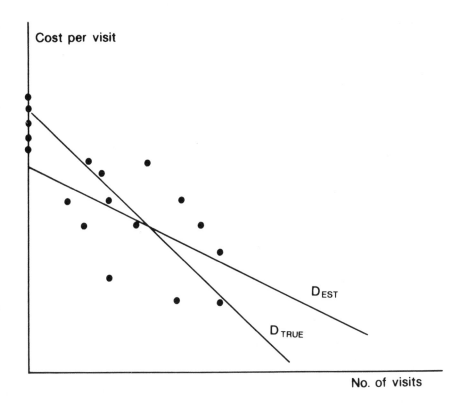

Figure 6.11 Truncation of non-visitors

Source: Pearce and Markandya (1989)

consumer surplus. In a TCM study of deer hunting quality, Balkan and Kahn (1988) found that OLS and ML estimates differed by relatively small amounts. On the other hand, Garrod and Willis (1991) found that, while some forest recreation sites produced relatively similar OLS and ML consumer surplus estimates, other sites produced very different results (one site differing by a factor of nearly 20).

Smith and Desvousges (1986) compared OLS and ML estimated TCM models for 33 water recreation sites. Estimates of mean variance obtained under both approaches were compared, and highly significant differences were taken as indicating high truncation effects. Using this approach, 11 of the 33 sites were identified as highly truncated and were omitted from further investigations.

Other work fundamentally questions the appropriateness of switching to ML estimation as a counter to truncation bias. Both Kling (1987;

Table 6.5 ZTCM–ITCM consumer surplus estimates for six UK forests

Forest	ZTCM		ITCM		CS:
	Travel cost coefficient	CS/visitor (£)	Travel cost coefficient	CS/visitor (£)	ZTCM/ ITCM
Brecon	− 0.384	2.60	− 0.358	1.40	1.86
Buchan	− 0.444	2.26	− 0.996	0.50	4.52
Cheshire	− 0.525	1.91	− 1.259	0.40	4.78
Lorne	− 0.694	1.44	− 0.327	1.53	0.94
New Forest	− 0.702	1.43	− 0.215	2.32	0.62
Ruthin	− 0.396	2.52	− 0.386	1.29	1.95

Notes: All coefficients produced via OLS techniques and significant at 5% level
Travel cost defined as full running costs
Consumer surplus estimates at 1988 prices
$N = 21$ for all forests

Source: Garrod and Willis (1991)

1988) and Smith (1988) suggest that while ML techniques are theoretically more appropriate, OLS techniques (once trimmed to remove predicted negative visits) may actually produce more accurate consumer surplus estimates. Future research appears necessary before any firm conclusions can be drawn regarding this problem.

Zonal versus individual TCMs
Throughout this chapter we have referred to both the zonal and individual variants of the TCM and, as Hanley (1990) points out, 'there is no consensus in the literature as to which option is preferable on theoretical grounds'. However, when both approaches are applied to the same data the two methods are capable of producing disturbingly different results. Table 6.5 illustrates this point with regard to a joint ZTCM–ITCM study of six UK forest sites. Using the same estimation procedure (OLS) and cost definition (full running costs) throughout, estimates of consumer surplus produced by the ZTCM ranged between 60% and 500% of those produced by the ITCM. As all cost coefficients produced by both methods are statistically significant, this points towards some serious methodological problems for one or both of these approaches.

There are a number of methodological problems associated with the use of an average value as a dependent variable. The use of a zonal visitor rate means that it is impossible to specify individual-specific explanatory variables. For example, membership of an environmental or

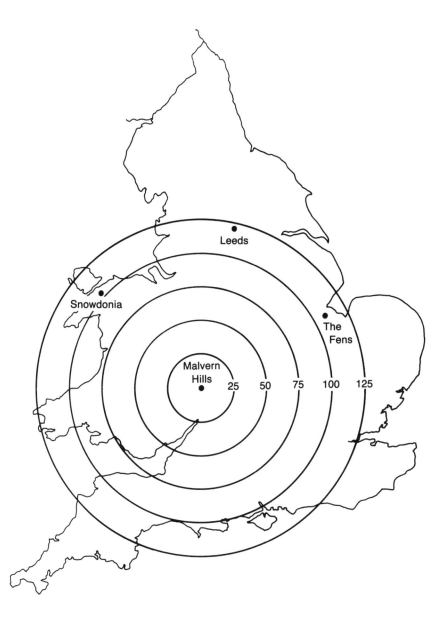

Figure 6.12   Concentric distance zones around the Malvern Hills

(a) ITCM and R^2

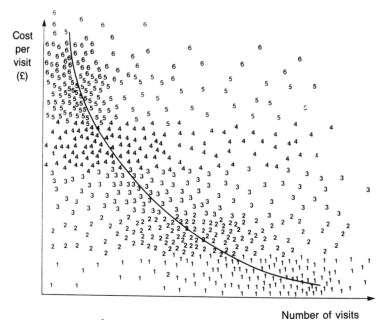

(b) ZTCM and R^2

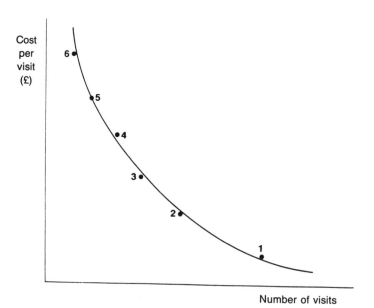

Figure 6.13 R^2 bias in TCM studies

outdoor pursuits association may well be a highly significant predictor of recreational visits. However, in the ZTCM such individual characteristics information cannot be used and a constructed zonal average for such a variable is likely to be meaningless. Similarly, intra-zonal variation is to a considerable degree lost in the ZTCM, as inter-zonal average effects dominate in curve fitting. An extreme case of this may occur where concentric zones are used: outer zones may encompass areas which are geographically very different from each other. For example, suppose that we were to carry out a ZTCM study estimating the recreation value of the Malvern Hills (Worcestershire, England) using distance bands 25 miles wide. Here the distance band between 100 and 125 miles from the Malvern Hills encompasses both the Snowdonia Mountains of North Wales and the flat Fenlands of eastern England (see Figure 6.12). It is likely, therefore, that anyone with a predisposition for hills (as the visitors to Malvern presumably have) would have far more substitute sites if he lived in Snowdonia than if he lived in the Fens. However ZTCM approaches can at best only construct comparisons of the attributes of the studied site with those of all sites perceived by the analyst as substitutes, irrespective of the distance individuals would have to travel to reach such substitutes. Such variables will always be weak compared to the individual-specific substitute variables which can be employed by the ITCM.

Figure 6.12 also highlights a problem with the ZTCM if straight-line distances are equated directly with both travel and time costs. Both Snowdonia and the Fens have relatively poor road links with Malvern, whereas Leeds (in the same zone) has a direct motorway link. Therefore both time and travel costs from Leeds will be considerably less than those for either of the others, a distinction which may be lost in any zonal average.

A further problem for the ZTCM, which again does not afflict the ITCM, is that R^2 statistics will always be upwardly biased. This arises as a natural consequence of aggregating individual responses across zones and so reducing the number of curve-fitting points to the number of zones. Figure 6.13 illustrates this point. Figure 6.13(a) shows the spread of individual observations recorded in a hypothetical TCM survey, each point being represented by a number which in turn is defined by a distance band away from the site. In fitting a demand curve the ITCM would employ all these observations as individual points. In Figure 6.13(b) these individual observations have been converted into zonal averages for use in a ZTCM. The number of observations has thus been reduced to the number of zones (here 6), which will in turn spuriously increase the R^2 of the fitted line.

Consequently the very high R^2 values recorded in many ZTCM studies should be treated with extreme caution. Their only real validity is as indicators of which model has relatively higher explanatory power within any particular functional form; their absolute value should be disregarded (and even not supported as it may well be misleading). This criticism does not apply to the ITCM, for which R^2 figures are unbiased.

A final criticism of the ZTCM approach arises from the methods by which zones are defined. Zones are conventionally concentric, although there is no reason why this should necessarily be so:[26] Bojö (1985), for example, uses county boundaries. The definition of the width and number of zones is typically either arbitrary or influenced by the availability of population data. In effect, each possible definition of zones implies a different aggregation of population and in practice almost certainly a different visitor rate. This in turn will imply changes in the estimated demand curve and thereby different consumer surplus estimates. Therefore, in practice, it is almost certain that an analyst could respecify zones so as either to inflate or to reduce valuation estimates as required. The extent to which such a change is possible is uncertain and the subject of ongoing research.[27]

In conclusion, the decision to use either zonal or individual TCM approaches is likely to have a significant impact upon the results obtained. While there appears to be no theoretical reason for preferring one approach to the other, this discussion highlights a number of methodological problems associated with the application of the ZTCM. However, this does not imply a clean bill of health for the ITCM and certainly all the application issues raised throughout this chapter would have to be satisfactorily addressed before we might begin to consider the adequacy of such an approach.

Non-use values

TCM measures only the 'use value' of recreation sites (see Chapter 5). Underestimation of site value due to the truncation of non-visitors would be made worse if the non-use value of both visitors and non-visitors were relevant. TCM is not capable of producing any total economic value estimate in that it cannot estimate non-use items such as existence value. This is because the basis of the technique is the level of use-based costs incurred by visitors visiting a site. If non-use values are thought to be significant then an appropriate methodology (e.g., CVM) must be employed to capture these values.

Comparison of TCM and CVM results may therefore be problematic. If we are using both techniques to assess the use value alone of a site then the fact that the TCM produces an uncompensated welfare measure while the CVM produces a compensated measure should lead us to expect a larger estimate from the TCM model than a CVM willingness-to-pay (for a welfare gain) survey. However, if we now repeat the experiment attempting to evaluate all aspects (use plus non-use) of a site (e.g., by having potential site destruction as our scenario) then we might well expect our CVM welfare measure (which includes use and non-use values) to exceed our TCM measure (particularly in the case of CVM willingness-to-accept scenarios). Such comparisons form the basis for convergent validity testing (see Chapter 5). However, uncertainty about which attributes of a site either technique is measuring make comparisons between the two methods a fairly weak test of estimate validity.

Variants of the TCM
Apart from those already discussed, a number of variants on the TCM have been developed. Pearse (1968), in a study of big game hunting in Canada, develops a variant of the ZTCM, stratifying his sample by income rather than distance zones. The individual with the highest visit costs is assumed to be the marginal user with no net benefits. All other users are assumed to have net benefits equal to the difference between their visit costs and those of the marginal user. Summing these net benefits across all visitors therefore gives us our consumer surplus estimate.

Burt and Brewer (1971) and Cicchetti *et al.* (1976) develop an 'extended TCM' approach to the valuation of a proposed new site. The first stage involves the estimation of a system of demand equations for existing sites. The existing site which is believed to be the closest substitute for the proposed new site is then isolated. Visitor patterns at this site are held constant while its travel costs are replaced with those which are expected to apply at the proposed new site. The entire system is then re-estimated. The introduction of the new site will cause a fall in net travel costs and this will provide our consumer surplus estimate of the benefit of introducing the new site. This approach, of course, assumes that the demand function for the new site can be adequately described by that of its closest substitute.[28]

Brown and Mendelsohn (1984) develop a 'hedonic TCM' approach in which the property prices of the hedonic price method are replaced with travel costs as below:

$$TC = f(Z_k, Y, X) \qquad\qquad (6.21)$$

where

TC = total travel cost (travel expenditure and time costs)
Z_k = site attributes ($k = 1, \ldots, n$)

Y = income
X = composite good.

The HTCM uses travel costs to estimate values for separate site attributes rather than whole site values. A two-stage process is employed. In stage one the independent variables of equation (6.21) (including characteristic levels for each of the k site attributes) are regressed, in separate estimations, against travel costs (i.e. distance from zone i to site j) and time costs.[29] Thus, for simplicity, setting $n=3$ and ignoring non-Z explanatory variables, we have

$$\underbrace{PTC.d_{ij}}_{A} = a_0 + a_1 Z_1 + a_2 Z_2 + a_3 Z_3 \qquad (6.22)$$

$$\underbrace{PTT_{ij}.TT_{ij}}_{B} = b_0 + b_1 Z_1 + b_2 Z_2 + b_3 Z_3$$

where

PTC = money expenditure on travel (petrol, etc.) per mile
d_{ij} = distance from zone i to site j (miles)
PTT_{ij} = opportunity cost per hour of travel time from zone i to site j
TT_{ij} = length of travel time from zone i to site j (hours.

To calculate the implicit prices of each Z_i attribute we now insert cost per mile and time costs. Hanley (1990) illustrates this using values of £0.20 per mile and £1.50 per hour, respectively. Using these figures and the amalgamated variables A and B in equation (6.22), the implicit price[30] of characteristic Z_i will be:

$$\frac{\partial P}{\partial Z_i} = 0.2 \frac{\partial A}{\partial Z_i} + 1.5 \frac{\partial B}{\partial Z_i} \qquad (6.23)$$

Stage two involves the estimation of demand curves for each attribute Z_i by regressing implicit price on the observed level of Z_i and other explanatory variables.[31] Summing over all observations gives an aggregate demand function for each attribute from which consumer surplus estimates may be obtained.[32]

Concluding comments

The TCM is a potentially useful evaluation tool producing uncompensated consumer surplus estimates of use value. It is best applied to the evaluation of well-defined recreation sites, or to the evaluation of a well-perceived, separable, environmental attribute within such a site.

This survey has highlighted several potential problems which may arise during the practical application of the TCM. These include:

(i)   the decision whether to use zonal and individual approaches and variation in results between these methods;
(ii) calculation of the cost elements and in particular determination of the opportunity cost of on-site and travel time;
(iii) multicollinearity between explanatory variables, especially site environmental characteristic levels;
(iv) choice of the appropriate functional form and its impact upon consumer surplus estimates;
(v)   truncation bias and the choice of appropriate estimation technique;
(vi) treatment of substitute sites;
(vii) accounting for potential congestion effects.

The validity of TCM welfare measures will be dependent upon the extent to which these problems can be minimised.

The hedonic pricing method

Introduction

The hedonic pricing method (HPM) attempts to impute a price for an environmental good by examining the effect which its presence has on a relevant market-priced good. The notion of land characteristics being reflected in land values can be traced back to Ricardo. However, it was Ridker (1967) who conducted the first recognisable HPM study, noting that 'if the land market were to work perfectly, the price of a plot of land would equal the sum of the present discounted stream of benefits and costs derivable from it' (i.e. environmental values are reflected in associated market prices); and, equally importantly, that 'since air pollution is specific to locations and the supply of locations is fixed, there is less likelihood that the negative effects of pollution can be significantly shifted onto other markets' (i.e. single markets can be identified as capturing these environmental values). While Rosen (1974) shows that the

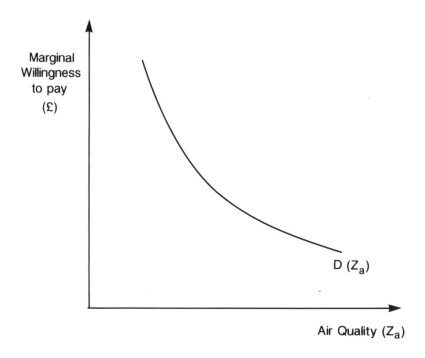

Figure 6.14   Demand curve for air quality

HPM can be applied to any market which can be fully estimated, the vast majority of HPM studies have looked at the property market as a reflection of surrounding environmental characteristics such as air quality, water quality and noise. By controlling for the structural (e.g., size), locational (e.g., access to workplace) and other characteristics of a house, we can isolate the effect which environmental characteristics have upon house price and thereby ascertain the implicit price of a specific public good such as air quality.

The objective of HPM studies is to define the (inverse) demand function relating the quality of the environmental good to individuals' marginal willingness to pay for that good. Figure 6.14 illustrates a typical demand curve for air quality.

Once the (inverse) demand function has been isolated we can evaluate a welfare change as the area under the demand curve between the initial and final environmental quality level. As this demand curve is obtained from property market-price data it includes income effects and therefore its integration produces uncompensated consumer surplus estimates of welfare change.

Theoretical issues: The method

This discussion will centre on a hypothetical property value application of the HPM to the evaluation of air quality benefits.[33]

The HPM is dependent on a number of assumptions for its operation. For the moment we will define these as follows:[34]

(i)   Willingness to pay is an appropriate measure of benefits.
(ii)   Individuals can perceive environmental quality changes; these changes affect the future net benefit stream of a property, and therefore people are willing to pay for environmental quality changes.
(iii) The entire study area can be treated as one competitive market with freedom of access across the market and perfect information regarding house prices and environmental characteristics.
(iv)   This housing market is in equilibrium, i.e. individuals continually re-evaluate their location such that their purchased house constitutes their utility-maximising choice of property given their income constraint.

We shall discuss the validity of these assumptions subsequently, but for the moment we assume that they hold.

We first define our property value or hedonic price function describing the house price (P_i) of any housing unit i as the function:

$$P_i = f(S_{1i}, \ldots, S_{ki}, N_{1i}, \ldots, N_{mi}, Z_{i1}, \ldots, Z_{ni}) \tag{6.24}$$

where

S = Structural characteristics $(1, \ldots, k)$ at house i, e.g., house size, number of rooms, type of construction;

N = neighbourhood characteristics $(1, \ldots, m)$ at house i, e.g., accessibility to work, quality of schools, local crime rates;

Z = environmental characteristics $(1, \ldots, n)$ at house i, e.g., air quality.

For simplicity we will assume that only one environmental variable is significant, namely air quality (Z_{ai}).

The functional form of equation (6.24) will be determined by the underlying utility function. For illustrative purposes let us suppose that this is linear, so that equation (6.24) can be respecified as

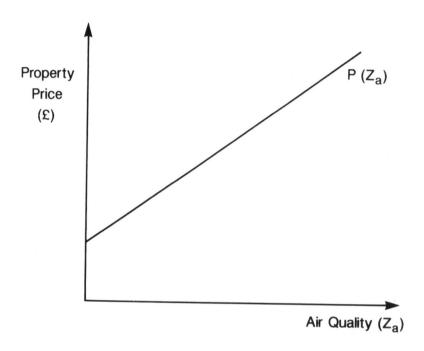

Figure 6.15   Rent function for air quality (linear hedonic price function)

$$P_i = \alpha_0 + \alpha_1 S_{1i} + \alpha_2 S_{2i} + \ldots + \alpha_k S_{ki} + \beta_1 N_{1i} + \beta_2 N_{2i} + \ldots$$
$$+ \beta_m N_{mi} + \gamma_a Z_{ai} \tag{6.25}$$

Estimating equation (6.25) we would expect $\gamma_a > 0$, i.e., house price increases with air quality improvement. This gives the total house price relationship (or rent function) for air quality as illustrated in Figure 6.15.

By differentiating equation (6.25) with respect to Z_a we can obtain the implicit marginal purchase price for air quality:

$$\frac{\partial P}{\partial Z_a} = \gamma_a \tag{6.26}$$

This shows the purchase price of each successive unit of air quality. Because of the linear form of equation (6.25) this is a constant, as shown by the line $R(Z_a)$ in Figure 6.16.

The constant marginal purchase price implied by the linear hedonic price function of equation (6.25) is unlikely to occur in reality. Air quality is likely to be a normal good and exhibit diminishing marginal

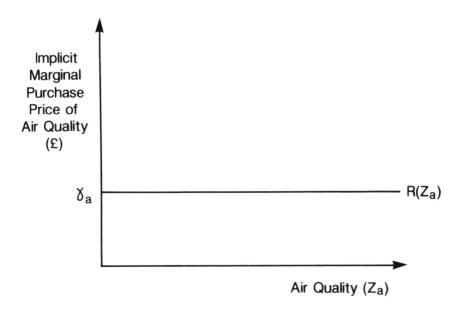

Figure 6.16   Implicit marginal purchase price of air quality (linear hedonic price function)

utility. The hedonic price function is therefore likely to be non-linear. If we have a multiplicative underlying utility function then a double-log hedonic price function such as the following may be appropriate (the number of non-environmental explanatory variables being reduced for simplicity):

$$\ln P_i = \alpha_i \ln S_{1i} + \ldots + \beta_i \ln N_{1i} + \ldots + \gamma_a \ln Z_{ai} \qquad (6.27)$$

This yields the rent function illustrated in Figure 6.17.

Differentiating equation (6.27) with respect to Z_a gives a new expression for the implicit marginal purchase price of air quality:

$$\frac{\partial P}{\partial Z_a} = \gamma_a \frac{P}{Z_a} \qquad (6.28)$$

Now notice that the implicit marginal purchase price of Z_a varies according to the ambient level of Z_a prior to the marginal change. This function is mapped out by the curve $R(Z_a)$ in Figure 6.18.[35]

The implicit marginal purchase price function describes the price paid

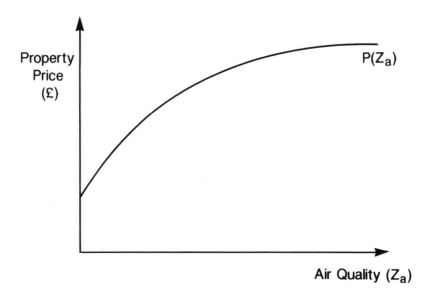

Figure 6.17   Rent function for air quality (double-log hedonic price function)

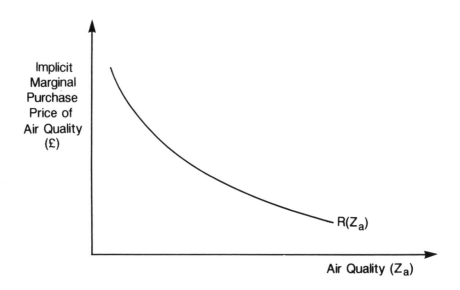

Figure 6.18   Implicit marginal purchase price of air quality (double-log hedonic price function)

for marginal increments of air quality (Z_a). However, it does not necessarily follow that it is the household demand curve for Z_a (see Figure 6.14) in the sense that it is unlikely to correspond to households' marginal willingness to pay (inverse demand curve) for Z_a. To demonstrate this, first consider household (inverse) demand for Z_a, which will itself be a function of the level of Z_a, income and other preference variables and can be determined by the following regression:[36]

$$W_{Z_{ai}} = f\,(Z_{ai},\ Y_i,\ X_i) \tag{6.29}$$

where

$W_{Z_{ai}}$ = household i's marginal willingness to pay for Z_a
Z_{ai} = air quality at household i
Y_i = income of household i
X_i = other explanatory variables.

The potential non-correspondence between this marginal willingness to pay (MWTP) function and the implicit marginal purchase price function (equation (6.26) or (6.28) depending upon functional form) can be demonstrated intuitively by considering an individual household's decision about whether or not to buy an additional unit of Z_a. At some low level of Z_a a particular household may be more than willing to pay the marginal purchase price of an additional unit of Z_a (MWTP greater than implicit marginal purchase price). However, at high levels of Z_a that same household may not be willing to purchase an additional unit (MWTP less than marginal implicit purchase price). Logically there will be some intermediate level of Z_a where the household is only just willing to pay for the incremental Z_a unit (MWTP equal to marginal implicit price). Figure 6.19 illustrates this relationship between household MWTP and implicit marginal purchase price for two households (i and j).

We can now show that, assuming all variables are continuous, then at its utility-maximising equilibrium, household MWTP for Z_a will be equal to this marginal implicit price. If we define utility (U) as a function of the consumption of the characteristics of housing (S, N and Z as per equation (6.24)) and of all other goods (X), then the conventional utility maximisation problem will be to

maximise $U = U\,(X,\ S,\ N,\ Z)$

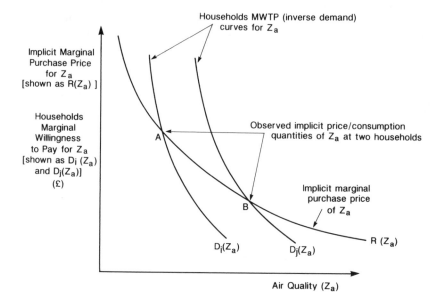

Figure 6.19 Relationship between implicit marginal purchase price of air quality and household marginal willingness to pay (inverse demand curve) for air quality for two households

subject to

$$Y = P_x.X + P$$

where

Y = income
P = price of property
P_x = vector of prices of all other marketed goods.

Then, considering the first-order utility-maximisation conditions, we can see that MWTP for the characteristic Z_a will be:

$$W_{Z_{ai}} = \frac{\partial U}{\partial Z_a} \tag{6.30}$$

This is the demand function for Z_a. At equilibrium, this will equal the marginal implicit price of Z_a, i.e.

$$\underbrace{\frac{\partial U}{\partial Z_a}}_{\substack{\text{MWTP} \\ \text{for } Z_a}} = \underbrace{\frac{\partial P}{\partial Z_a}}_{\substack{\text{implicit marginal} \\ \text{price of } Z_a}} \tag{6.31}$$

Each household therefore chooses a position where its own MWTP for Z_a is equal to the marginal implicit price of Z_a.

Therefore, we can see that empirical observations of implicit marginal purchase price and corresponding level of air quality (as mapped out by $R(Z_a)$) only tell us about single points on each household's inverse demand curve for air quality. Thus the implicit marginal purchase price curve can normally only be used to approximate the benefit of marginal changes in air quality. In the case of such small changes the implicit marginal purchase price curve is an acceptable approximation of the household's inverse demand curve. However, as soon as we consider non-marginal changes these curves begin to diverge markedly and the implicit marginal purchase price curve will give a biased estimate of the benefits of air quality change.

Freeman (1979c) shows that there is one case in which the implicit marginal purchase price curve correctly estimates the benefits of non-marginal air quality changes. This occurs when all households have identical utility functions and incomes.[37] In such a case all household inverse demand curves will lie on top of each other along the marginal implicit price curve which can then be integrated to determine consumer surplus welfare measures. Figure 6.20 illustrates such a case for an air quality improvement from Z_{a1} to Z_{a2}, with resultant benefits shown as the shaded area.

Such an assumption is obviously weak,[38] therefore we are left with two broad courses of action: either we can make certain assumptions about the shape of household inverse demand curves; or we can attempt to directly estimate the shape of these curves. Freeman (1979c) and Hufschmidt et al. (1983) consider both options.

One assumption considered by Freeman is that households might have a constant marginal benefit for air quality improvements, i.e. they have a horizontal inverse demand curve, $D_i(Z_a)$, as in Figure 6.21. If this is so and the magnitude of the air quality change (ΔZ_a) is also known, then the benefits of that change are calculable as the rectangle $Z_{a1}A_1A_2Z_{a2}$.

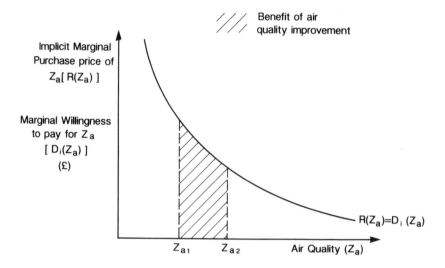

Figure 6.20   Estimating the benefits of air quality improvement assuming identical household utility functions and incomes

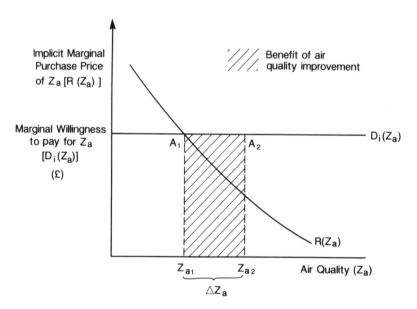

Figure 6.21   Estimating the benefits of air quality improvement assuming constant marginal household benefits for air quality

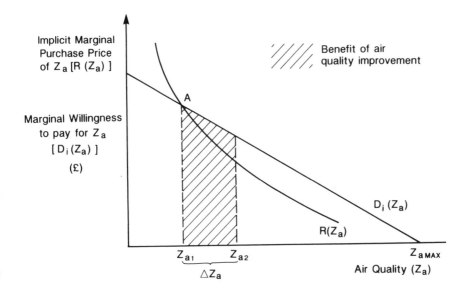

Figure 6.22 Estimating the benefits of air quality improvement assuming linearly declining household benefits for air quality

A second assumption considered by Freeman (1979c) is that household marginal willingness to pay might decline linearly from its observed level (point A in Figure 6.22) to reach zero where air quality reaches a maximum (i.e. zero pollution level) at Z_{aMAX}. Given such an assumption the slope of $D_i(Z_a)$ can be estimated and, knowing ΔZ_a, we can calculate the benefit of air quality improvement.

Despite Freeman's (1979c) assertions that all of the assumptions underlying Figures 6.20–6.22 are 'plausible', in reality all three are highly improbable. Given that (as Figure 6.19 shows) a utility-maximising explanation of household behaviour requires that household inverse demand curves be steeper than the implicit marginal purchase price curve, the three assumptions discussed above are all liable to result in overestimation of the benefits of air quality improvement (or underestimation of the costs of air quality loss). Furthermore, both of the assumptions of Figures 6.21 and 6.22 would fail to result in equilibrium solutions as increases in air quality above the initial ambient level (Z_{a1}) result in a widening excess of household MWTP over implicit marginal purchase price.

One alternative assumption which we suggest is to reverse the previous approach and assume that household MWTP for the first unit of air

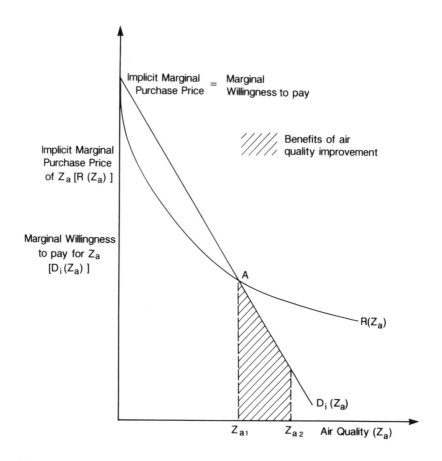

Figure 6.23 Estimating the benefits of air quality improvement, assuming initial marginal willingness to pay equals implicit marginal purchase price

quality is equal to the implicit marginal purchase price of that unit.[39] Assuming also that we have linear inverse demand curves, we can evaluate the slope of these curves via observations of point A and thereby estimate the benefits of air quality improvement. The assumption of linearity is likely to lead to underestimation of benefits, but the approach would have the advantage of producing defensible lower boundary benefit estimates. Figure 6.23 illustrates such an approach.

All these assumptions regarding the shape of the inverse demand curve are highly questionable. Therefore many commentators have focused upon direct estimation of this curve, estimating its slope via regression techniques and using this information in conjunction with observed household consumption levels.

Household demand for air quality will be a function of socioeconomic level, income and other relevant explanatory variables. However as Freeman (1979c) notes, an important issue here is the speed with which the supply side of the property market can adjust to demand for housing characteristics such as air quality. Three permutations are possible. The first of these is that supply may adjust immediately to demand (i.e. a demand-constrained system with perfectly elastic supply at a given price). Since we can observe the level of Z_a for any given household and we can use our implicit price function to find the marginal willingness to pay for Z_a (denoted W_{Z_a}) at that level of provision, a regression of observed quantities of Z_a on W_{Z_a} and other independent variables should identify the demand function for Z_a thus:

$$S(Z_a) \equiv D(Z_1) \Rightarrow Z_{ai}$$

where

$$
\begin{aligned}
S(Z_a) &= \text{supply of air quality} \\
D(Z_a) &= \text{demand for air quality}
\end{aligned}
$$

i.e. perfectly elastic supply producing a demand constrained system. Therefore, under such an assumption, we can use the implicit marginal purchase price curve to estimate household MWTP for air quality (W_{Z_a}). Thus we can directly estimate the inverse demand curve:

$$Z_{ai} = f(W_{Z_{ai}}, Y_i, X_i) \tag{6.32}$$

for example, the linear form:

$$Z_{ai} = A_0 + A_1 W_{Z_{ai}} + A_2 Y_i + A_3 X_i + e \tag{6.33}$$

where

$$
\begin{aligned}
Z_{ai} &= \text{air quality at household } i \\
Y_i &= \text{income at household } i \\
X_i &= \text{other explanatory variables e.g., socioeconomic factors at household } i \\
e &= \text{error term.}
\end{aligned}
$$

Rearranging the estimated form of equation (6.33) gives us the slope of the household inverse demand curve, $-1/\hat{A}_1$. Integrating under this curve between the limits of a given air quality increase (decrease) will give our estimate of household benefits (costs), while summing this across all households gives our estimate of total benefits (costs).

An assumption of perfectly elastic supply is, however, difficult to justify. House building often exhibits lagged response and there does not appear to be a vast surplus of housing. An alternative scenario, therefore, is to postulate a completely supply-constrained system with a supply of Z_a fixed irrespective of individual households' demand. In this context, households bid for fixed amounts of housing with desirable environmental characteristics.

In such a case we have

$$D(Z_a) \equiv S(Z_a) \Rightarrow Z_{ai} \qquad (6.34)$$

i.e. perfectly inelastic supply dictating a supply constrained system. We can now regress observed household MWTP for Z_{ai} on observed quantities of Z_{ai}, household income and other explanatory variables thus:

$$W_{Z_{ai}} = f(Z_{ai}, Y_i, X_i) \qquad (6.35)$$

for example, the linear form:

$$W_{Z_{al}} = B_0 + B_1 Z_{ai} + B_2 Y_i + B_3 X_i + e \qquad (6.36)$$

Now the slope of the household inverse demand curve is directly given by the estimated coefficient $\hat{B}_1$.

Compared to the elastic supply assumption, the assumption of a fixed supply of environmental characteristics is far more defensible. Freeman (1979c) feels that the speed at which market supply can adjust to demand will be sufficiently slow to allow an assumption of fixed supply to hold and asserts that 'in general it seems reasonable to treat air quality as exogenous, that is, independent of its implicit price, and to assume that ordinary least squares estimation of [the individual household's MWTP function] identifies the inverse demand curve for [air quality]'.

Harrison and Rubinfeld (1978a; 1978b)[40] adopted such a position in their HPM study of air pollution in Boston. The authors estimated household demand curves with two structural, eight neighbourhood, two access and one environmental (air quality) variables, the latter being the concentration of nitrogen oxide at households measured in parts per hundred million. In estimating the hedonic price function, the air quality variable was shown to be highly significant. The implicit marginal purchase prices could then be estimated giving observations on household MWTP for air quality (denoted W_{Za} in equation (6.37) below) and the level of household air pollution (denoted by the variable *NOX* below). Assuming a supply-constrained system, the household inverse demand curve was estimated by regressing W_{Za} on *NOX*, household income (*INC*) and other explanatory variables as per equation (6.35). Harrison and Rubinfeld fitted a variety of functional forms to their data, the double-log form being reported as follows (all significant variables given):

$$\text{Ln } W_{Za} = 1.08 + 0.87 \ln NOX + 1.00 \ln INC \qquad (6.37)$$

Notice that because the air quality variable *NOX* measures increasing air pollution (decreasing air quality) it is assigned a positive coefficient, i.e. MWTP for a unit increase in air quality increases as levels of air pollution rise. Therefore, we do have the expected negative relationship between increased air quality (decreased *NOX*) and MWTP for that air quality. Notice also that equation (6.37) indicates, as expected, a positive relationship between income and marginal willingness to pay. Figure 6.24 illustrates the derived inverse demand curves for three income groups, $Y_1 < Y_2 < Y_3$. In practice, Harrison and Rubinfeld (1978b) used average income and *NOX* figures to calculate the average household benefits arising from cuts in emissions.

While the fixed-supply assumption may be empirically attractive, in reality both the quantity supplied of the air quality characteristic Z_a and the quantity demanded may be a function of its implicit price. In such a case the market is simultaneously determined and equations for both the demand and supply sides need to be specified. The demand side can be expressed as

$$D_i (Z_a) = f(W_{Z_{ai}}, Y_i, X_i) \qquad (6.38)$$

and the supply side as

$$S_i (Z_a) = f(W_{Z_{ai}}, Y_i, X_i) \qquad (6.39)$$

Clearly, as it stands we would have problems estimating such a system as we cannot identify whether shifts in MWTP are the result of demand- or supply-side pressures. Nelson (1978a) attempts to address such a situation by separately specifying the supply side as

$$OP = f (Z_a, d, K) \qquad (6.40)$$

where

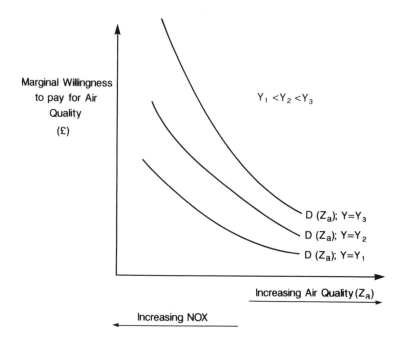

Figure 6.24   Inverse demand curves for air quality for three income groups

OP = offer price (from house vendors)
Z_a = level of air quality
d = distance from central business district
K = other relevant variables

Nelson argues that the two-stage least-squares estimation of this system identifies both the demand and supply functions for air quality. The validity of such approaches should be judged by the statistical validity of the estimated functions.

Assuming for the moment that the individual household's inverse demand curve has been estimated, the household's welfare gain or loss engendered by some change in the provision of an environmental good can be estimated by integrating under this curve between the initial and subsequent provision levels. Total benefit is then estimated by summing individual household benefits thus:

$$B = \sum_{i=1}^{n} \int_{Z_{ai1}}^{Z_{ai2}} D_i(Z_a) \, dZ_a \qquad (6.41)$$

where

B = uncompensated consumer surplus change aggregated over all households

i = households $(1, \ldots, n)$

Z_{ai1} = initial level of air quality

Z_{ai2} = final level of air quality

$D_i(Z_a)$ = household's inverse demand for Z_a function (MWTP for Z_a).

As stated, this procedure produces measures of uncompensated consumer surplus. We can test the specific validity of such a measure as an estimator of true (compensated) welfare change using the Willig (1973; 1976) formulae. Freeman (1979c) uses such a procedure to test figures from Harrison and Rubinfeld's (1978a; 1978b) HPM study of nitrogen oxide and particulates in Boston. However, while Freeman uses the results obtained using Harrison and Rubinfeld's linear form, we can repeat this test using their preferred single-log (dependent) functional form. Using mean values for all explanatory variables, this function estimates a mean capitalised consumer surplus value for air quality of $1,613 which (following Freeman's approach) we can annualise to about $160 p.a. using an assumed 10% discount rate. Reported values for average income and income elasticity were $11,500 p.a. and 1.0 respectively. Willig shows that for consumer surplus to be a valid estimator of compensated welfare change, two conditions must be satisfied. Following Varian's (1984) simplification, the first condition is that:

$$\frac{CS}{Y} \cdot \frac{\epsilon^Y}{2} \leqslant 0.5 \qquad (6.42)$$

where

CS = annual consumer surplus estimate

Y = income

ϵ^Y = income elasticity

Here we have

$$\frac{160}{11,500} \cdot \frac{1.0}{2} = 0.007$$

therefore the first condition is well satisfied. Willig's second condition is that:

$$\frac{CS}{Y} \leqslant 0.9 \tag{6.43}$$

Here we have

$$\frac{161}{11,500} = 0.014$$

therefore the second condition is again well satisfied. We can conclude that, in this study at least, consumer surplus estimated by the HPM can be taken as a valid approximation of the compensated welfare measure.

The HPM does have a consistent theoretical basis which allows for the valid estimation of consumer surplus measures of welfare change. We now turn to consider methodological aspects and applications issues specific to the method.

Application issues

Assumptions
The HPM relies upon a number of restrictive assumptions, the most fundamental of which are those of perfect information and perception regarding the environmental characteristics which define a particular housing bundle. In the case of air quality change, this assumes that individuals can perceive such change as a continuous variable. However, it is quite feasible that individuals will have perception thresholds for pollutants such that concentrations below this threshold are not perceived. This leads to the possibility of a non-continuous implicit marginal purchase price function and associated estimation and inter-pretation problems. A similar problem arises where pollution-related diseases are subject to long time delays, or only arise due to a critical accumulation of pollutants. The degree to which individuals perceive environmental quality (or its absence) is therefore a crucial determinant of the validity of any particular HPM study.

A further aspect of this problem is whether individuals' perceptions and consequent property purchase decisions are based upon actual or historic levels of pollution and environmental quality. If expectations are not the same as actuality (as measured by present pollution readings) then there are clearly problems in relating this to values derived from purchases. Similarly, Mäler (1977) points out that expectations regarding future environmental quality may bias present purchases away from that level dictated by present characteristic levels.

Concern can also be raised regarding the extent to which an assumption of utility maximisation can be maintained within the rigidity of the property market (see subsequent discussion). Mäler (1977) also criticises the HPM for making the implicit assumption that households continually re-evaluate their choice of location.

Furthermore, there is considerable doubt that such an assumption can hold in the context of spatially large study areas (see Garrod and Willis 1992). If people cluster for social or transportation reasons (for example, if intra-urban transport is poor) then HPM results will be biased.

These criticisms regarding the dubious nature of the assumptions underpinning the HPM (compounded by the severe difficulties surrounding the construction of satisfactory assumptions regarding the form of the inverse demand curve, discussed previously) constitute the most fundamental problem facing the practical application of the HPM.

Statistical problems

Functional form As with CVM and TCM studies, HPM researchers face a fundamental problem in that theory provides us with no particular expectation regarding the nature of the functional form for the implicit marginal purchase price and inverse demand function (other than that they are unlikely to be linear). The most commonly adopted forms are the double and semi-log.[41] However, Freeman (1979c) discusses eight potential functional forms, all of which have been applied.

Apart from the linear form, most functional forms generally produce the expected result that the implicit marginal purchase price of air quality is dependent upon its absolute level. However, the nature of that dependence (diminishing or increasing) varies according to the specific functional form. We have previously examined a double-log form, and we can contrast this with the exponential form discussed by Hanley (1991). Here the environmental characteristic is measured as the air pollution level variable *POL* rather than air quality. A simple exponential form (ignoring all other explanatory variables) is:

$$P_i = aPOL^b \tag{6.43}$$

where P_i = property price. Setting $b < 0$, we obtain the hedonic and implicit marginal purchase price functions shown in Figure 6.25. Notice

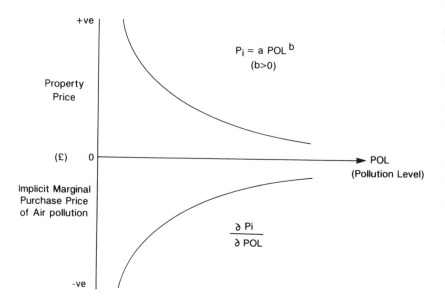

Figure 6.25   Hedonic price and implicit marginal purchase price of air pollution, assuming an exponential functional form

that, as we are dealing with increasing levels of air pollution (decreasing levels of air quality) along the horizontal axis, we now have a downward-sloping hedonic price function (upper graph), in contrast to the upward-sloping curve used in our previous discussions (increasing air quality). However, such a function provides the implicit marginal purchase price curve shown in the lower graph which now diminishes with absolute pollution level (i.e. the hedonic price function has a positive second derivative) indicating, as we would expect, that initial doses of pollution impose a greater marginal cost than do subsequent units of pollution. This seems plausible; however, in this example the choice of functional form has been imposed, i.e. it is not data-determined. Anderson and Bishop (1986) discuss the adoption of a Box–Cox transformation procedure in which relationships are not imposed from the outset but instead functional form is derived using tests of best statistical fit (the linear, semi-log and double-log forms all being special cases of the Box–Cox linear model). Such an approach could adopt a function form such as:

$$\frac{P_i^c - 1}{c} = a + bPOL \tag{6.45}$$

where c is an unknown parameter. Such a functional form gives the following second derivative.[42]

$$(1 - c)b^2 P_i^{1-2c} \tag{6.46}$$

this will be signed as:

positive for $c < 1$ (i.e. diminishing implicit marginal purchase price of pollution)

zero for $c = 1$ (i.e. constant implicit marginal purchase price of pollution)

negative for $c > 1$ (i.e. increasing implicit marginal purchase price of pollution).

Measurement and multicollinearity   Pollution measurements may well be subject to temporal variation; however, there is no clear rationale for aggregation. In particular, as previously noted, individuals may be more perceptive of peak intensities rather than averages. Even if averages are appropriate the periodicity is again uncertain.

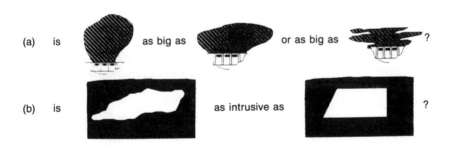

Figure 6.26   Landscape evaluation problems for the HPM

Source: Price (1991)

Price (1991) comments upon HPM measurement problems in the context of landscape valuation, noting that 'a view is not measured only by the surface area covered, and may as easily be related to the edge perimeter with other components of the view, as to the solid angle subtended at the viewpoint'. Figure 6.26(a) illustrates this point.

Multicollinearity is a similarly endemic problem for HPM studies. Explanatory variables are very likely to be collinear – for example, sulphates and particulates; noise and dust. Ignoring such problems by using one pollution variable will lead to biased OLS estimators (usually upwardly biased as the chosen single pollutant picks up variation due to other pollutants) unless it is made clear that the chosen environmental characteristic is a flag variable for all those it is collinear with (see Turner and Bateman 1990).

Again Price (1991) highlights the relevance of such concerns in HPM landscape evaluation exercises, noting that multicollinearity invalidates the assumption that different attributes affect utility separably: 'given that landscape quality is strongly determined by composition of features, this assumption is unsound'. Figure 6.26(b) illustrates this point.

Other variables   Given that estimation of a valid hedonic price function involves the inclusion of a variety of variables alongside environmental characteristics, omission of any of the structural, locational or neighbourhood variables will lead to biased estimators. Freeman (1979c) makes an interesting argument for the potential inclusion of income as a neighbourhood explanatory variable. Generally a hedonic price function would not include income, it being a characteristic of households rather than houses. However, it can be argued that the general income level of households in the area may be perceived as a socioeconomic characteristic of a particular house, so that, for example, some median income figure might defensibly be included.

However, researchers, should also be wary of 'kitchen-sink' type approaches in which a plethora of spurious explanatory variables is included in order simply to boost overall explanatory power statistics. All explanatory variables should have a valid theoretical basis for their inclusion.

A major problem for HPM researchers has been the gathering and measurement of the large volume of explanatory variable data necessary to operationalise models – for example, measuring access distances to a variety of amenities (workplace, leisure centres, etc.). A new tool which holds great promise for such applications is the Geographical Information System (GIS) which is capable of manipulating a number of digitised maps simultaneously and can, for example, computerise the measurement of access distances so as to facilitate a vast improvement in the collection and analysis of data. A second application of GIS would be the simulation of 3-D maps (viewsheds) around a house to allow the use of actual views to amenity/disamenity sites rather than just using straight-line distances which may be less relevant. There is good reason to suppose that such an approach would provide a very considerable enhancement of HPM studies. In an HPM study of amenity values, McLeod (1984) used a visual inspection of the view from each sample property related to whether or not that property had a river view. The resultant variable was found to be not only highly significant but to exert the largest influence upon house price. However, we are currently not aware of any studies which have yet taken advantage of the specific advantages afforded by GIS.[43]

The property market

Data Property value data may be obtained directly from sales (see, for example, Garrod and Willis 1992) or estimated using professional or homeowners' estimates (Button and Pearce 1989). Clearly there are potential errors and/or biases inherent in the latter options, and sales price is always to be preferred (however, the availability of such data may be restricted). Thus in England and Wales sales price tends to be known only by the vendor, purchaser and mortgage lender.[44] Data in Scotland are far more accessible via the sales register, while in the USA availability varies by state but can be very good (see Brookshire *et al.* 1982). Australian data are typically excellent, with database records of house valuations for most urban properties (see McLeod 1984). The cost of obtaining data is likely to rise inversely with its availability, and this may be an important factor limiting the applicability of HPM studies.

As already indicated, property prices are just as likely to reflect supply conditions as demand. This is particularly true of the UK, where large-scale state intervention, both directly through provision of public housing and indirectly via tax concessions to homeowners, has considerably distorted the market. Therefore, the analyst must consider both sides of the market and the adjustments necessary to avoid the biasing of benefit values.

Another consideration in the treatment of property price data is the possibility of averting behaviour (defensive expenditure). For example, a houseowner may mitigate against local pollution levels by, say, fitting double glazing to reduce noise, or installing air purifiers to combat air pollution. Such behaviour may enhance the value of the house above that defined by local environmental characteristics. Therefore, where such factors are significant, the hedonic price function should be fitted with a specific variable to account for this; however, such a variable may be difficult to operationalise.

Market rigidities The HPM assumes that the property market is both fluid and in equilibrium, i.e. that individuals are free to move to utility-maximising positions subject only to their income constraints and the housing market clearing in each period. Violation of these assumptions means that individuals will no longer be able to equate their MWTP curves with the implicit marginal purchase price function and, in some circumstances, this will cause the HPM to be invalid.

Further complication arises where the property market studied is shown not to be a single whole but a segmented collection of regional markets (Hyman 1981). This will occur where there are certain barriers to mobility between regions (for example, due to ethnic discrimination) and at the same time demand and/or supply conditions vary between these regions.[45]

Is the property market sufficient?   The HPM only examines households' willingness to pay for environmental quality at the home. It therefore ignores both the value of improvements which occur away from the home but which directly affect the household (i.e. HPM does not capture the totality of use values) and those which do not directly affect the household but are still valued (i.e. HPM does not capture non-use values). For example, air quality both at the home and at the workplace may affect the decision to locate. Therefore, both property prices and wage differentials may need examination. However, we do run a risk of double counting if we add all benefit estimates obtained from various individually valid sources. For example, if, for a reduction in air pollution, we added soiling benefits to property value benefits and wage differentials, then we are likely to double-count and overestimate the value of environmental improvements (see Cummings *et al.* 1986; Freeman 1979c).

Conclusions

The HPM is founded upon a coherent theoretical base. As such, it is theoretically capable of producing valid uncompensated consumer surplus estimates. However, the method can be strongly criticised according to the number and strength of assumptions which need to be made in order that valid results be obtained. Pearce (1978) feels that these stringent requirements invalidate the practical use of HPM, while Price (1991) feels that any results obtained are at very best highly site-specific and non-transportable. These difficulties are compounded by the need to consider both the supply and demand side of the property market, dictating that the use of HPM is likely to remain restricted to a few expert users.

Optimal applications for the HPM are defined by the requirements of the method itself. The environmental characteristic under evaluation must be well perceived by householders, and the impacts of that characteristic must be well contained within the market being studied. Typical applications include the evaluation of visible particulates or well-perceived aircraft noise in areas where the property market is fluid and can reasonably be said to be in equilibrium. In such conditions the HPM may provide defensible evaluations of those environmental characteristics.

Notes

1. The author is grateful to Ken Willis and Guy Garrod at the University of Newcastle upon Tyne and to Ian Langford and Kerry Turner at the University of East Anglia for comments regarding this chapter. Errors remain the responsibility of the author. A fuller, more technical version of this analysis can be found in Bateman (1993).

2. Several texts make the simplifying assumption that consumer surplus for the marginal user (here the most remote zone) is zero (Sinden and Worrell 1979; Hufschmidt *et al.* 1983). This will typically lead to an underestimate of true consumer surplus.

3. Many empirical studies employ a variety of simplifications of equation (6.5) – for example, on-site time costs may be ignored. All these variables will be considered subsequently.

4. See Willis and Garrod (1991) for an application of this latter approach to both ZTCM and ITCM models.

5. For example, if individual i states that 60% of the day's enjoyment was due to the on-site experience then $E_{ij} = 0.60$.

6. Care has to be taken in the aggregation procedure as data may well have been gathered in the form of household or party visits whereas total annual visitor data is usually held as numbers of individuals. Household data must be converted to individual visit data to avoid underestimation (or, on occasion, double counting).

7. Unfortunately, an ambiguity in the questionnaire for this study made it impossible to determine whether responses represented visitors' perceptions of marginal (petrol only) or average (petrol plus standing charges) cost.

8. This analysis follows McConnell (1975), Wilman (1977), Freeman (1979c) and Johansson (1987).

9. The entrance of priced and non-priced recreation goods into the household production function is considered in Smith and Desvousges (1986).

10. Note that the term $t_{ij}D_{ij}$ (the length of travel time, in hours, from individual i's home (or zone i in ZTCM) to site j) in equation (6.14)) equates to the term TT_{hj} in equation (6.5) (TT_{ij} in the ITCM version of equation (6.5)).

11. A further practical problem arises when attempting to estimate a model which contains both a wage-related time cost variable and an income variable as the two are likely to be highly or perfectly collinear.

12. A reverse scenario could also be plausibly constructed.
13. In addition, Smith and Desvousges (1986) consider the site attribute lake size relative to whole-site size and two measures of water quality. However, for simplicity, these are ignored here.
14. An alternative two-stage approach is to employ factor analysis. However, such an approach has been used more in the context of urban planning and is not pursued here.
15. Smith and Desvousges (1986) actually use a semi-log (dependent variable) form of equation (6.17); however, for simplicity a linear form is reported here. This study used both travel and time costs in its definition of C_{ij}, with the full wage rate being taken as the opportunity cost of time.
16. In estimating a semi-log (dependent variable) version of equation (6.17) for 22 US lakes, Smith and Desvousges (1986) report R^2 values between 0.02 and 0.54 with an unweighted mean of 0.22.
17. Approaches such as cluster analysis and multi-level modelling (Jones 1991) are not discussed here. For further discussion, see Bowes and Loomis (1980; 1982), Christensen and Price (1982) and Vaughan et al. (1982).
18. Interestingly, in a simulated model test of this hypothesis, Connolly and Price (1991) found that curve-fitting errors could more than outweigh the impacts of substitute sites (see discussion of functional form).
19. The impact of improvements in the environmental quality of a single and substitute sites is discussed in Freeman (1979c), Burt and Brewer (1971) and Cicchetti et al. (1976).
20. See, for example, Ziemer et al. (1980); Vaughan et al. (1982); Desvousges et al. (1983); Smith and Desvousges (1986); Hanley (1989); and Benson and Willis (1990).
21. See subsequent comments concerning R^2 figures for ZTCM studies.
22. Tested by the standard Breusch–Pagan test.
23. See Maddala (1977, 265).
24. Wilcoxon (1945); see Mendenhall et al. (1986, 806).
25. Mann and Whitney (1947); see Kazmier and Pohl (1987, 496).
26. Furthermore, zones may either be cut off at some finite distance or the outer band may be infinite. Englin and Mendelsohn (1991), in their study of rainforest tourism, analyse visits from all countries.
27. The author is currently examining this problem.
28. See Hof and King (1982) for further discussion of this assumption.
29. Brown and Mendelsohn (1984) only consider travel time, i.e. they assume that on-site time is a constant. A separate regression (on site time cost) equation would be required if this were not the case.
30. The implicit price tells us the value of a marginal improvement in attribute i.
31. These include income, other socioeconomic variables and the predicted number of trips from each zone, the latter being derived from a standard TCM tgf.
32. Further examples of the HTCM approach include Loomis et al. (1986), Bell and Leeworthy (1990) Bowes and Krutilla (1989), and Hanley and Ruffell (1992). The latter two studies both describe applications to forestry.

33. Air quality has been the main focus for empirical HPM studies – see Anderson and Crocker (1971); Waddell (1974); Pearce (1978); Pearce and Edwards (1979); Freeman (1979a; 1979b); Brookshire *et al.* (1982); Pearce and Markandya (1989); Pennington *et al.* (1990); Turner and Bateman (1990).

34. See also Mäler (1977) and Hufschmidt *et al.* (1983).

35. Ignoring non-environmental explanatory variables, we have the hedonic price function

$$\ln P = \gamma_a \ln Z_a$$

Using the integral log rule:

$$\int \frac{1}{x} \, dx = \ln x + c$$

we can rewrite the function as:

$$\int \frac{\partial P}{P} = \gamma_a \int \frac{\partial Z_a}{Z_a}$$

$$\frac{\partial P}{P} = \gamma_a \frac{\partial Z_a}{Z_a}$$

$$\frac{\partial P}{\partial Z_a} = \gamma_a \frac{P}{Z_a}$$

i.e. the implicit marginal purchase price of air quality (Z_a).

36. Necessary assumptions are that households only buy one housing bundle and that household utility functions are weakly separable.

37. Strictly speaking, this correspondence would also occur if the (hedonic) implicit price function were linear as this would imply a constant marginal implicit price, i.e. there would be no price–quantity relationship. However, empirical studies have found non-linear functional forms to be consistent with data (Freeman 1979c).

38. Although this is unlikely to be the case, some studies have adopted an assumption of identical household inverse demand curves (i.e. identical utility curves) so as to use the marginal implicit price function as a demand curve for the purposes of welfare measure estimation (see, for example, Brown and Pollakowski 1977).

39. In reality, this is also weak as the implicit marginal purchase price of the first unit of air quality is a completely abstract concept and is arguably equal to infinity.

40. See also the excellent discussion in Hufschmidt *et al.* (1983).

41. Freeman (1979c, 156–60) presents and discusses empirical results for some 15 case studies.

42. Note that the inclusion of P_i in the second derivative of the Box–Cox form implies that the implicit price of pollution is also dependent upon the levels of the other non-environmental characteristics in the housing bundle.
43. Current work by the author and others at CSERGE, UEA, is examining the application of GIS to HPM studies.
44. As Garrod and Willis (1992) show, mortgage companies can prove good sources of house price data.
45. Straszheim (1974) showed this to be significant in San Francisco, whereas Nelson (1978b) could not show segmentation to be a significant factor in Washington, DC.

References

Anderson, G. and Bishop, R. (1986). The valuation problem. In Bromley, D. (ed.), *Natural Resource Economics*. Kluwer-Nijhoff, Boston.

Anderson, R.J. and Crocker, T.D. (1971). Air pollution and residential property values. *Urban Studies* 8, 171–80.

Balkan, E. and Kahn, J.R. (1988). The value of changes in deer hunting quality: A travel-cost approach. *Applied Economics* 20, 533–9.

Bateman, I.J. (1993). *Valuation of the Environment: A Survey of Revealed Preference Methods*, CSERGE GEC Working Paper. University of East Anglia, Norwich, and University College London.

Bateman, I.J. *et al.* (1993). Recent experiments in monetary evaluation of the environment. Environmental Appraisal Group, University of East Anglia, forthcoming.

Bell, F. and Leeworthy, V. (1990). Recreational demand by tourists for saltwater beach days. *Journal of Environmental Economics and Management* 18, 189–205.

Benson, J.F. and Willis, K.G. (1990). *The Aggregate Value of Non-Priced Recreation Benefits of the Forestry Commission Estate*. Report to the Forestry Commission. Dept of Town and Country Planning, University of Newcastle, Newcastle.

Bockstael, N.E., Strand, I.E. Jr and Hanemann, W.M. (1984). Time and income constraints in recreation and demand analysis. Dept of Agriculture and Resource Economics, University of Maryland, College Park, Maryland.

Bojö, J. (1985). *A Cost–Benefit Analysis of Forestry in Mountainous Areas: The Case of Valadalen*. Stockholm School of Economics, Sweden.

Bowes, M.D. and Loomis, J.B. (1980). A note on the use of travel cost models with unequal zonal populations. *Land Economics* 56, 465–70.

Bowes, M.D. and Loomis, J.B. (1982). A note on the use of travel cost models with unequal zonal populations: Reply. *Land Economics* 58, 408–10.

Bowes, M. and Krutilla, J. (1989). *Multiple-Use Management: The Economics of Public Forestlands*. Resources for the Future, Washington, DC.

Brookshire, D., Thayer, M., Schulz, W. and d'Arge, R. (1982). Valuing public goods: A comparison of survey and hedonic approaches. *American Economic Review* 72, 165–71.

Brown, G. and Mendelsohn, R. (1984). The hedonic travel cost model. *Review of Economics and Statistics* 66, 427–33.

Brown, G.M. and Pollakowski, H.O. (1977). Economic valuation of shoreline. *Review of Economics and Statistics*, 59, 272–8.

Brown, W.G. and Nawas, F. (1973). Impact of aggregation on the estimation of outdoor recreation demand functions. *American Journal of Agricultural Economics* 55(2), 246–9.

Burt, O.R. and Brewer, D. (1971) Estimation of net social benefits from outdoor recreation. *Econometrica* 39, 813–27.

Button, K. and Pearce, D.W. (1989). Infrastructure restoration as a tool for stimulating urban development. *Urban Studies* 26, 559–71.

Cesario, F.J. (1976). Value of time in recreation benefit studies. *Land Economics* 55, 32–41.

Cesario, F.J. and Knetsch, J.L. (1970). Time bias in recreation benefit estimates. *Water Resources Research* 6, 700–4.

Cesario, F.J. and Knetsch, J.L. (1976). A recreation site demand and benefit estimation model. *Regional Studies* 10, 97–104.

Cheshire, P.C. and Stabler, M.J. (1976). Joint consumption benefits in recreational site 'surplus': An empirical estimate. *Regional Studies* 10, 343–51.

Christensen, J.B. (1983). An economic approach to assessing the value of recreation with special reference to forest areas. PhD thesis, University College of North Wales, Bangor.

Christensen, J.B. and Price, C. (1982). Weighting observations in the derivation of recreation trip demand regression – a comment on Bowes and Loomis. *Land Economics* 58, 395–9.

Cicchetti, C.J., Fisher, A.C. and Smith, V.K. (1976). An econometric evaluation of a generalized consumer surplus measure: The Mineral King controversy. *Econometrica* 44, 1259–76.

Clawson, M. (1959). *Methods of Measuring the Demand for and Value of Outdoor Recreation*. Reprint no.10. Resources for the Future, Washington, DC.

Clawson, M. and Knetsch, J.L. (1966). *Economics of Outdoor Recreation*. Resources for the Future and Johns Hopkins University Press, Baltimore, MD.

Common, M. (1973). A note on the use of the Clawson method. *Regional Studies* 7, 401–6.

Conger, A.J. (1974). A revised definition for suppressor variables: A guide to their identification and interpretation. *Educational and Psychological Measurement* 34, 35–46.

Connolly, D.S. and Price, C. (1991). The Clawson method and site substitution: Hypothesis and model. Manuscript, University College of North Wales, Bangor.

Cummings, R.G., Cox, L.A. and Freeman, A.M. III (1986). General methods for benefit assessment. In Bentkover, J.D., Corello, V.T. and Mumpower, J. (eds), *Benefit Assessment: The State of the Art*. Reidel, Dordrecht.

Desvousges, W.H., Smith, V.K. and McGivney, M.P. (1983). *A Comparison of Alternative Approaches for Estimating Recreation and Related Benefits of Water Quality Improvements*, EPA-230-05-83-001. Office of Policy Analysis, US Environmental Protection Agency, Washington, DC.

Englin, J. and Mendelsohn, R. (1991). A hedonic travel cost analysis for valuation of multiple components of site quality: the recreation value of forest management. *Journal of Environmental Economics and Management* 21, 275–90.

Fisher, A.C. and Krutilla, J.V. (1972). Determination of optimal capacity of resource-based recreation facilities. In Krutilla, J.V. (ed.), *Natural Environments: Studies in Theoretical and Applied Analysis*. Resources for the Future and Johns Hopkins University Press, Baltimore, MD.

Freeman, A.M. III (1979a). The hedonic price approach to measuring demand for neighbourhood characteristics. In Segal, D. (ed.), *The Economics of Neighborhood*. Academic Press, New York.

Freeman, A.M. III (1979b). Hedonic prices, property values and measuring environmental benefits: A survey of the issues. *Scandinavian Journal of Economics* 81(2), 154–73.

Freeman, A.M. III (1979c). *The Benefits of Environmental Improvement: Theory and Practice*. Johns Hopkins University Press, Baltimore, MD.

Garrod, G. and Willis, K.G. (1991). *Some Empirical Estimates of Forestry Amenity Value*, Working Paper 13. Countryside Change Unit, University of Newcastle upon Tyne.

Garrod, G. and Willis, K.G. (1992). The environmental economic impact of woodland: A two stage hedonic price model of the amenity value of forestry in Britain. *Applied Economics*, 24, 715–28.

Gibson, J.G. (1978). Recreation land use. In Pearce, D.W. (ed.), *The Valuation of Social Cost*, Allen & Unwin, London.

Greig, P.J. (1977). Forecasting the demand response to changes in recreational site characteristics. In Elsner, G.H. (ed.), *State of the Art Methods for Research, Planning and Determining the Benefits of Outdoor Recreation*. USDA Forest Service, Washington, DC, pp.11-12.

Gum, R.L. and Martin, W.E. (1975). Problems and solutions in estimating the demand for and value of rural outdoor recreation. *American Journal of Agricultural Economics* 57, 558–66.

Hanley, N.D. (1989). Valuing rural recreation benefits: An empirical comparison of two approaches. *Journal of Agricultural Economics* 40, 361–74.

Hanley, N.D. (1990). *Valuation of Environmental Effects: Final Report – Stage One*. Industry Department of Scotland and the Scottish Development Agency, Edinburgh.

Hanley, N.D. (1991). *Valuation of Environmental Effects*, ESU Research Paper No.22. Industry Department for Scotland/Scottish Development Agency.

Hanley, N.D. and Common, M.S. (1987). *Evaluating the Recreation Wildlife and Landscape Benefits of Forestry: Preliminary Results from a Scottish Study*. Papers in Economics Finance and Investment no.141. University of Stirling, Scotland.

Hanley, N.D. and Ruffell, R. (1992). *The Valuation of Forest Characteristics*, Working Paper 849. Institute for Economic Research, Queens University, Kingston, Ontario.

Harrison, D. Jr and Rubinfeld, D.L. (1978a). Hedonic housing prices and the demand for clean air. *Journal of Environmental Economics and Management* 5, 81–102.

Harrison, D. Jr and Rubinfeld, D.L. (1978b). The distribution of benefits from improvements in urban air quality. *Journal of Environmental Economics and Management* 5, 313–32.

Hof, J.G. and King, D.A. (1982). On the necessity of simultaneous demand estimation. *Land Economics* 58, 547–52.

Hufschmidt, M.M., James, D.E., Meister, A.D., Bower, B.T. and Dixon, J.A. (1983). *Environment, Natural Systems and Development: An Economic Valuation Guide*. Johns Hopkins University Press, Baltimore, MD and London.

Hyman, E.L. (1981). The valuation of extramarket benefits and costs in environmental impact assessment. *Environmental Impact Assessment Review* 2, 227–64.

Johansson, P.-O. (1987). *The Economic Theory and Measurement of Environmental Benefits*. Cambridge University Press, Cambridge.

Jones, K. (1991). *Multi-Level Models for Geographical Research*, CATMOG 54. Institute of British Geographers, Norwich.

Kazmier, L.J. and Pohl, N.F. (1987). *Basic Statistics for Business and Economics*. McGraw-Hill, Singapore.

Kling, C.L. (1987). A simulation approach to comparing multiple site recreation demand models using Chesapeake Bay survey data. *Marine Resource Economics* 4, 95–109.

Kling, C.L. (1988). Comparing welfare estimates of environmental quality changes from recreation demand models. *Journal of Environmental Economics and Management* 15, 331–40.

Loomis, J., Sorg, C. and Donnelly, D. (1986). Economic losses to recreational fisheries, due to small-head power development. *Journal of Environmental Management* 22, 85–94.

Maddala, G.S. (1977). *Econometrics*. McGraw-Hill, New York.

Mäler, K.G. (1977). A note on the use of property values in estimating marginal willingness to pay for environmental quality. *Journal of Environmental Economics and Management* 4, 355–69.

Mann, H.B. and Whitney, R. (1947). On a test of whether one of two random variables is stochastically larger than the other. *Annals of Mathematical Statistics* 18, 50–60.

McConnell, K. (1975). Some problems in estimating the demand for outdoor recreation. *American Journal of Agricultural Economics* 57, 330–4.

McConnell, K.E. and Strand, I. (1981). Measuring the cost of time in recreation demand analysis: An application to sport fishing. *American Journal of Agricultural Economics* 63, 153–6.

McLeod, P.B. (1984). The demand for local amenity: an hedonic price analysis. *Environment and Planning A* 16, 389–400.

Mendenhall, W., Reinmuth, J.E., Beaver, R. and Dunhan, D. (1986). *Statistics for Management and Economics*, 5th edn. Duxbury Press, Boston.

Morey, E.R. (1981). The demand for site-specific recreational activities: A characteristics approach. *Journal of Environmental Economics and Management* 8, 345–71.

Morey, E.R. (1984). Confuser surplus. *American Economic Review* 74, 163–73.

Morey, E.R. (1985). Characteristics, consumer surplus and new activities: A proposed ski area. *Journal of Public Economics* 26, 221–36.

Nelson, J.P. (1977). Accessibility and the value of time in commuting. *Southern Economic Journal* 43, 3, 1321–9.

Nelson, J.P. (1978a). Residential choice, hedonic prices and the demand for urban air quality. *Journal of Urban Economics* 5, 357–69.

Nelson, J.P. (1978b). *Economic Analysis of Transportation Noise Abatement.* Ballinger, Cambridge, MA.

Parson, G.R. (1991). A note on choice of residential location in travel cost demand models. *Land Economics* 67, 360–4.

Pearce, D.W. (ed.) (1978). *The Valuation of Social Cost.* Allen & Unwin, London.

Pearce, D.W. and Edwards, R. (1979). The monetary evaluation of noise nuisance: Implications for noise abatement policy. In O'Riordan, T. and d'Arge, R.C. (eds), *Progress in Resource Management and Environmental Planning*, Vol. 1. Wiley, New York.

Pearce, D.W. and Markandya, A. (1989). *The Benefits of Environmental Policy.* Organisation for Economic Cooperation and Development, Paris.

Pearse, P.H. (1968). A new approach to the evaluation of non-priced recreation resources. *Land Economics* 44, 87–99.

Pennington, G. *et al.* (1990). Aircraft noise and residential property values adjacent to Manchester International Airport. *Journal of Transport Economics and Policy* 24, 49–59.

Prewitt, R.A. (1949). *The Economics of Public Recreation – An Economic Survey of the Monetary Evaluation of Recreation in the National Parks.* National Park Service, Washington, DC.

Price, C. (1978) *Landscape Economics.* Macmillan Press, London.

Price, C. (1979) Interpreting the Clawson demand curve: some philosophical problems in evaluating additional facilities. *Proceedings of IUFRO Meeting: Economics of Recreation*, March 11–15, Washington DC.

Price, C. (1991). Landscape Valuation and Decision Making. School of Agricultural and Forest Sciences, University College of North Wales, Bangor.

Price, C. *et al.* (1986). Elasticities of demand for recreation site and recreation experience. *Environment and Planning A* 18, 1259–63.

Ravenscraft, D.J. and Dwyer, J.F. (1978). *Reflecting Site Attractiveness in Travel Cost-Based Models for Recreation Benefit Estimation*, Forestry Research Report 78-6. Dept of Forestry, University of Illinois at Urbana-Champaign.

Ridker, R.G. (1967). *Economic Costs of Air Pollution: Studies and Measurement.* Praeger, New York.

Rosen, S. (1974). Hedonic prices and implicit markets: Product differentiation in perfect competition. *Journal of Political Economy* 82, 34–55.

Sellar, C., Stoll, J.R. and Chavas, J.-P. (1985). Validation of empirical measures of welfare change: A comparison of non-market techniques. *Land Economics* 61, 156–75.

Sinden, J.A. and Worrell, A.C. (1979). *Unpriced Values: Decisions without Market Prices.* John Wiley, New York.

Smith, R.J. and Kavanagh, N.J. (1969). The measurement of benefits of trout fishing: Preliminary results of a study at Grafham Water, Great Ouse Water Authority, Huntingdonshire. *Journal of Leisure Research* 1, 316–32.

Smith, V.K. (1988). Selection and recreation demand. *American Journal of Agricultural Economics* 70, 29–36.

Smith, V.K. and Desvousges, W.H. (1986). *Measuring Water Quality Benefits.* Kluwer-Nijhoff, Boston.

Stankey, G. (1972). A strategy for the definition and management of wilderness quality. In Krutilla, J.V. (ed.), *Natural Environments: Studies in Theoretical and Applied Analysis*, Resources for the Future and Johns Hopkins University Press, Baltimore, MD.

Straszheim, M. (1974). Hedonic estimation of housing market prices: A further comment. *Review of Economics and Statistics* 56, 404–6.

Talheim, D.R. (1978). A general theory of supply and demand for outdoor recreation, Manuscript, Department of Agricultural Economics, Michigan State University.

Turner, R.K. and Bateman, I.J. (1990). *A Critical Review of Monetary Assessment Methods and Techniques.* Environmental Appraisal Group, University of East Anglia, Norwich.

Vaughan, W.J. and Russell, C.S. (1982). *Freshwater Recreational Fishing: The National Benefits of Water Pollution Control.* Resources for the Future, Washington, DC.

Vaughan, W.J., Russell, C.S. and Hazilla, M. (1982). A note on the use of travel cost models with unequal zonal populations: Comment. *Land Economics* 58, 400–7.

Varian, H.R. (1984). *Microeconomic Analysis.* Norton, New York.

Vaux, H.J. and Williams, N.A. (1977). The costs of congestion and wilderness recreation. *Environmental Management* 1, 495–503.

Waddell, T.E. (1974). *The Economic Damage of Air Pollution*, report no. EPA-600/5-74-012. US Environmental Protection Agency, Washington, DC.

Wilcoxon, F. (1945). Individual comparisons by ranking methods. *Biometrics* 1, 80–3.

Willig, R.D. (1973). *Consumer's Surplus: A Rigorous Cookbook*, Technical Report No.98. Institute for Mathematical Studies in the Social Sciences, Stanford.

Willig, R.D. (1976). Consumer's surplus without apology. *American Economic Review* 66, 589–97.

Willis, K.G. and Benson, J.F. (1988). A comparison of user benefits and costs of nature conservation at three nature reserves. *Regional Studies* 22, 417–28.

Willis, K.G. and Garrod, G. (1991). An individual travel-cost method of evaluating forest recreation. *Journal of Agricultural Economics* 42, 33–42.

Wilman, E.A. (1977). A note on the value of time in recreation. Mimeo referred to in Freeman (1979).

Ziemer, R., Musser, W.N. and Hill, R.C. (1980). Recreational Demand Equations: Functional form and consumer surplus. *American Journal of Agricultural Economics* 62, 136–41.

Part 2

Practice

Chapter 7

Valuing Environmental Assets in Developed Countries

Kenneth G. Willis and John F. Benson

Introduction

This chapter considers the value of environmental assets when they are 'public goods' or externalities: where there is zero opportunity cost of consumption, or where exclusion is not possible by the producer or the consumer. Clearly, it is impossible to review all studies valuing environmental assets, hence this chapter concentrates on four areas: nature conservation; forests; landscape; and green belts. A number of common perspectives have been used to value these types of environmental asset.

First, laws and regulations can be enacted through a political choice framework prohibiting economic activity which threatens the environment. Such an approach either assumes the asset has an infinite value; or places a value implicitly determined by policing costs and fines to enforce the regulations multiplied by the probability of being caught – which often indicates a very small value (Amacher *et al.* 1976). Majority voting rules suggest that only by chance will the correct amount of resources be devoted to environmental protection through a public choice framework (Willis 1980). However, one of the few advantages of political decisions and general taxation to finance environmental protection is the prevention of free-riding by those expressing conservation preferences. On the other hand, analysis of past political decisions on environmental resource allocation has shown how irrational such decisions are in terms of efficiency and distributional impacts. The political choice mechanism smacks too much of intuitive judgement, with little foundation in social science.

Second, assets can be valued through financial opportunity cost, measured in terms of either

(i)   compensation payments for profits forgone, e.g., management agreements under the Wildlife and Countryside Act (WCA) 1981 or payments to farmers in Environmentally Sensitive Areas to farm in a less intense way (Turner and Brooke 1988; Bowers and O'Riordan 1991);

(ii)   the replacement value of the asset, creating an equivalent environmental asset elsewhere;

(iii) the cost of diverting the 'polluting activity', e.g., whether it be a road such as the proposed Winchester bypass to avoid an area of outstanding natural beauty and historical interest, or a reservoir such as Cow Green reservoir in Upper Teesdale (Gregory 1975);

(iv)   the capital cost of site purchase to preserve it into perpetuity, e.g., nature reserves purchased by wildlife trusts;

(v)   negotiating planning gain (see Bowers 1990), e.g., by agreeing to development on part of a site, English Nature may secure a long-term agreement on the remainder owned by a landowner, or secure ownership or a management agreement on another site elsewhere.

None of these approaches explicitly considers the benefits produced by an environmental asset, to set alongside the opportunity costs of conservation.

Third, benefit–cost analysis can be employed, with estimates of the benefits conferred by environmental assets set alongside the opportunity costs (benefits forgone) of some alternative economic activity. For example, nature conservation benefits are typically not priced by the market: wildfowl and other trusts often charge entry fees, and wildlife trusts requiring membership fees exist, but much 'consumption' of wildlife is of an open access type. This means that some mechanism must be used of either inferring values through expenditure on related private goods (e.g., travel costs to gain access to a site), or asking individuals directly through contingent valuation techniques how much they would be willing to pay for the conservation of particular sites and species (see Chapters 5 and 6).

Fourth, value may be determined in terms of sustainable development: a concept of development which meets the needs of the present without compromising the ability of future generations to meet their own needs. Sustainable environmental development requires maximising the net benefits of economic development, subject to maintaining the services and quality of natural resources (Barbier 1989). However, a gallery of definitions exist on the concept (see Pearce *et al.* 1989). An accounting framework would suggest taking account of changes in the stock of an environmental asset, just as conventional accounting takes into account changes in stocks held by a firm in assessing its overall economic position.

This chapter concentrates on studies which have attempted to measure benefits conferred by environmental assets through travel cost methods

(TCM), contingent valuation methods (CVM), and hedonic price methods (HPM) – see Chapter 5 – rather than alternative valuation methods. It views values from a willingness-to-pay (WTP) framework. Beneficiaries are assumed to have to buy property rights to the asset to secure user and non-user benefits, a practice common to many environmental assets; for example, English Nature pays compensation through management agreements to protect SSSIs from intensive agriculture and other activities and to secure access rights to the land; while the RSPB, wildlife trusts and other bodies purchase properties to secure archaeological, historical, or wildlife sites for their members and the general public.

However, these WTP values are contrasted on occasion with those derived from ascribing rights to consumers and determining how much individuals would be willing to accept to relinquish those rights. The whole question of property rights is crucial to determining outcomes in situations where transactions costs and income effects are significant; but not otherwise, in theory (Coase 1960).

Nature conservation

Financial compensation payments for opportunities forgone dominate in valuing nature conservation in Britain, in *ex-ante* appraisals (Benson and Willis 1988) and *ex-post* evaluations (Willis and Benson 1988); although in the USA more emphasis is placed on valuing benefits (see, for example, Walsh 1986). Under the WCA 1981, financial compensation is payable to landowners and/or occupiers, where, to preserve wildlife, agricultural output cannot be expanded. Compensation is based on the difference between the proposed improvement in agriculture and the existing pattern regarded as compatible with wildlife habitats. The net annual profit forgone, i.e. the annual value of compensation payable, is

$$(c - a) - (b - d) - k$$

where

c = extra revenue from agricultural improvements
a = extra variable and operating costs incurred in the improvement
b = revenue forgone from existing agricultural pattern
d = variable and operating costs saved on existing agricultural pattern
k = additional capital expenditure to effect the agricultural improvement (annuitised).

The full financial cost of conservation to English Nature, in addition to the above, would also include administrative costs and legal fees incurred in the process, as well as labour (wardens, etc.) and material (fencing, etc.) costs in maintaining the habitats. Annual payments are also made to farmers under other schemes, for example, to maintain dry stone walls, hedges, traditional buildings, hill flocks, footpaths, etc., under the North York Moors Farm Scheme operated by the North York Moors National Park.

Because of a variety of factors, principally government intervention in agriculture engendering protective tariffs and subsidies, financial prices do not reflect social opportunity costs. There are a number of methods by which output from agricultural land can be revalued on social terms, among which are: comparing domestic and world prices; calculating producer subsidy equivalents (PSEs); and assessing effective protection rates (EPRs) (Willis *et al.* 1988). The first two methods concentrate on output only. In his theory of effective protection, Corden (1966) also included the effective protection on inputs to the final level of production on outputs. Protection can, therefore, be more accurately measured in terms of an EPR:

$$EPR = \frac{VA_m - VA_w}{VA_w}$$

where

VA_m = value added at domestic prices
VA_w = value added at world prices.

The terms on the right-hand side of this equation may be expanded as follows:

$$VA_m = P_j [(1 + T_j) - A_{ij} (1 + T_i)]$$

$$VA_w = P_j (1 - A_{ij})$$

where

P_j = nominal price of commodity j in free trade
A_{ij} = share of input i in cost j at free trade prices
T = nominal tariff (on i or j as the case may be).

The inclusion of inputs is important in valuing the social value of forgone agricultural output as it implies that the inputs would have alternative uses if not devoted to increasing output on SSSI land. Variables such as fertilisers and fuel will have alternative uses; but other inputs such as agricultural labour may not over the short run (because of the need to retrain), and some capital and other fixed costs (e.g., specific farm buildings and field drains) will be immobile over the long run.

Whatever method is adopted, the basic assumption is that the social value of agricultural output lost through nature conservation is the value of the output on the free market, since this is the price which would have to be paid to replace the lost output; and the opportunity cost is this social value minus the social value of inputs, since it is assumed that these could be used elsewhere.

The important questions in this methodology are whether world price is the appropriate marginal social valuation, and what actually happens to the agricultural output produced: whether it can be sold on the world market, enters storage, is destroyed, or is sold below world market price. If the EC's Common Agricultural Policy (CAP) were abandoned, it could be argued that Britain would not adopt a free trade position in agriculture but move to some (intermediate) form of intervention. The empirical problem here is specifying what alternative position society would adopt. If stock could not be sold on the world market, even with export subsidies at current world prices, then the social value of an increase in agricultural output would be less than the world price: indeed the social value would be zero if such output were simply destroyed. If the output were stored instead, then additional storage costs would be incurred, which might be quite high.

Table 7.1 presents some empirical results for three nature reserves and

Table 7.1 Some costs and benefits of nature conservation (£ per ha per year, 1986 prices)

	Derwent Ings	Skipwith Common	Upper Teesdale
Opportunity costs			
financial: WCA 1981	175	181	165
social: without CAP	127	64	97
Benefits			
TCM users (wildlife)	23	65	32
TCM all users	32	80	264
CVM user + non-user	504	2290	440

TCM = travel cost method; CVM = contingent valuation method.

SSSIs, in terms of financial compensation to farmers and the social cost of conservation based upon world price as a measure of the opportunity cost of replacing agricultural output from restrictions on SSSI land (see Willis *et al.* 1988). These results could, therefore, be viewed as upper bounds on the social cost of nature conservation: the actual social cost of lost agricultural output may be less than these estimates. In any case, changes in CAP support for particular commodities over time would require estimates to be continually updated: these results should not be generalised over time or space. Nevertheless, in this example, the social cost of conservation is only a fraction of the financial cost.

The estimated consumer surplus (CS) per visit at the three SSSIs in Table 7.1 was £1.02 at Skipwith Common, £1.15 at Derwent Ings, and £2.29 at Upper Teesdale, based upon a zonal travel cost model (ZTCM) (Willis and Benson 1988). For these sites wildlife benefits to visitors were much less than the financial compensation payable under the WCA 1981, assuming 1986 price support levels for the agricultural products. Benefits also tended to be lower than social opportunity costs of agricultural output forgone. Benefits only began to exceed social and financial costs when CS to all visitors, wildlife as well as general recreation, was enumerated, or when non-use values were also included.

For many nature conservation sites visitors are not specifically encouraged: large numbers of visitors would degrade wildlife interest. Thus for this environmental good, total economic value (TEV) (comprising CS from actual use plus option use value plus existence or preservation value plus bequest value) is perhaps a more relevant concept. Studies estimating TEV use CVMs, since existence and option values are not readily accessible via revealed preference (e.g., TCM) techniques. Studies in the USA have shown that CS is typically only 25–50 per cent of TEV. Indeed, Walsh *et al.* (1984) and Boyle and Bishop (1987) argue that option, existence and bequest values ought to be added to CS from recreation use to determine the TEV of wilderness and endangered species to society: in the absence of information on WTP for preservation values, insufficient land may be allocated to wilderness protection and misleading project or policy decisions made, especially in cases where other competing developments, such as mining in Colorado, may irreversibly degrade natural environments.

The lack of markets for nature conservation has engendered concern about whether respondents can give an accurate WTP valuation in CVM studies. This has prompted the use of simulated markets in the USA for wildlife, by linking payments for licence permits to hunt. Some licences are free, e.g., goose hunting; some are free but limited in availability, e.g., in order to manage bighorn sheep on a sustained yield basis; others have recently experienced a moratorium, e.g., hunting grizzly bears in

Wyoming; or require an annual fee. Respondents (hunters) are familiar with these situations, and, in addition, this framework readily lends itself to the estimation of option price and value. Where conservation closely resembles a public good, referendum (voting) elicitation methods have been advocated, in which a single discrete response is sought to a take-it-or-leave-it type WTP question – that is, a yes/no response is sought to a specific payment which varies randomly between respondents. Such a method is easier for respondents to answer, and simpler to implement in a mail survey. The participant may regard the referendum method as being in accord with some plurality decision rule, with everyone having to abide by it; and since the price is exogenous this may remove some element of strategic bias in the response. Referendum models have been successfully applied by Boyle and Bishop (1987) in valuing the TEV of the bald eagle in Wisconsin and the less well-known striped shiner; and by Bowker and Stoll (1988) in evaluating whooping cranes, a rare species of around 110 birds which migrate between Aransas National Wildlife Refuge in Texas and Wood Buffalo National Park in Canada. However, caution is necessary in applying discrete choice models in CVM studies. The results are sensitive to the functional form of the model adopted, truncation, and the WTP estimator (mean or median) adopted. Of course, revealed preference models (e.g., TCMs) are also sensitive to functional form (Hanley 1989; Willis and Garrod 1991a).

Respondents to CVM questions on preservation values will rationally provide quite different estimates of benefits depending upon the size of conservation area provided (Walsh *et al.* 1984); the degree of environmental damage likely to be sustained (Imber *et al.* 1991); or the probability of encountering wildlife (Brookshire *et al.* 1983). In addition, the amount of information provided to respondents influences WTP values: Samples *et al.* (1986) recorded increases in preservation values for whales after an information film was seen by respondents; and valuations of different species changed with the presence or absence of information on endangered status and the species of animals. However, the issue of how much and what kind of information should be provided to respondents remains unresolved.

Even with fairly complete information many respondents sampled on a random basis across a nation will not be WTP for a specific site, such as small local nature reserves. Respondents may have genuinely zero preferences for many sites, especially those at a distance, quite independent of any strategic bidding considerations.

Most empirical studies reveal a relationship between WTP and income or education (Loomis 1987); or a decrease in WTP with increasing distance from a site (Willis 1990); and this provides a sensible way of aggregating preservation values across the whole population. The decrease in perceived benefits with distance from a site probably reflects an information effect. However, the Kakadu National Park and Conservation Zone study in Australia shows that this relationship does not necessarily hold: Northern Territories residents were WTP much less than other residents in the national sample (Imber *et al.* 1991). Aboriginal cultural values may be an important factor in preservation values nationally; while respondents in the Northern Territories may have taken account of possible financial and personal gains from mining and netted these out of the benefits of environmental preservation. Thus the framing of the CVM question and the scenario can be very instrumental in determining the CVM responses.

Where individuals have rights to a public good, WTA estimates for decrements in the good are required in theory in any cost–benefit analysis (CBA), rather than WTP measures. However, where compensation is not customarily paid to those who experience decrements in natural and environmental amenities, iterative acceptance formats which collect willingness-to-accept (WTA) bids do not appear to collect reliable data. Thus Brookshire *et al.* (1980) suggest that WTA measures be estimated by collecting WTP data and using economic theory to derive WTA values from the relationship:

$$WTA - WTP = \frac{uM^2}{Y}$$

where

Y = income
M = Marshallian consumer surplus
u = price flexibility of income for the good.

Much effort has been devoted to valuing the benefits of nature conservation, but fewer studies report a full CBA on particular site or policy. Hanley and Craig (1991) attempt a CBA using a Krutilla–Fisher model (Porter 1982):

$$NPV_\text{D} = -C + \int_0^\infty D_t\, \mathrm{e}^{-(r+b)t}\, \mathrm{d}t - \int_0^\infty P_t\, \mathrm{e}^{-(r-g)t}\, \mathrm{d}t$$

where

NPV_D = net present value of the development decision
C = capital cost, assumed to occur in year 0
D_t = positive net benefits of the development (revenue minus labour and other annual costs)
P_t = annual preservation benefits.

Both D and P may be growing or falling in current value terms, e.g., due to changing world timber prices, rising incomes and increasing preferences of preservation. These decay and growth rates are defined as b and g, and the social discount rate as r. Plausible values for the terms in the model suggest that there are no efficiency grounds for subsidising further afforestation in the Flow Country of Scotland; and that this rare habitat ought to be preserved.

The loss of wildlife on a site through either development or excessive use may be irreversible: species and habitats may be lost forever. Ecological arguments for preserving wildlife point to the need to preserve a gene bank. Some wild species have proved valuable in developing disease resistant plants and providing new drugs and pharmaceutical products (Dixon and Sherman 1990; *The Economist* 1991). This suggests the need for caution in making irreversible decisions where there is uncertainty about the future: for a risk-neutral decision-maker under uncertainty, the expected value of the benefits from an irreversible decision is less than the value of the benefits under certainty. A benefit–cost ratio of 1.0 as a criterion for a development decision is then inappropriate. Hodge (1984) documents the implications of the effects of different growth rates in preservation values and discount rates, and suggests that a ratio greater than 1 ought to be applied to these development decisions.

Forests

Forestry is a distinct land-use because of its long rotation periods. While cost calculations in agriculture are complicated by government interventions in the market, in forestry there is free world trade and limited intervention, so that simple investment appraisal and resource cost calculations are more straightforward. However, the effects of discounting in long rotations means that much forestry is uneconomic in strict financial terms; while social benefit arguments for import savings are invalid, and benefits in terms of rural job creation are equivocal (HM Treasury 1972; National Audit Office 1986). As well as timber, forests can produce many other benefits (Table 7.2), though most are unmarketed and many are public goods. Recent detailed work has estimated the value of forests for informal recreation (Willis and Benson

Table 7.2   Potential benefits and costs from forests

Benefits	Costs
Timber	Production costs: land, labour, capital, etc.
	Opportunity costs: the alternative output value which the land could produce if not under timber, e.g. crops
Employment	Employment displacement
Security from reliance on imports	
Recreation and sporting	Loss of prior recreation
Landscape/amenity values	Amenity loss/visual intrusion
Wildlife/ecological benefits	Habitat loss/ecological costs
Biodiversity protection	Biodiversity loss
Watershed protection, storage	Overprotection, water shortage
Prevention of soil erosion	Soil acidification, degradation
Removal of pollutants	Concentration of pollutants
Microclimatic regulation	
Shelter	
Fuelwood, berries, nuts, etc.	

Source: modified after Bateman (1991).

1989; Benson and Willis, in press) using the zonal TCM to produce use values in the range of £1.43 to £3.31 per visit, with a weighted mean of £2 (1988 prices), while further analysis using the individual TCM (Willis and Garrod 1991a) estimates lower values (in the range £0.40 to £2.32), though the values in each method are very sensitive to assumptions made about travel costs and the functional form of the model used. However, the figures are all comparable to a number of related studies at nature reserves (Willis and Benson 1988) and other work in forests (Hanley 1989). Earlier studies of recreation in UK forests (see, for example, Grayson *et al.* 1975; Everett 1979; Christensen *et al.* 1985) have shown similar results when comparable assumptions are made. Despite the need for care in the application of the method (see Chapter 6), the results quoted above imply that each visitor would, on average, be prepared to pay an extra £1 to £2 for each (forest) visit over and above his existing costs; typical entry fees at rural sites where charges are made (e.g., National Trust properties, ancient monuments) are of the same order. There is also work from elsewhere in Europe – for example, Merlo and Signorello's (1990) study of forests and mountain areas in Italy – which estimates consumer surplus of the same magnitude. Valuations for other environmental benefits (Table 7.2) are available in part (Bateman 1991), and in aggregate suggest that the environmental benefits would be much greater than the direct financial returns from timber production.

Forestry is one land use where valuations have been used to examine the policy implications of non-marketed benefits. For example, the distribution of forests in the UK is variable (Figure 7.1(a)), and successive governments have supported investment by both the public (Forestry Commission) and private sectors. The distribution of visitors to the Forestry Commission estate is shown in Figure 7.1(b), together with an estimate of the recreational use benefits per hectare based on the travel cost calculations referred to above. In remote areas, the benefits are negligible (£1 per hectare per year), while in lowland areas, as might be expected, the benefits are more significant, typically around £200 per hectare per year and up to £400 in exceptional cases like the New Forest. If it is assumed that demand for recreation will grow in future, Figure 7.1(c) indicates where new forests should be planted (and where redesign and restructuring is a priority) in order to maximise use benefits; that is, in lowland areas, close to large centres of population, with below average forest cover at present. It is significant that new policy initiatives by the Countryside Commission (and Forestry Commission), approved by the British government, for a major new Midlands forest and for several local community forests in the urban fringe (Countryside Commission 1985), are almost exclusively located in this indicative zone. Note, however, that the environmental economic calculation was not an explicit influence on the new policies: CAP budget problems, surpluses in agricultural land, low internal rates of return (IRR) estimated for upland afforestation on marginal land, and other factors, were more important; but the results do show a convergence on a policy which appears to make both political and economic sense. There are a number of caveats, however; in particular, it is essential that any new forests are designed and managed for multiple benefits. Also, the consequences for investment appraisal are not straightforward. Table 7.3 shows the impact of adding recreational use benefits to some typical commercial forest rotations (but ignores any cost impacts of redesign for multiple use). First, note that the IRR is very sensitive to when the benefits are assumed to start; from year 0, the change in the IRR is very significant where the benefits are large, but from year 26 the effect is more modest but still positive. Is it reasonable to assume that even in the best-designed forest, recreational benefits can start so early in a rotation? Second, the table shows the dramatic effect of land costs, especially in lowland areas where land coming out of agriculture is unlikely to be cheap, even after the effect of CAP support is removed. The IRR may therefore still be low and may not pass a test discount rate of 8% or even 6%. Third, a large question remains for private

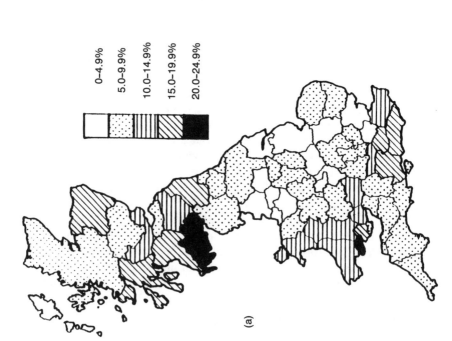

0–4.9%

5.0–9.9%

10.0–14.9%

15.0–19.9%

20.0–24.9%

(a)

Figure 7.1 Spatial implications of informal recreational use for forestry expansion (Forestry Commission estate only)

(a) Woodland area as a percentage of land and inland water area (counties and regions).

(b) Variation in visitor numbers (carborne) and consumer surplus (CS) per hectare per year on Forestry Commission land (forest districts).

(c) Indicative areas for:

(1) new planting: below average woodland cover, highest use values in existing FC forests and high accessibility to large populations;

(2) existing forests: where further development of recreational use could produce high use values, but woodland cover is above average;

(3) other areas: above or below average woodland cover but low to medium use values in existing FC forests

Note: that the maps are generalised from large administrative units and therefore mask local variations and opportunities for new planting

Sources: (a) Locke (1987); (b) Benson and Willis (in press)

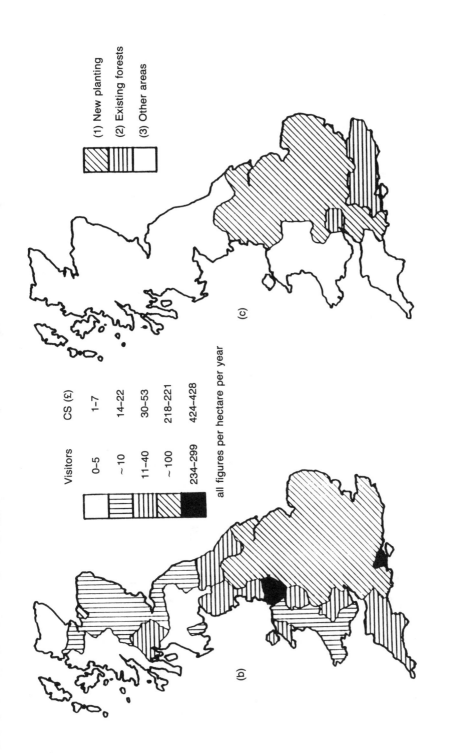

Visitors	CS (£)
0–5	1–7
~10	14–22
11–40	30–53
~100	218–221
234–299	424–428

all figures per hectare per year

(b)

(c)

(1) New planting
(2) Existing forests
(3) Other areas

Table 7.3 Some discounted (6%) cash flow forest rotation models including recreational benefits and land costs

	1. Lowland High CS (£220)	*	2. Lowland Low CS (£50)	*	3. Lowland Very high CS (£424)	*	4. Upland Low CS (£3)	*	5. Upland Low CS (£30)	*
Without land costs										
Wood										
NPV	−156	−156	320	320	−1618	−1618	−719	−719	−719	−719
IRR	5.7	5.7	6.7	6.7	1.5	1.5	3.8	3.8	3.8	3.8
Wood + Rec 26										
NPV	302	491	475	540	22	917	−711	−707	−638	−603
IRR	6.6	6.9	7.0	7.1	6.0	7.2	3.9	3.9	4.1	4.3
Wood + Rec 16										
NPV	794	1091	638	739	1325	2506	−702	−696	−551	−496
IRR	7.5	8.0	7.3	7.5	8.4	9.6	3.9	3.9	4.4	4.5
Wood + Rec 0										
NPV	2507	2918	1206	1343	5867	7346	−669	−661	−218	−140
IRR	18.9	20.6	9.2	9.6	107.7	111.4	4.0	4.0	5.2	5.5
With land costs										
Wood										
NPV	−2604	−2604	−2128	−2128	−4066	−4066	−1138	−1138	−1138	−1138
IRR	3.2	3.2	3.8	3.8	0.8	0.8	3.2	3.2	3.2	3.2

Table 7.3 contd

	1. Lowland High CS (£220)	*	2. Lowland Low CS (£50)	*	3. Lowland Very high CS (£424)	*	4. Upland Low CS (£3)	*	5. Upland Low CS (£30)	*
Wood + Rec 26										
NPV	−2146	−1957	−1973	−1908	−2426	−1531	−1130	−1127	−1058	−1023
IRR	3.9	4.1	4.0	4.1	3.9	4.9	3.2	3.2	3.5	3.6
Wood + Rec 16										
NPV	−1654	−1357	−1810	−1709	−1123	58	−1122	−1116	−971	−916
IRR	4.3	4.7	4.2	4.3	5.0	6.0	3.2	3.3	3.7	3.8
Wood + Rec 0										
NPV	60	470	−1242	−1105	3419	4898	−1089	−1081	−637	−560
IRR	6.1	6.6	4.6	4.8	12.5	13.8	3.3	3.3	4.3	4.5

See text for further explanation; Rec 0, 16, 26 means that recreation benefits are assumed to begin in those years after planting and continue throughout the rotation; * in these columns indicate that recreation benefits increase at 1% per annum compound.

CS = consumer surplus for recreation. The models are (1) Corsican pine YC14; (2) Douglas Fir YC15; (3) Oak YC6; (4), (5) Sitka Spruce YC12. NPV = net present value; and IRR = internal rate of return.

sector forests where the use benefits, if generated by open access, cannot easily be recovered by the landowner.

These use benefits (for recreation), measured using the TCM, aggregate a range of forest attributes: the value to people of wildlife, landscape, peace and quiet and other environmental features of the forest (Benson, in press). Non-use (or preservation) values are also expected to be significant and positive. For example, Oosterhuis and van der Linden (1987) estimated, using CVM, that Dutch households are on average willing to pay approximately 21 guilders per month to preserve forests and heaths (1987 prices). With around 6 million households and over 300,000 ha of forest and heath in the Netherlands, the figures represent non-use benefits of approximately 5000 guilders (£1500) per hectare per year, compared to the use-benefits calculated for the UK Forestry Commission estate (approximately 1,000,000 ha) of £47 per hectare per year on average. In North America, Walsh *et al.* (1990) have estimated the values of national forests in Colorado using CVM. WTP per resident household was $47 per year in total, but the researchers were able to disaggregate this into an allocation of $13 for recreation use value (27.4%), $10 for option value (21.9%), $10 for existence value (21.1%) and $14 for bequest value (29.6%). With over 1 million households in the state, the aggregate benefit estimate is $55.7 million, but only approximately $10 per hectare per year for the national forests in Colorado. These comparisons show significant scale effects, but the fact that the direct use benefits may represent under 30% of the total public valuation is significant. There is a range of evidence to confirm this rather typical result that the non-use or preservation values of many natural resources are as large as or larger than the current use values. Robust figures for UK forests are not available yet, but the effect of doubling the non-marketed benefits attributed to new forests in the rotations shown in Table 7.3 would be very significant, especially if the benefits accrue from the first year, and assuming, of course, that the design and management of the forest is truly multipurpose and in accordance with people's wishes.

Landscape

There has been a large body of work on the evaluation of landscape quality, often involving ordinal approaches, despite a common view that it is difficult or impossible to say anything quantitative about such a subjective topic as natural or scenic beauty, landscape and amenity, or however the idea is expressed. There has been much work on valuation (Price 1978). Part of the problem lies in the difficulty of defining a unit for consumption or measurement, for landscape strictly has no boundaries, only clines of change. There is therefore a serious problem, whether using HPM, TCM or CVM or other techniques, in defining the quantum of landscape, or the specific change to be valued. Possible routes to valuation depend on how the landscape is experienced. One of the earliest applications of the CVM (Randall *et al.* 1974) examined aesthetic environmental improvements occurring with the abatement of air pollution associated with power plant and mining development in a scenic area of New Mexico, while Rowe *et al.* (1980) used photographs in a study of the visibility and impact of a geothermal power plant in the Four Corners region of south-western USA. Other work in North America has also examined the effects of air quality on visibility (see, for example, Brookshire *et al.* 1976; Thayer 1981), although whether this is better regarded as valuing the deleterious effect of pollution rather than the value of landscape change may be debatable.

Inhabited landscape quality, where people live or spend their working lives, may or may not be a reason for choice of location. HPMs have been used to value a number of environmental attributes or use benefits (and non-use benefits?) associated with land or property; for example, air quality (Brookshire *et al.* 1982), aircraft noise levels (Nelson 1980), neighbourhood parks (More *et al.* 1988) and groups of trees (Anderson and Cordell 1988); only the latter are likely to be regarded strictly as part of the landscape of an area. A recent application in a rural area of the UK (Garrod and Willis 1991) suggests that a positive effect on house prices can be attributed to several countryside characteristics, including proximity to a canal, river or woodland, but with a negative effect attributable to views over urban areas and proximity to areas of wetland. This evidence begins to show the value of several factors, some of which would be included within the generic idea of landscape.

If the landscape is visited, specifically for recreational or leisure purposes, then the TCM can be used. One major difficulty with the method (see Chapter 6) is multipurpose trips: given a valuation for the whole trip, on what basis can this be disaggregated between visits to particular sites, the pleasure of driving, the enjoyment of specific viewpoints and the overall quality of the landscape in which the trip takes place? For visits to forests, Benson (in press) invited respondents to value the attributes of the site, defined as wildlife, landscape, active recreation, informal recreation and interpretation facilities, which allowed a disaggregation of the overall consumer surplus. However, this would pose considerable difficulties in the case of a very diverse trip, involving the definition and overlap of attributes, as well as synergistic and substitution effects. Combinations of TCM and CVM are likely to be useful. In many circumstances, the landscape is incidental to the main purpose of the trip, whether for work, pleasure or other domestic purposes. In such cases the difficulties of valuation are more formidable.

The CVM is perhaps a more appropriate tool to value landscapes, and appears to have more to offer than TCMs in this area, in being better able to distinguish more sharply the aesthetic dimensions of a policy-induced change. The CVM can also value the landscape in its entirety, taking into account the whole bundle of varying attributes in a spatial area.

Landscapes can alter over time as a consequence of economic demands, technological innovation and policy changes in agriculture. Willis and Garrod (1991b) assessed the preferences and values of different agricultural landscapes which might arise in the future in the Yorkshire Dales National Park, as a consequence of policy changes in agricultural support subsidies. The landscapes assessed comprised images of a range of possible future agricultural landscapes: today's landscape; abandoned; semi-intensive agriculture; intensive agriculture; planned; conserved; sporting; and wild landscapes. Pictorial displays can be used with the CVM. These permit an existing landscape to be modified in a montage according to likely impacts of policy changes. Such a procedure was adopted in the Dales study. The results of the study revealed some differences between visitors' and residents' preferences for particular attributes in the landscape. However, a majority of both visitors and residents of the Dales preferred today's landscape, although the conserved landscape was also valued highly. A comparison of the public costs of maintaining each landscape, with the respective benefits of each landscape estimated by the CVM, indicated that more public expenditure should be devoted to protecting and enhancing environmental attributes such as dry stone walls and stone barns, wild flowers and hay meadows, and small broadleaved woodland. Methodological tests on the CVM underpinning the study suggested that the results are reliable and robust.

Landscape has option, existence and bequest values, in addition to consumer surplus from existing use. Walsh *et al.* (1984) used the CVM to value wilderness in Colorado, and disaggregated households' total WTP into separate use and non-use valuations, as well as examining the influence of a wide range of variables on the results. Evidence suggests that preservation values are likely to be as large as use values.

Progress in valuing landscape faces both conceptual and methodological challenges. Conceptually a better understanding is required of landscapes and how to characterise them, together with better methods of measuring landscapes and their attributes. Relatively little research in CVM has explored WTP aggregation over attributes. Part–whole bias is one issue here, but there are many other aggregation problems to be explored.

Green belts

Green belts analysed from an Alonso bid-rent model have been described
as failing both tests of microeconomic policy and being inefficient and
inequitable (Evans 1973). Green belts can also be analysed from an
environmental economics framework, particularly in relation to the
outputs they produce:

(i)   amenity and aesthetic benefits: a more pleasant environment and
        the maintenance of the character of towns and villages by preven-
        ting them from merging;
(ii)   avoidance of the diseconomies of scale of urban growth;
(iii)   agricultural output from green belt land;
(iv)   recreational benefits where there is public access to the land.

Green belt policy has the effect of causing resources in the economy
to be used in a different way than would ensue in their absence. The
difference between these two positions, the net social benefit (NSB),
therefore represents a measure of the value or cost of the policy. The
NSB comprises the difference between the value of factors under a green
belt policy and their social opportunity cost:

$$NSB = \sum_{t=0}^{T} \frac{\sum_{j=1}^{n} Q_{tj}(P_{tj} - P_{tj}^*)}{(1 + i)^t}$$

where

Q_{tj} = quantity of factor j used at time t
P_{tj} = value of factor or good j at time t under green belt policy
P_{tj}^* = social opportunity cost of factor or good j at time t
T = time period of the policy
i = discount or time preference rate.

This section concentrates on the amenity of green belts in the NSB
calculation, as being the most important benefit produced by green belts.
Amenity value can be estimated by HPM or CVM. The HPM assesses
how much people are prepared to spend to live near a green belt. A
considerable amount of research has focused on ways in which amenities
and disamenities such as neighbourhood parks (Weicher and Zerbst
1973), lakes and reservoirs (Knetsch 1964; Darling 1973), air pollution
(Ridker and Henning 1967), and aircraft noise (Walters 1975), have
affected urban property values. In these models housing is taken to

represent a bundle of services comprising structural characteristics (e.g., number of bedrooms, garage), locational advantages (e.g., in relation to town centre, schools) and amenity (e.g., nearness to green belt). If green belts are a public good conferring benefits, one would expect the price of neighbouring houses to include a value reflecting this green belt externality *ceteris paribus*, given the competitive nature of house purchase and assignment of properties in accordance with maximum utility conditions.

Both Correll *et al.* (1978) and Wabe (1970) estimated the effect of green belts on neighbouring property values. In the Correll study in Boulder, Colorado, distance from the green belt was a continuous variable, in the Wabe study a dummy variable. Regression results revealed that proximity to a green belt had a statistically significant positive impact on the price of residential property. In Boulder the effect appeared to be dramatic, with price decreasing by $13.78 for every 1 metre increase in distance from the green belt; thus, holding other variables constant, the average value of properties adjacent to the green belt was 32% higher than that of properties only 1 km away.

After standardising for structural characteristics of the house, its location (time and price of reaching central London), Wabe estimated that for sites within 2 miles of London's green belt, the average plot value increased by £276 (at 1968 prices) or 4.9 per cent of the price of the house.

The values derived by Wabe can be used to estimate the value of green belt land from assumptions about development around the green belt, density of development, and area of London's green belt (Willis and Whitby 1985). This produced an amenity value of £2275 per hectare. Given some extreme assumptions (unlikely to hold in practice) an upper bound estimate of amenity value was calculated to be £62,368 per hectare. This environmental value was much less than the alternative use (housing) value of green belt land. Indeed the average land plot value of the house in Wabe's study itself implied a development value of £97,050 per hectare at 1968 prices, with 30 houses per hectare. Thus one might reasonably conclude, assuming for the moment that amenity is the only output attached to green belts, that the benefits of London's green belt land are less than its opportunity cost, at the present scale and configuration of designation.

The CVM can also be used to value green belt land, in terms of both compensating and equivalent variation methods: WTA the loss of the green belt, and WTP to avert the loss. The payment instruments used in the past have included property taxes paid to the authority, which is also responsible for green belt designation. A CVM study on the Tyneside green belt produced plausible results: no respondent suggested the green belt had infinite amenity or aesthetic value; nor did any respondent suggest that 100% reduction in taxes was required as compensation for its loss. The maximum reduction expressed or desired was 60%; 26% of respondents said they would not require any compensation for the loss of the green belt; and for all but two households WTA was greater than WTP. The average WTA value was £104 per house, i.e. a 22 per cent reduction in property taxes; and for WTP £35. Again such values can be aggregated over all households and the value of green belt land estimated.

There is, however, the basic question of which set of property rights should be used to base the value. WTA values are much more likely to approach the opportunity cost of green belt designation (the alternative value as housing land). WTA has been accepted as the principle for valuing existing possessions: things are valued at the minimum people are prepared to accept as just compensation for their loss (Commission on the Third London Airport 1970; Department of the Environment 1974). But amenity rights are not recognised in English planning law, not subject to compensation, and planning authorities would probably argue that designating a green belt does not confer rights to neighbouring residents.

Environmental management

The benefits which environments confer can be valued with varying degrees of accuracy. How such information is used in environmental management decisions and whether better outcomes emerge is another issue. Some of these issues are discussed in different countries in Europe by Barde and Pearce (1991). Environmental management is a large field in its own right, so this section will only briefly touch on three issues in the use of environmental valuations.

Even though accurate environmental benefits and costs have been estimated, the practical incorporation of these in decision aids such as CBA may be biased. It is generally recognised that analysts and decision-makers should not have a prior commitment to the outcome of a CBA. However, Bowers (1988) has eloquently demonstrated that in land drainage projects the set of rules and conventions used in CBA to value costs and benefits and the administrative framework in which appraisals are carried out are intrinsically intertwined. Consequently, pressures exist to interpret the economic rules and modify them in ways favourable to land drainage. Similar institutional capture of Environmental Impact Assessments (EIAs) has also been detected.

There may be a failure of intervention in the environment. Turner (1991a; 1991b) cites a number of examples of intervention failure through intersector policy inconsistencies. For example, even though the benefits of wetlands have been recognised by one arm of government, the British Ministry of Agriculture, Fisheries and Food continued to subsidise farming which led to their destruction. Similarly, both public and private forestry received grants and favourable tax treatments which resulted in afforestation of globally scarce blanket bog in Scotland, for a marginal 1 per cent return on forestry investment. In addition, degrading the peatland resulted in a net increase in carbon in the atmosphere (loss from peat is greater than that fixed from trees; Adger *et al.* 1991); at the same time the Department of the Environment was incurring public expenditure to implement costly regulation elsewhere to reduce global warming.

The sectoral division of responsibilities and the value derived from economic goods might also lead to a sub-optimal management of the resource. For example, in the USA green belt case cited earlier, Correll *et al.* (1978) point out that although additional taxes, as a consequence of increased property values because of the green belt, would greatly exceed the cost of green belt land purchase, 86 per cent of the additional property tax would accrue to units of local government other than the city, the unit of government responsible for establishing the green belt. Thus a major policy and environmental management issue is the extent to which the concentration of cost to one authority and the dispersal of tax benefits to other units of government will result in the non-optimal provision of an environmental resource.

Conclusions

A multitude of applications of CVM, HPM, TCM and other methods now exist, valuing a variety of environmental assets, many more than can be mentioned in this short text (for a recent survey, see Navrud 1992). The methods now in widespread use provide reasonable estimates of the value of specific environmental assets: comparable ball-park estimates with one another and with similar environmental goods where these are marketed or an entrance charge is made.

However, environmental economics is far from placing a single value on a particular environmental asset, or a value correct to the last penny. The greater understanding of the theoretical issues underlying environmental economic valuation (WTP, WTA, expected, option, and existence values and risk preferences, etc.), and the technicalities of the methods employed to obtain a value, have led to a greater understanding of how values can vary. The way questionnaires are framed, the nature of the sample, and the functional forms of the models can all affect the results. Indeed, since TCM studies now run into hundreds in the USA, meta-analysis is beginning to explore how results very depending upon the assumptions in the models (Walsh *et al.* 1989; Smith and Kaoru 1990). In this way further insights are being provided into the factors influencing environmental value estimates.

The fact that precise values are not derivable should not come as any great surprise. The tremendous stock market financial crashes of the 1970s and 1980s, the fluctuation in the value of firms even though their output and revenue remained constant, and the significant rise and fall in house prices in the late 1980s in Britain emphasise the conditional nature of price. Asset values are conditional, depending upon institutions, demand, supply and expectations about these conditions. Price at any one point in time reflects all the information available at that point in time. Price is conditional upon information unique to that point in time, which is unlikely to be replicated. So in a sense each environmental valuation is a unique experiment which cannot be exactly replicated. Thus CVM, HPM and TCM values are reflections of information perceived by individuals, usually unknown to the researcher. Hence, over time values may vary. But it is important to remember that this value is infinitely better than ignoring the value of benefits in making decisions.

It is heartening to observe that environmental economics and benefit estimates are now being used much more by government in Britain in environmental management decisions (Markandya *et al.* 1991). Its acceptance by regulatory agencies, government and the courts is still far short of its standing in the USA. But its increasing acceptance also places pressure on environmental economists to be concerned much more about the accuracy of their valuations when people take notice of them, to reflect on their results, as well as leading to a greater maturity of the discipline.

Finally, distributional issues rarely receive much comment in environmental economics, despite the early observation by Baumol and Oates (1975) that environmental regulation may be regressive, benefiting the rich and penalising the poor. The whole distributional issue of who gains and who loses in the provision and subsidisation of specific environmental commodities deserves a much higher research profile.

References

Adger, N., Brown, K., Sheil, R. and Whitby, M. (1991) *Dynamics of Land Use Change and the Carbon Balance*, Countryside Change Working Paper Series, WP15. Department of Agricultural Economics and Food Marketing, University of Newcastle upon Tyne.

Amacher, R.C., Tollison, R.D. and Willett, T.D. (1976). The economics of fatal mistakes: fiscal mechanisms for preserving endangered species. In R.C.

Amacher, R.D. Tollison, and T.D. Willett (eds), *The Economic Approach to Public Policy*. Cornell University Press, Ithaca, NY.

Anderson, L.M. and Cordell, H.K. (1988). Influence of trees on residential property values in Athens, Georgia (USA): a survey based on actual sales prices. *Landscape and Urban Planning* 15, 153–64.

Barbier, E. (1989). *Economics, Natural Resources, Scarcity and Development*. Earthscan, London.

Barde, J.-P. and Pearce, D.W. (eds) (1991). *Valuing the Environment: Six Case Studies*. Earthscan, London.

Bateman, I. (1991). Placing money values on the unpriced benefits of forestry. *Quarterly Journal of Forestry* 85, 152–65.

Baumol, W.J. and Oates, W.E. (1975). *The Theory of Environmental Policy*. Prentice Hall, Englewood Cliffs, NJ.

Benson, J.F. (in press). Public values for environmental features in commercial forests. *Quarterly Journal of Forestry*.

Benson, J.F. and Willis, K.G. (1988). Conservation costs, agricultural intensification and the Wildlife and Countryside Act 1981: a case study and simulation on Skipwith Common, North Yorkshire, England. *Biological Conservation* 44, 157–78.

Benson, J.F. and Willis, K.G. (in press). Valuing informal recreation on the Forestry Commission estate. *Forest Bulletin*.

Bowers, J.K. (1988) Cost–benefit analysis in theory and practice: Agricultural land drainage projects. In R.K. Turner (ed.), *Sustainable Environmental Management: Principles and Practice*. Belhaven, London.

Bowers, J.K. (1990). *The Economics of Planning Gain: a re-appraisal*, Working Paper No.2, ESRC/NCC Project on Planning Gain: a strategy for conservation?, Department of Geography, University of Bristol.

Bowers, J.K. and O'Riordan, T. (1991). Changing landscapes and land-use patterns and the quality of the rural environment in the United Kingdom. In M.D. Young (ed.), *Towards Sustainable Agricultural Development*. Belhaven Press, London.

Bowker, J.M. and Stoll, J.R. (1988). Use of dichotomous choice nonmarket methods to value the whooping crane resource. *American Journal of Agricultural Economics* 70, 372–81.

Boyle, K.J. and Bishop, R.C. (1987). Valuing wildlife in benefit–cost analysis: a case study involving endangered species. *Water Resources Research* 23, 943–50.

Brookshire, D.S., Ives, B.C. and Schulze, W.D. (1976). The value of aesthetic preferences. *Journal of Environmental Economics and Management* 3, 325–46.

Brookshire, D.S., Randall, A. and Stoll, J.R. (1980). Valuing increments and decrements in natural resource service flows. *American Journal of Agricultural Economics* 62, 478–88.

Brookshire, D.S., Thayer, M., Schulze, W. and d'Arge, R. (1982). Valuing public goods: a comparison of survey and hedonic approaches. *American Economic Review* 72, 165–77.

Brookshire, D.S., Eubanks, L.S. and Randall, A. (1983). Estimating option and existence values for wildlife resources. *Land Economics* 59, 1–15.

Christensen, J.B., Humphreys, S.K. and Price, C. (1985). A revised Clawson method: one part solution to multidimensional disaggregation problems in recreation evaluation. *Journal of Environmental Management* 20, 333–46.

Coase, R. (1960). The problem of social cost. *Journal of Law and Economics* 3, 1–44.

Commission on the Third London Airport (1970). *Papers and Proceedings Vol 7 (Parts 1 and 2) – Stage 3 Research and investigation – Assessment of Shortlisted Sites* (Roskill Commission). HMSO, London.

Corden, W.M. (1966). The structure of a tariff system and the effective protection rate. *Journal of Political Economy* 74, 221–37.

Correll, M.R., Lillydahl, J.H. and Singell, L.D. (1978). The effects of green belts on residential property values: some findings on the political economy of open space. *Land Economics* 54, 207–17.

Countryside Commission (1985). *Forestry in the Countryside*, Countryside Commission Publication 245. Cheltenham.

Darling, A.H. (1973). Measuring benefits generated by urban water parks. *Land Economics* 49, 22–34.

Department of the Environment (1974). *Environmental Evaluation: The Cost–Benefit Approach*. R. Travers Morgan and Partners, Urban Motorways Project Team, Report Paper No.1. DoE, London.

Dixon, J.A. and Sherman, P.B. (1990). *Economics of Protected Areas: A New Look at Benefits and Costs*. Earthscan, London.

The Economist (1991) Loving yew. *The Economist*, 9 February, p.51.

Evans, A.W. (1973). *The Economics of Residential Location*. Macmillan, London.

Everett, R.D. (1979). The monetary value of the recreation benefits of wildlife. *Journal of Environmental Management* 8, 203–13.

Garrod, G.D. and Willis, K.G. (1991). *The Hedonic Price Method and the Valuation of Countryside Characteristics*, Countryside Change Working Paper Series, WP14. Department of Agricultural Economics and Food Marketing, University of Newcastle upon Tyne.

Grayson, A.J., Sidaway, R.M. and Thompson, F.P. (1975). Some aspects of recreation planning in the Forestry Commission. In G.A.C. Searle (ed.), *Recreational Economics and Analysis*. Longman, London.

Gregory, R. (1975). The Cow Green Reservoir. In P.J. Smith (ed.), *The Politics of Physical Resources*. Penguin, Harmondsworth.

Hanley, N.D. (1989). Valuing rural recreation benefits: an empirical comparison of two approaches. *Journal of Agricultural Economics* 40, 361–74.

Hanley, N.D. and Craig, S. (1991). The economic value of wilderness areas: an application of the Krutilla–Fisher model to Scotland's 'Flow Country'. Department of Economics, Stirling University.

HM Treasury (1972) *Forestry in Great Britain: An Interdepartmental Cost Benefit Study*. HMSO, London.

Hodge, I. (1984). Uncertainty, irreversibility and the loss of agricultural land. *Journal of Agricultural Economics* 35, 191–202.

Imber, D., Stevenson, G. and Wilks, L. (1991). *A Contingent Valuation Survey of the Kakadu Conservation Zone*, Research Paper No.3. Resource Assessment Commission, Canberra.

Knetsch, J.L. (1964). The influence of reservoir projects on land values. *Journal of Farm Economics* 46, 231–43.

Locke, G.M.L. (1987). Census of Woodland and Trees 1979–82. *Forest Bulletin* 63.

Loomis, J.B. (1987). Expanding contingent value sample estimates to aggregate benefit estimates: Current practices and proposed solutions. *Land Economics* 63, 396–402.

Markandya, A., Pearce, D.W. and Turner, R.K. (1991). United Kingdom. In J.-P. Barde and D.W. Pearce (eds), *Valuing the Environment: Six Case Studies.* Earthscan, London.

Merlo, M. and Signorello, G. (1990). Alternative estimates of outdoor recreation benefits: some empirical applications of travel cost method and contingent valuation. Paper presented to the International Conference Environmental Cooperation and Policy in the Single European Market, Venice.

More, T.A., Stevens, T. and Allen, P.G. (1988). Valuation of urban parks. *Landscape and Urban Planning* 15, 139–52.

National Audit Office (1986). *Review of Forestry Commission Objectives and Achievements.* Report by the Comptroller and Auditor General. HMSO, London.

Navrud, S. (ed.) (1992). *Pricing the European Environment.* Oxford University Press, Oxford.

Nelson, J.P. (1980) Airports and property values: a survey of recent evidence. *Journal of Transport Economics and Policy* 14, 37–52.

Oosterhuis, F.H. and van der Linden, J.W. (1987) Benefits of preventing damage to Dutch forests: An application to the contingent valuation method. Paper presented to the International Conference Environmental Policy in a Market Economy, Wageningen.

Pearce, D., Markandya, A. and Barbier, E.B. (1989). *Blueprint for a Green Economy.* Earthscan, London.

Porter, R.C. (1982). The new approach to wilderness preservation through cost–benefit analysis. *Journal of Environmental Economics and Management* 9, 59–80.

Price, C. (1978). *Landscape Economics.* Methuen, London.

Randall, A., Ives, B. and Eastman, C. (1974). Bidding games for valuation of aesthetic environmental preferences. *Journal of Environmental Economics and Management* 1, 132–49.

Ridker, R.G. and Henning, J.A. (1967). Determinants of residential property values with special reference to air pollution. *Review of Economics and Statistics* 49, 246–57.

Rowe, R.D., d'Arge, R.C. and Brookshire, D.S. (1980). An experiment on the economic value of visibility. *Journal of Environmental Economics and Management* 7, 1–19.

Samples, K.C., Dixon, J.A. and Gowan, M.M. (1986). Information disclosure and endangered species. *Land Economics* 62, 306–12.

Smith, V.K. and Kaoru, Y. (1990). Signals or noise? Explaining the variation in recreation benefit estimates. *American Journal of Agricultural Economics* 72, 419–33.

Thayer, M.A. (1981) Contingent valuation techniques for assessing environmental impacts: further evidence. *Journal of Environmental Economics and Management* 8, 27–44.

Turner, R.K. (1991a). Economics and wetland management. *Ambio* 20, 59–63.

Turner, R.K. (1991b). The United Kingdom. In R.K. Turner and T. Jones (eds), *Wetlands: Market and Intervention Failures*. Earthscan, London.

Turner, R.K. and Brooke, J. (1988). Management and valuation of an Environmentally Sensitive Area: Norfolk Broadland, England. Case study. *Environmental Management* 12, 193–207.

Wabe, J.S. (1970). *A Study of House Prices as a Means of Establishing the Value of Journey Time, the Rate of Time Preference and the Valuation of Some Aspects of the Environment in the London Metropolitan Region*. Research Paper 11. Department of Economics, University of Warwick.

Walsh, R.G. (1986). *Recreation Economic Decisions*. Venture Publishing, Boulder, CO.

Walsh, R.G., Loomis, J.B. and Gillman, R.A. (1984). Valuing option, existence and bequest demands for wilderness. *Land Economics* 60, 14–29.

Walsh, R.G., Johnson, D.M. and McKean, J.R. (1989). Issues in non-market valuation and policy application: a retrospective glance. *Western Journal of Agricultural Economics* 14, 178–88.

Walsh, R.G., Bjonback, R.D., Aiken, R.A. and Rosenthal, D.H. (1990). Estimating the public benefits of protecting forest quality. *Journal of Environmental Management* 30, 175–89.

Walters, A.A. (1975). *Noise and Prices*. Clarendon Press, Oxford.

Weicher, J.C. and Zerbst, R.H. (1973). The externalities of neighbourhood parks: an empirical investigation. *Land Economics* 49, 99–105.

Willis, K.G. (1980). *The Economics of Town and Country Planning*. Collins (Granada), London.

Willis, K.G. (1990). Valuing non-market wildlife commodities: an evaluation and comparison of benefits and costs. *Applied Economics* 22, 13–30.

Willis, K.G. and Benson, J.F. (1988). A comparison of user benefits and costs of nature conservation at three nature reserves. *Regional Studies* 22, 417–28.

Willis, K.G. and Benson, J.F. (1989). Recreational values of forests. *Forestry* 62, 93–110.

Willis, K.G. and Garrod, G.D. (1991a). An individual travel-cost method of evaluating forest recreation. *Journal of Agricultural Economics* 42, 33–42.

Willis, K.G. and Garrod, G.D. (1991b). *Landscape Values: A Contingent Valuation Approach and Case Study of the Yorkshire Dales National Park*, Countryside Change Working Paper Series WP21. Department of Agricultural Economics and Food Marketing, University of Newcastle upon Tyne.

Willis, K.G. and Whitby, M.C. (1985). The value of green belt land. *Journal of Rural Studies* 1, 147–62.

Willis, K.G., Benson, J.F. and Saunders, C.M. (1988). The impact of agricultural policy on the costs of nature conservation. *Land Economics* 64, 147–57.

Chapter 8

'Heritage Landscapes': A New Approach to the Preservation of Semi-Natural Landscapes in Canada and the United States

Norman Henderson

Introduction

This paper is ideological, it is descriptive, and it is prescriptive. The *ideological* component runs throughout its content, though it shows most strongly in the conclusion. This conclusion emphasizes the role of protected nature conservation landscapes *not* as preservationist islands of refuge for man, nor as storehouses of natural biota, nor as global engines essential to the maintenance of world ecostability, but rather as educational tools whose main objective should be the altering of man's behaviour *outside* of conservation landscapes. This interpretation is not to everyone's taste.

The paper's *descriptive* component dominates the two sections which follow this introduction. These provide a brief description of the inadequacies of the current North American (here meaning Canadian and American) approach to nature conservation and the possibilities of expanding the remit of nature conservation beyond publicly controlled wilderness landscapes and onto privately controlled semi-natural ones.

This descriptive component is followed by three sections which are *prescriptive*. The first two of these prescribe what the objectives of nature conservation should be on semi-natural landscapes in North America; they introduce the concept of a *Heritage Landscape*. The third examines a study site on the Canadian grasslands just north of the American frontier to see how the Heritage Landscape approach could work in practice. Some detailed management recommendations result from applying a Heritage Landscape philosophy to the study site, the Alberta-Saskatchewan Cypress Hills.

Readers may reasonably disagree with my ideological standpoint. They may not accept my description of current nature conservation efforts on semi-natural lands in North America as wholly inadequate. Or they may find fault with the prescriptions for the Cypress Hill study site. But I do expect them to be challenged to look at North America's vast semi-natural landscapes in a new light.

The need for a new nature conservation approach in settled landscapes

Right at the start it must be said that even to recognize the need for the preservation of semi-natural landscapes in North America implies a radical shift in nature conservation thought. Traditionally, North American nature conservation has focused on the preservation of wilderness landscapes. Such landscapes, the national parks, constitute the well-known 'crown jewels' of conservation in North America. The wilderness emphasis is historically explicable and for the regions of the continent where wilderness landscapes are still to be found it is sensible – even if a careful scrutiny of the underlying philosophy behind wilderness preservation thought reveals some logical inconsistencies.[1] But a wilderness emphasis fails to address those extensive areas in North America where sometimes almost entire ecosystems have passed largely into private hands. Such areas are subject to strong development and population pressures.

In western Canada, for example, most of the grassland ecoprovince has been appropriated (with the full encouragement of governments) by private landowners for agricultural production. Here the original vegetation of the tall-grass and medium-grass ecosystems has been almost totally extinguished. Significant remnants of the short-grass ecosystem alone have survived, thanks only to their low agricultural utility. The traditional conservation strategy of protecting some of those short-grass remnants by removing them from private ownership and creating a new park reserve has proven extremely difficult on an occupied landscape. This is exemplified by the protracted efforts made by the federal government to create a national park in the short-grass country of southern Saskatchewan. Although relatively few landholders are involved it has still proven extremely difficult for the government to acquire land title to lands within the future park. Expropriation is politically not feasible and paying ranchers sometimes inflated asking prices can be too expensive. The federal government must also compensate the provincial government for the loss of subsurface mineral rights, as mining or oil extraction are not allowed in a national park. Although the legal status of a Grasslands National Park appears finally to have been ensured, it is estimated that it will take up to 40 years before actual land assembly is completed (Mondor and Kun 1984). Trying to restore ploughed fields to something like native prairie (i.e. grassland prior to European impacts) conditions could take even longer (Jim Masyk, pers. comm. 1989).[2] With specific reference to the 20-year struggle for Grasslands, Alan Appleby, Assistant Deputy Minister of the Saskatchewan Department of Parks, commented:

another agreement has just been signed between Canada and Saskatchewan to establish Grasslands National Park. It will probably be the last major park area to be designated in southern Saskatchewan in our lifetimes, so that alternative means for landscape preservation need to be explored. (pers. comm. 1988)

The difficulties of acquisition as a strategy are also recognized in the United States. According to the Conservation Foundation, based in Washington, DC, 'an increasingly settled nation will need to find imaginative ways to protect far more land than it will ever be possible to buy'. The Foundation notes:

In many of America's special places . . . massive federal land acquisition would be inappropriate or unacceptable, not only because of its cost but also because of its disruption of landowners and their communities. Only [non-purchase protective techniques] . . . hold promise of accommodating and protecting these 'living landscapes', shaped in part by the continuing choices of private residents.

The present problem status of some privately controlled semi-natural landscapes of nature conservation interest in Canada and the United States is similar to situations long encountered in Europe. Given increasing populations and increasing demands on resources, parts of Canada and the United States will continue to approach more and more closely the dominant European situation.

Agriculture's key role

As farmers (or ranchers) are the major private landholders in North America, affecting farming practice is the greatest single opportunity for nature conservation on private lands. Farmers, particularly those engaged in rearing stock, are the key group controlling semi-natural landscapes with high conservation value.

Marginal agricultural land is usually of higher nature-conservation and recreation value than prime agricultural land. The latter is usually farmed intensively and its suitability for wildlife declines accordingly. Marginal land may often have received less intensive agricultural use to begin with, so that a relatively natural environment may be extant, or be fairly easily reproducible. Conveniently, compensatory costs of reducing agricultural production or restricting agricultural management in ways favourable to nature conservation are much less on marginal than on prime agricultural land. Many intensively farmed prime agricultural landscapes undoubtedly also have high inherent potential nature-conservation value, but the opportunity costs of their conversion would be high.

A web of government regulation and subsidy has so altered the agricultural economic environment that only complex and contentious economic modelling experiments can suggest what farming (and the results for nature conservation) would be like in a non-interventionist, non-protectionist, subsidy-free world. However, the results of government price support and other subsidies in North America and elsewhere have been fairly clear: increasingly intensive farming and the cultivation or grazing of areas which it would be uneconomical to exploit without support – see, for Britain, Body (1984) and Bowers and Cheshire (1983); for Canada, Henderson (1987a). This has clearly been to the detriment of nature conservation. In some cases (typically on agriculturally marginal land) it might actually be cheaper for governments to pay farmers to produce less and to do this in ways that benefit nature conservation, than to pay the usual production subsidies.

There exists in North America a widespread conviction that farmers manage their land to the general benefit and satisfaction of society as a whole. Deeply ingrained in many people is a related belief in a kind of divine agrarian right that empowers farmers to do as they see fit with their land. These two beliefs are, naturally enough, carefully nurtured by agricultural organizations and by farmers' representatives in government ministries of agriculture. Yet in North America agricultural history has been characterized by soil exploitation, exhaustion and abandonment from early European settlement onwards (see, for example, Seymour and Gidardet 1986). The presumption that farmers will by nature necessarily be a positive landscape force is false. But most politicians and policy-makers have adapted to two realities of modern agriculture: first, that it is politically unproductive to criticize agricultural landholders; and second, that the broad societal will to effect major environmental policy changes which would enhance nature conservation does not exist.

Farmers are more convinced even than the public that they are good stewards of the land. Appreciation of this conviction is critical when dealing with landholders. The difficulties involved are well illuminated by Appleby (pers. comm. 1989), again describing the Grasslands National Park case:

the ranchers want to see the park, and yet do not want to give up ranching . . . they have been stewards of the land for as long as they can remember . . . They see it as it *is*, not as it was, covered in bison, or as it might be, a prairie wilderness. They think they've done a marvellous job of maintaining the natural prairie . . . and they see fewer changes needed to make a park acceptable to them than the park professionals do.

This attitude is characteristic of landowner thinking. In a soil-conservation sense even intensive crop farmers can be good managers, but promoting nature conservation is another question entirely. Certainly monoculture grain production has almost no nature-conservation value. On the other hand, the ranchers described by Appleby undoubtedly *have* preserved many elements of the original ecosystem through their maintenance of native range. The key management problem in such landscapes is how to guarantee that the natural elements do not disappear in future and, ideally, to reintroduce some of the natural elements that have been eliminated.

The role of a 'Heritage Landscape'

Let us apply the name 'Heritage Landscape' to a region of semi-natural ecology and high conservation value. Such a landscape includes some settlement, is typically privately controlled, and shows significant human impacts (i.e. it is not a wilderness area). A Heritage Landscape could serve the following nature-conservation functions:

Greenbelt protection

A Heritage Landscape could act as a belt of protection around a more highly controlled jurisdiction. The Conservation Foundation (1985) reported that in 1980 more than half of the reported threats to American national parks originated largely or wholly outside park boundaries. A Heritage Landscape might be valuable for gaining some control over some of these threats – although some problems, such as air pollution, which may have its source hundreds of kilometres away, cannot be so dealt with.

It may in certain cases be possible to extend positive protection to particular habitats or species which extend outside of park boundaries and into the surrounding landscape. This is particularly important in the case of large mammals which require extensive ranges to be viable. Problems with 'depredation' by animals moving and feeding on either side of the boundary between a park and private land are well known. Biogeography island theory has demonstrated that the larger a reserve area is, the greater the number of species that can survive in it (Scanlon 1981). Problems vary among individual parks and reserves, but a

Heritage Landscape might offer one way to extend the range for specific species.

Protection of representative environments

A Heritage Landscape could focus more strongly on everyday plants, animals and landscapes than do existing reserve designations in North America, which often seem to emphasize the unusual, the unique or the bizarre more than 'representative' environments. This bias to the unusual has been belatedly recognized. Government park system plans now emphasize the protection of representative environments, but however much they are reclassified after the fact we cannot alter the fact that almost all existing national parks have been originally selected and designated on singularity grounds.

Heritage Landscapes could help to redress the bias to the unusual. It is, in fact, accepted by many people that not only rare species and spectacular environments are worthy of protection, but also everyday nature. Many Canadians and Americans enjoy seeing deer, beaver, lilies or puffball mushrooms in the general landscape, but it is only the occasional connoisseur that finds pleasure in discovering a rare western painted turtle or a lady's slipper orchid. Conservation focused on rarity is vulnerable to charges of elitism. It is also intrinsically a bad strategy since species often become rare through the destruction of their requisite habitat, by which time the real conservation battle (holistic ecosystem protection) has effectively been lost. Heritage Landscapes offer a way of extending some protection to more of our valuable, everyday environments.

Education

The most important objective of Heritage Landscape should be education – the education of people as to their responsibilities to the land. Fred Halliday, past chairman of the British Nature Conservancy Council, comments: 'To be effective nature conservation policy must call upon a wide range of approaches and methods . . . it entails conserving species and their habitats, but also influencing people's attitudes and actions and capturing their interest and imagination' (Mabey 1980, 12).

No type of conservation designation in itself guarantees protection. Individual reserves of various types will continue to be created and delisted dependent on future competing land-use demands and valuations. David Lowenthal (1990) is correct when he concludes: 'Social habit and attitudes are more important for preserving landscape than designation.' Heritage Landscapes, by virtue of being privately controlled and inhabited systems, can contribute to fighting a fundamental error in our approach to nature – our tendency to ghettoize it. We do this whenever we zone off one area of landscape (primarily one of those increasingly few areas we have discarded as of not significant alternative economic use) for the protection and/or promotion of species other than ourselves.[3] This division perpetuates modern man's dangerous and relatively recent view of himself as something apart from and independent of the rest of the natural world (Mumford 1967).

The mere existence of public conservation reserves has been used to justify a lack of concern for natural values in privately controlled landscapes, personal responsibility having been abdicated to the state. Richard Westmacott (1983) describes farmers' attitudes in the United States: 'few farmers would take steps to improve [their farms] for . . . wildlife. The immense national parks, state parks, wilderness areas and other federal lands have encouraged an attitude among farmers that these are the appropriate places for wildlife conservation, and not the farm'. In Britain, Brian Goodey (1990) describes Sites of Special Scientific Interest as 'pernicious lines on the map. If they exist in your proposed development site you prepare yourself for the planning commission accordingly; otherwise you are free to ignore conservation concerns.' It would not be an exaggeration to say that current nature-conservation reserve designations have completely failed to instil a sense of responsibility in individuals making their day-to-day land-management decisions.

Postulating a need to expand an appreciation of nature-conservation values onto private lands where they are not currently considered relevant is not a new idea. It was already well expressed by Aldo Leopold in many of his writings as far back as the 1940s (see, for example, Leopold 1949). But we have not progressed. Leopold's land ethic has failed to develop; the abuse of land has not become socially contemptible as he had hoped it would. James Shaw (1985, 272) provides a powerful summary of the current North American nature conservation situation:

the fact that his writings of long ago ring true today is less a tribute to Leopold's visionary abilities than it is an indictment of the lack of progress in wildlife conservation. Very little has changed especially on private lands where habitat conditions have grown steadily worse.

There is even now no widely accepted land ethic. Land is still regarded as mere property, and property rights, including rights to abuse and destroy, remain sacred.

Many conservation professionals have failed to understand the reason why we continue to shuffle our feet at the gates of the preservation road: the general public's undervaluation of the importance of nature conservation. This undervaluation occurs because nature-conservation objectives are often long-term (genetic diversity and habitat preservation, for example) and threats complex and often apparently intractable (for example, acid rain, oceanic pollution or soil degradation). Norman Moore (1987) compares nature conservation against agriculture and medicine. These latter fields have goals which are ancient, obvious, often immediate and generally understandable. But that which is difficult to understand is typically underrated in importance – nature conservation's fate to date. Meeting this perceptual challenge is the most important task facing conservationists everywhere. If we want to preserve species, habitats and biotic diversity we *must* first educate.

The fundamental objectives of a Heritage Landscape

To fulfil its role as described in the previous section, a Heritage Landscape must satisfy four objectives:

The philosophical and educational objective

Canadians and Americans traditionally are inclined to take a polarized stance to conservation issues. This results from operating within a wilderness-versus-development conceptualization of natural values (Henderson, 1992). In *some* ecoregions of North America we need to be able usefully to reconcile nature-conservation objectives with existing human impacts. We need to recognize that such areas exist, recognize their nature-conservation and cultural value, and begin to preserve ecological components within an existing and legitimate cultural framework. A Heritage Landscape must be a conceptual tool to aid in the adaptation of our philosophical approach to nature conservation in non-wilderness areas.

The cultural objective

A Heritage Landscape must aim to maintain valued traditional cultural values. This is desirable for two reasons: first, the lifestyle of the inhabitants and the landscape that results from their way of life have a symbiotic attractiveness and importance to Canadians or Americans that legitimizes supporting the cultural heritage; and second, these inhabitants will be, at least potentially, the natural allies of most nature conservation goals in the Heritage Landscape. The second point should hold in any Heritage Landscape where residents have evolved some kind of stable working relationship with their immediate natural world.

The aesthetic objective

Preservation of aesthetic landscape integrity is perhaps the vaguest of the four basic objectives. Taste and aesthetics vary among people and change over time. It is not inconceivable, for example, that under certain conditions oil pumpjacks (today viewed as negative landscape intrusions) one day could be viewed as positive components of some cultural landscape. But this proves only that a variety of opinions need to be considered in arriving at aesthetic judgements and that aesthetic views of a cultural landscape will evolve over time just as landscapes themselves do. The importance of aesthetic factors to people is not denied by the fact tastes vary and change.

The physiographic and ecological objective

The ecological objective for a Heritage Landscape must centre on the maintenance of the vegetative component of the landscape in as 'natural' a state as possible. A useful mental aid to this end is to view natural and altered environments as two ends of a spectrum. For practical purposes most major policy questions are clear at least in the direction of movement on the nature versus controlled environment spectrum that possible resolutions of a given management issue imply.

The protection of the fauna component of a Heritage Landscape's ecology should parallel the preservation of native vegetation. All existing wildlife species should be seen as legitimate. Significant interference with natural population fluctuations should be avoided.

Normally the impacts of human settlement and activity will be more evident on the ecology of an area than on its basic physiography, but specific sites (such as easily erodible badlands, deserts, or cliff walls) sometimes may need special attention. But aside from the case of extreme environments ecological protection generally implies physiographic stability as well.

The Heritage Landscape in practice: an interpretation for a Plains grassland test site

Why a site in the grasslands?

Potential Heritage Landscapes can be found throughout settled North America. To apply the theory of Heritage Landscapes in practice, I have chosen a study site on the North American grasslands. One reason for this choice is that the Great Plains represent the largest ecoregion in North America dominated by private control. And, as we have already seen, they are resistant to traditional landscape protection through the accumulation of land and the designation of a park reserve by government.

There is another important and more subtle reason why the western grasslands are an appropriate place to abandon classic North American wilderness thought: no one has any definitive answer to the question of what the grasslands were as ecosystems prior to the arrival of the Europeans. Nor are we ever likely to know, as Jan Looman (1983, 182) states in summarizing a long discussion on this question: 'In how far natural grasslands in their present floristic composition resemble the vegetation prior to man's [native Americans'] influence cannot be determined.' The wilderness conservation paradigm that may still be applicable to boreal forest, tundra, or other natural environments is inapplicable to the grasslands. The existing 'natural grassland' or 'native prairie' is a derivative of human activity dating back thousands of years to the widely presumed use of fire by native Americans.[4]

The grasslands may well not have been in any state of equilibrium even prior to first European contact. Since contact, man's influences have obviously increased (the suppression of fire, the extirpation of larger grazers and predators, the introduction of domestic cattle and controlled grazing regimes, fencing, roads and controlled hunting), but the effects of even these relatively recent impacts are not always clear.

Why the Cypress Hills site?

It was desirable that the Plains study site be under some 'threat', that is to say, that environmentally undesirable developments might occur in the foreseeable future. The 'threat' attribute both makes the case study of more immediate practical applicability and can serve to point out any inadequacies of the status quo approach. On the other hand, the existence of a single, overwhelming threat would be less than ideal; the intention of the study was not to focus attention too narrowly on a single management issue. Among various potential study areas the Alberta-Saskatchewan Cypress Hills offered a good balance of environmental threats. Oil and gas exploration and development, new road construction, long-term overgrazing, new land breakage, and expanded tourism and development were all topical issues, yet no single concern was paramount.

Although this is not clear to all residents or resource managers, the

Cypress Hills are far from a wilderness landscape and are thus a suitable candidate for Heritage Landscape philosophy. It is possible to maintain that the present relationship between forest (present on the highest parts) and grassland in the Hills is the result of their development over a relatively short time period since the most recent glacial ice-sheet retreat and that the present vegetation 'has matured under the influence of a combined ranching–agriculture–recreation economy. The extent and significance of this influence will probably never be known' (Scace 1972, 229). Opinions can be ventured as to man's historical impacts, as Scace points out, but even hypotheses about man's effects during the European historical era often must remain at best educated guesses. The uncertainty surrounding pre-man environmental conditions in the Hills has direct parallels in much of Europe. In Britain, for example, one can make only educated guesses about the pre-man environment over most of the island. In such situations the value of using theoretical antecedent 'pristine' ecosystems as reference points for modern ecosystem management is badly handicapped.

The Cypress Hills: a brief discussion

For a much fuller exposition of Cypress Hills climate, physiography, flora and fauna, pre-European environment and post-European settlement history than will be given here the reader is referred to Henderson (1991).

The Hills cover an area of about 4500 square kilometres straddling the Alberta–Saskatchewan border and centred about 70 km north of the Montana boundary. Technically they are the eroded remnants of a strongly dissected plateau which is elevated some 600 m above the surrounding plain. The climate is continental, sub-humid on the higher slopes and becoming increasingly drier with descent to the surrounding semi-arid plains. On the humid north- and east-facing slopes of the higher hills Lodgepole Pine, White Spruce and Aspen Poplar are the dominant vegetation. Rough Fescue grass dominates in slightly drier areas and short-grass prairie associations characterize still drier lower slopes.

Compared with the surrounding short-grass plains, the Hills are rich in fauna diversity and numbers. Wolf, bison, grizzly bear and black bear were all historically present in the Hills but extirpated by white settlers. Elk, moose, pronghorn, whitetailed deer and mule deer are still found in relatively high concentrations. Smaller mammals have generally survived settlement well. Bird and insect densities and diversity are also high in what has been termed 'a plains oasis'.

The Cypress Hills are dominated by a single land use, ranching. Land tenure is not predominantly private, or 'freehold' as local terminology puts it, even if largely controlled by private operators. A typical western Canadian rancher owns part of 'his' land outright but more often than not the majority of his holding is on a long-term (in Saskatchewan typically 33-year) agricultural lease from the relevant provincial government. Historically the leases have been such long-term affairs and tenure so secure that *de facto* ownership rights may in some senses be imputed to the lessee. Certainly the lessees themselves typically think of leased land as their own, to do with as they wish. But the provinces do retain title to the land and they have reserved to themselves ultimate development and disposition control. However, the leverage available to government as to how these lands are to be managed (ultimately including the refusal to renew a lease), may only be exercised at peril owing to the political danger in agrarian provinces of appearing to harass agriculturalists.

A long-standing problem in the Hills is the issue of wildlife management. In the Cypress Hills most of the wooded top land is administered as provincial park land, with strong game management objectives. Unfortunately, elk and deer ranges encompass both the parks and parts of the surrounding privately operated ranchland.

The main reason why the Hills have retained high conservation interest is the *combination* of semi-humid forested uplands and surrounding grasslands. The forested uplands are largely government-administered in the form of provincial parks and the federal Fort Walsh National Historic Park, while the grasslands are largely privately owned or leased. The key feature of these grasslands is that they are largely 'native prairie', i.e. they have not been ploughed and sown to an exotic forage crop like Russian wild rye or crested wheat grass. Essentially it is this balance of native range and wooded toplands, together with their associated fauna, that any nature conservation designation should seek to ensure and promote.

I interviewed individually about 50 Hills ranchers as well as government resource managers in the Hills. Landholders notably failed to note the ecological and landscape changes wrought by agricultural settlement and consistently underplayed their role in changing the Hills. Extirpations of species, the fencing of the open range, road construction, the suppression of fire, and the introduction of controlled grazing regimes with exotic grazers are major environmental changes (to name only the most obvious), but they are not perceived as such by ranchers. Both Hills ranchers and government resource managers have difficulty in seeing the Hills for what they are: a heavily man-modified semi-natural ecosystem.

Most of the major changes noted above that were wrought by or as a consequence of the ranching community in the Hills occurred in the first 50 years of settlement. The ranching lifestyle now seems well entrenched and environmentally stable. A rough equilibrium between ranchers and the land has been established in this latter half of the first century of European occupation.

Although ranchers are generally conservative land managers, my enquiry found that younger ranchers and landholders with smaller landholdings are more likely to pursue innovative or aggressive land management, such as breaking new land. The likeliest practitioners of conservative traditional land management would appear to be the older ranchers with larger spreads. This group is generally under less financial pressure, is more risk-averse, and is able to pursue satisficing behaviour at lower levels of stock production per unit area. Older ranchers and rangers with larger holdings are therefore the natural allies of conservation in the area.[5]

The misapplication of 'Eastern' conceptions of environmental controls

Prior to European settlement many of the environmental controls in the Hills took the form of 'macro-events', cyclical but unpredictable phenomena of great, often catastrophic, impact. Fire is an obvious example, completely changing the landscape in a few hours. Buffalo and associated grazers likely made a similarly dramatic impact through irregular flood grazing. Historical locust plagues or the current infestation of pine beetles in the lodgepole pine forests are other examples of macro-events. Severe drought is still another natural, recurring, but essentially unpredictable phenomenon that shapes the Hills. The harsh Hills environment differs radically, and in principle, from the relatively benign, constant and predictable natural environments of north-western Europe, or the dominant hardwood ecoprovince of heavily populated eastern North America. Catastrophic events do occur in these latter environments too, of course, but as environmental controls they are there the exception rather than the rule. In these 'benign' environments environmental change most often proceeds via micro-events (such as the fall of a tree and gradual succession processes). This presents a sharp contrast to the situation in the Hills.

A resident in the Hills environment is someone perpetually 'living on the edge'. This condition is not something European settlers seem to accept or enjoy on its own terms.[6] Everywhere on the North American Plains they have struggled to impose constancy and predictability on the environment. Since European settlement of the Hills fires have been stopped, water supplies dammed and regulated, grazing stabilized, and insect infestations controlled. Even relatively minor and somewhat predictable cyclical fluctuations, such as the natural expansion and contraction of wildlife populations, are not accepted. Feeding stations are maintained to prevent elk and deer population crashes in severe winter, while hunting pressure is increased if populations are 'too high'. Those environmental risk factors that defy effective human control may be downplayed or suppressed subconsciously, as Thomas Saarinen (1966) reports in his studies of Plains farmers who, he found, tended to underestimate the risks of drought.

Many human manipulations in the Hills increase the eventual severity of macro-events or, if they succeed in controlling one macro-event, substitute the risk of a different macro-event variant. For example, the beetle infestation of lodgepole pine in the Hills forests results directly from the suppression of regenerative fires. In the grasslands substituting native range associations with exotics increases the susceptibility of pasture to mismanagement and particularly to drought damage.[7]

The failure to appreciate the macro nature of ecosystem controls in the Hills is a direct result of an inappropriate approach to the local environment. The misunderstanding underlying this approach has its roots in the cultural baggage brought by European immigrants via eastern North America. Efforts to squeeze out the cyclicality of natural systems on the semi-arid grasslands, and the downplaying of the risks and consequences of interference, create management problems in the Hills as fast as they appear to be solved.[8]

An interpretation of the Heritage Landscape for the Hills

This subsection details what the objectives of a Heritage Landscape would be for the specific case of the Cypress Hills.

Education

Residents and resource managers need to understand better the extent of past environmental manipulation and the implications for the future of the Hills environment. It is not generally understood that:

(i) Natural controls in the Hills are by their nature often catastrophic.
(ii) Efforts to control this seem a natural (European) human response.
(iii) The partially controlled environment that follows from (ii) inevitably deviates greatly from any 'natural' state.
(iv) Control efforts can only be partially successful. Macro-events remain possible, and sometimes more likely and of increased

eventual intensity as time passes. New macro-event variants may be a direct result of man's interference.

Culture

The cultural objective of a Hills Heritage Landscape should focus on promoting the traditional values of the Hills. These are mostly associated with the dominant economic activity of the past 100 years, ranching. Efforts should be made to identify physical cultural elements of note in the Hills that contribute to heritage quality. Examples include tepee rings, abandoned homestead sites, and some instances of rancher settlement architecture. Any especially noteworthy examples of the above should be formally recognized. The development of a heritage ranch site, along the lines of the existing Fort Walsh National Historic Park, could be useful as a representative both of local history and development and of modern life and practice. Local minority groups, too, particularly residents of the area's only Indian reserve, should be given input into the interpretation of the Hills.

Aesthetics

An interpretation of the aesthetic objective for the Hills is fairly straightforward, if difficult to implement. Most landscape threats to the Hills are visual. In particular, the danger exists that the vistas and the impression of wide open space that characterize the Hills could be impaired by roads, fences or gas wells. But aural and olfactory integrity are important landscape components too. The Hills should not be the site of screeching oil well pumpjacks, or stinking gas flare stations.

Landscape policy should centre on discouraging the provision of new access. Access is the 'soft underbelly' by which the landscape can be undermined by subsequent undesirable use. Not only should governments not undertake road improvements that increase effective access, but also the gas industry should be limited and restricted so far as is politically possible. New landscape intrusions such as transmission lines, industry, gas and oil development, or excessive fencing should be discouraged.

Realistically, economic development pressures make landscape protection difficult, especially as it relates to gas exploration and development. But new developers can be obliged to make their impacts on the landscape as sympathetic as possible. For that oil and gas development which is unavoidable policy should demand the sensitive design of access roads and their effective blockage and post-exploration reclamation, including reseeding with an appropriate native seed mix. Drilling pad sites should also be reclaimed. Permanent service access roads should be absolutely minimized. Control of usage during exploration activity is also important, since driving heavy equipment over roads softened by rain can erode land severely.

Ecology

Of the four Heritage Landscape objectives the physiographic and ecological objective is the most complex to interpolate for the Hills. But the key is simple enough: to maintain the ecological integrity of the Hills policy must centre on discouraging the breaking of new land. The amount of broken land in the higher Hills currently is low at less than 10% (Saskatchewan Department of Tourism and Renewable Resources 1976), an amount easily acceptable in a cultural landscape.[9]

Many of the driving economic forces encouraging the breaking of new land are national or provincial in nature. Key programmes that indirectly or directly encourage breaking include the Wheat Board quota system, crop insurance, and various provincial cultivation incentives (Henderson 1991). All agricultural programmes should be redesigned so as to be at worst cultivation-neutral. Politically and economically this is an enormous task and, realistically, unachievable in the short term; it is none the less a crucial element for nature-conservation policy and must be pursued.

Some land systems need more government attention than others. In the North American Plains the level of public awareness and concern is much higher for wetlands than for grasslands. This is almost entirely because wetlands produce a commercially valuable by-product, game wildfowl. Because of the size and wealth of the hunting lobby millions of dollars have been raised privately for 'wetland conservation' (in truth, for waterfowl production) on the Plains. An argument could therefore be made that native range grassland should have priority over wetlands for government conservation attention, since the former lacks comparable commercial attraction for private groups. In practice, the very lack of effective interest-group concern for threatened privately controlled environments on the grasslands other than wetlands has encouraged a government focus on wetland preservation.

One positive step would be to free native rangeland from land taxation. Indeed, most desirable would be a system that eliminates taxation on any 'natural' lands. In the Canadian prairie provinces, for example, northern forest, some parkland belt areas, badlands, wetlands, and native rangeland might all qualify as exempt from land tax under an appropriate programme. Acceptable criteria defining the term 'natural' for each ecosystem would need to be devised. That appropriate programmes can be practical is clearly demonstrated by the Minnesota property tax credit programme for wetlands and native grasslands (Peterson and Madsen 1981).

Keeping native range 'native' through judicious control of fertilizer and pesticide use is a more subtle issue than discouraging breaking. Moderate applications of fertilizer probably do not pose a problem if the intent is the retention of native range or its regeneration after overgrazing. The key test of fertilization is whether it would change the basic species components of native range. Eventually it might be necessary to calculate desirable limits to fertilization (as is done, for example, in designated Environmentally Sensitive Areas in Britain). For pesticide use the bottom-line criterion should again be whether or not native range flora and fauna would be significantly damaged by pesticide application (although obviously the intent of a pesticide is severely to impair one or more species!). Initial presumption should be against pesticide use, although exceptions are conceivable where pesticides could be used as an aid to restoring native range.

To restore damaged range areas native seed mixtures should be promoted by government. The availability of native range seed mixtures is not well known among ranchers, or even government agricultural experts. Consideration should be given to subsidizing native range seeding.

Protection needs to be given to the minority flora environments in the Hills as well as to the majority grassland components. Minority environments include the wooded top lands and coulées and the wetland vegetation complexes in the sloughs and ponds of the rolling interplateaux and lower slope country. Bush clearing should be constrained, especially in creek bottoms prone to erosion and of high wildlife habitat value. Constraining bush clearing is not so clear-cut a policy issue as constraining breaking of native range, as it may be demonstrated that bush has greatly expanded since the elimination of fire. But, so long as there is doubt as to its pre-European prevalence and role, our preference should be to retain this stabilizing and minority natural element in the Hills.

Research has not determined with certainty the modern-day effects of fire in the general North American landscape. The only certainty about fire's historical role in the Hills is that it was important. Culturally the use of fire is not traditional in the Hills among ranchers, though it was among American Indians. Government should encourage study and experimentation that would help to clarify fire's past and current effects on Hills forest, bush and grassland. In the meantime government should maintain a neutral stance towards fire in the Hills.

Native fauna should be protected holistically. The mainstream approach, which tries to stabilize and maximize commercial game production, is not acceptable in a Heritage Landscape whose objective is to preserve ecosystems for their own sakes.[10] Lure crops and wildlife feeding stations are therefore not preferred options to deal with wildlife depredation as they are an increase in interference with the natural feeding hierarchy. Ranchers who suffer depredation should be compensated or, in the case of feed stacks, supplied with protective fencing.

Major work is needed to establish the carrying capacities of the Hills as a whole for deer, elk, antelope, moose, bison, cattle and horses, and the competitive interrelationships between these species. Ranchers must be major participants in the necessary studies, as it is important they accept the results as credible. Waterton Biosphere Reserve in Alberta offers some useful examples of cattle–wildlife interaction studies (Lieff 1985; 1987). Decisions about desirable cattle numbers in the Hills should partly reflect wildlife numbers.

Horses and sheep were both major elements of the early Hills cultural landscape. Either one would probably be acceptable additions to the current cattle-dominated landscape. Sheep are not economically viable owing to adverse climate and cheap foreign imports. Free-running horses already contribute to the atmosphere of Fort Walsh National Historic Park, and having horses as well as cattle in the provincial parks could contribute greatly to the heritage feel of the Hills landscape. Possible competitive effects on wildlife would have to be considered if a significant number of horses were introduced.

Although desirable, the reintroduction of wolf and bear, spectacular and emotive animals, would not be practical in the Hills owing to conflicts with the ranching business.[11] But it may be feasible to run some bison in parts of the provincial parks. This would definitely be a positive contribution to the Heritage Landscape.

General cattle grazing policy should be the promotion of continuous grazing systems as currently practised in the Hills. Intensification should not be encouraged. In effect, this means no support for systems such as the Savory which purport to allow double or triple the headage on the same land as traditional grazing methods. If, however, research eventually demonstrates that more intensive systems are indeed sustainable in the Cypress Hills specifically, this judgement may need to be changed. But besides the obvious risks of overgrazing and land degradation, the increased fencing associated with intensive grazing rotations and the potential for increased disruption of wildlife are major negatives.

Overgrazing itself is *not* as important an issue in the Hills as one might at first suspect. While overgrazing is not desirable, its consequences are not nearly as drastic as those of breaking. Overgrazing was likely a natural and recurring phenomenon in the pre-European landscape. Early European explorers frequently traversed areas of the Plains grazed to the ground by bison and associated grazers (Nelson 1973; Howard 1943). The patchily overgrazed landscape of the Plains today may simply be a reflection of the pre-European past, albeit on a much finer scale.

I cannot conclude whether or not overgrazing in the Hills study area has increased in the long term since European settlement. In the ranching economy overgrazing appears to be cyclical, driven by climate and by cattle and grain price fluctuations. But government can limit the prevalence of overgrazing through intelligent lease policy.

Unfortunately for the conservationist, government agricultural policy specifically targets special support to the younger rancher or farmer, who is more likely to pursue aggressive management than the established agriculturist.[12] While this bias cannot effectively be countered in the Hills, it would aid conservation if lease land was at least not assigned to the younger rancher under financial pressure. Most importantly, lease land blocks should not be broken into smaller holdings that would imply the need for a rancher to intensify production to make his desired living.

Concluding notes

Many of the strategies involved in meeting the objectives of a Heritage Landscape are dependent on cooperation between private landholders and government. This would be an essential characteristic of any Heritage Landscape. In the Hills, for example, research on the nature of original environments in the Hills, on native range productivity and its role as a 'heritage ecosystem', on the role of fire, on cattle–wildlife relationships, and on strategies to accommodate wildlife depredation are all necessary cooperative ventures.

A Heritage Landscape approach can be a useful tool for many semi-natural areas of North America. Our successful future may in fact depend more on evolving appropriate and balanced relationships between man and nature in settled areas than on preserving wilderness tracts. To promote nature conservation values on semi-natural lands requires a shift in approach than can lead us beyond the current North American paradigm, a paradigm that polarizes issues into a nature conservation versus development debate.

But we must be cautious. In some regions of North America wilderness is and remains *the* valid conservation goal. By postulating Heritage Landscapes as a legitimate nature-conservation tool on occupied landscapes I absolutely do not intend to be understood to be supporting the expansion of human economic activity or the hypothesis of 'sustainable economic growth' into wilderness areas. We are faced with many environmental choices where it is dishonest to suggest anything other than that the development values proposed are fundamentally incompatible with the natural values threatened. Rather than reduce the perception of the stark incompatibilities between development and wilderness values, my intent here is to propose an approach that allows for expanding the remit of conservation and preservation into North America's settled areas as a cautious complement to a healthy wilderness ethic.

By shifting nature-conservation issues and responsibilities out of the strictly public administration domain and into the everyday business of the landholder and the public Heritage Landscapes *may* represent a small step forward in our conceptualization of our relationship to land and nature. If so, then the strategy's incremental effect is valuable and correct. But if, despite positive intent, it can be demonstrated that a 'Heritage Landscape' as it is herein understood will be interpreted by people as a licence to continue or intensify the abuse of land (either within a designated landscape or in the majority of the continent that will always be outside any conservation designation) then the designation will have failed its most important objective, that of helping to change our misperception of ourselves as divorced from the natural world. The Heritage Landscape approach will then have to be discarded, even regardless of any apparent conservation benefits within any designated landscape. Our goal reaches beyond the land itself: we must seek to make people as environmentally sensitive as landscapes are.

Notes

1. For a discussion of the origins and logic of the preservation philosophy in North America and a comparison with an alternative conservation conceptualization demanding intervention and manipulation of the environment see Henderson (1992).
2. For a discussion of the problems of establishing what indeed 'native prairie' really was prior to human influences and what the term can usefully mean in a modern context, see Henderson (1991).
3. There are striking parallel objectives pursued through the restriction by European settlers of Amerindian tribes to small unwanted parcels of land. These Indian reserves, like nature reserves today, served to assuage the consciences of the time at a minimal resource cost to the invaders. As with nature conservation, the only path to full equal status in Western society for Amerindians appears to be through a process of deghettoization.
4. Grassland fires may have been started to drive game in specific directions, to promote new grass growth that would attract grazers, to make travel easier, as a signalling tool, as an instrument of war (Looman 1983; Nelson 1973), or to control insects around campsites (Canadian Wildlife Service, n.d.). Fires may have been started for still other reasons as well and some grassland fires will have been started accidentally.
5. The converse group, i.e. the natural antagonists of nature conservation, are young agriculturalists under pressure. An excellent example of such a group are Israeli farmers of the 1950s and 1960s who responded with innovative and successful agricultural techniques to unpromising environmental

conditions and economic uncertainty. Early Israeli farmers had the rare advantage of usually *not* being the sons and daughters of farmers themselves and were therefore open to all sorts of experimental approaches. 'Making the desert bloom' may be an impressive technical achievement, but it is the antithesis of a Heritage Landscape's objectives.

6.   In a manager interview Tom Galimberti described his disorientation and shock at moving from his 'benign' home environment in southern Ontario to the Canadian Plains. Here for the first time he visualized nature as a serious external force: 'At 40 below you can die.' In Saskatchewan he has a feeling of 'living on the edge', where any day a minor environmental shift might 'blow everyone away'.

7.   Native range is extremely resilient and withstands overgrazing and mismanagement more successfully than exotics like crested wheat grass or Russian wild rye. Even after abusive management, given a few years of proper management, native range can almost always recover. Native range is also more resistant to drought than the exotics (Manske and Conlan, n.d.).

8.   John Livingston (1981, 33) notes we have a natural (and, in his view, fatal) tendency to embrace any approach which 'tends to shore up the ancient human hallucination having to do with the predictability and thus the controllability of all things'.

9.   Patches of exotic feeds probably allow for a higher number of some big game species in the Hills than would be the case in a pure native grass environment. Similarly, new water point development by ranchers undoubtedly helps some wildlife species. But as the proposed fauna goal is not the production of game species, but rather the preservation of whole (semi-)natural ecosystems, this is not a justification for the planting of exotics, nor for water development.

10.   Intensive wildlife production techniques are increasingly prevalent in North America. Fish stocking, waterfowl production, and ungulate management are all major government enterprises.

  The factory production approach to wildlife may be on its own terms both acceptable and efficient in many areas, but it has little in common with the holistic nature-conservation objective of protecting ecosystems in their natural entirety. When government wildlife departments say they are promoting 'wildlife conservation', as when grain farmers practise 'soil conservation', one must beware of confusing these uses of the term 'conservation' with nature conservation itself. Sometimes this confusion is created deliberately.

11.   Ranchers would simply shoot or poison wolves to protect their cattle. An interesting comparison is with the strategy of the Maasai in the Ngorongoro Conservation Area in Tanzania. The Maasai seasonally retreat their cattle in the face of advancing wildebeest to protect their stock from malignant catarrh fever, a fatal disease that can be passed on from wildebeest to cattle (Perkin 1990). European origin ranchers would doubtless not retreat!

12.   It is fair to say that the younger agriculturalist is more likely to need assistance than the older, so a greater individual need often does exist. It

does not follow, though, that a *societal* need or justification for this support exists. Latest census data in fact show a fall in the average age of Saskatchewan farmers, despite an increasing national average age profile (Henderson 1987b).

References

Body, Richard 1984. *Farming in the Clouds*. Maurice Temple Smith, Hounslow, Middlesex.

Bowers, J.K. and Paul Cheshire 1983. *Agriculture, the Countryside and Land Use: An Economic Critique*. Methuen, London.

Canadian Wildlife Service (n.d.). *Last Mountain Lake Grasslands Trail Guide*. Saskatoon.

Conservation Foundation 1985. *National Parks for a New Generation*. Donnelly and Sons, Harrisonburg, VA.

Goodey, Brian 1990. Market factors in the production of heritage landscapes. Landscape, Heritage and National Identity Conference, Nottingham, 7–9 September.

Henderson, Norman 1987a. *A Critique of the Conservation Reserve Component of the Soil Conservation Strategy*. Prairie Farm Rehabilitation Administration, Regina, Sask.

Henderson, Norman 1987b. *Farms and Farmers: The Outlook for Saskatchewan 1987–92*. Saskatchewan Chamber of Commerce, Regina.

Henderson, Norman 1991. Nature conservation on private and lease lands in Canada: British experience and the Alberta-Saskatchewan Cypress Hills as a test case semi-natural landscape. PhD thesis, University of East Anglia, Norwich.

Henderson, Norman 1992. The nature conservation ideal: Canada, the United States, and Britain contrasted. *Ambio* 21, 394–9.

Howard, Joseph 1943. *Montana High Wide and Handsome*. Yale University Press, New Haven, CT.

Leopold, Aldo 1949. *A sand county almanac*. Oxford University Press, Oxford.

Lieff, Bernie 1985. Waterton Lakes biosphere reserve: developing a harmonious relationship. *Parks* 10(3), 9–11.

Lieff, Bernie 1987. Case study: Waterton biosphere reserve. In *Proceedings of the Symposium on Biosphere Reserves — 4th World Wilderness Congress*. Estes Park, Colorado, pp. 134–41.

Livingston, John (1981). *The Fallacy of Wildlife Conservation*. McClelland and Stewart, Toronto.

Looman, Jan 1983. Grassland as natural or semi-natural vegetation. In W. Holzner *et al.* (eds), *Man's Impact on Vegetation*. Dr. W. Junk B.V. Publishers, The Hague, pp. 173–84.

Lowenthal, David 1990. Keynote address. Landscape, Heritage and National Identity Conference, Nottingham, 7–9 September.

Mabey, Richard 1980. *The Common Ground*. Hutchinson, London.

Manske, L. and Conlan, T. (n.d., but post 1984). *Complementary Rotation Grazing system in Western North Dakota*. Dickinson Experimental Station, Dickinson, ND.

Mondor, Claude and Kun, Steve 1984. The Lone Prairie: protecting natural grasslands in Canada. In J. McNeely and K. Miller (eds), *National Park Conservation and Development*. Smithsonian Institution Press, Washington, DC, pp. 508–17.

Moore, Norman 1987. *The Bird of Time*. Cambridge University Press, Cambridge.

Mumford, Lewis 1967. *The Myth of the Machine*. Secker and Warburg, London.

Nelson, J.G. 1973. *The Last Refuge*. Harvest House, Montreal.

Perkin, Scott 1990. Land degradation, pastoral development and wildlife conservation in the Ngorongoro conservation area, Tanzania. Unpublished MS, School of Development Studies, University of East Anglia, Norwich.

Peterson, C. and Madsen, C. 1981. Property tax credits to preserve wetlands and native prairie. Transactions of The North American Wildlife Natural Resources Conference, 46, 125–9.

Saarinen, Thomas 1966. *Perceptions of Drought Hazard on the Great Plains*. University of Chicago Press, Chicago.

Saskatchewan Department of Tourism and Renewable Resources (B. Hart) 1976. *Present Land Use* (map sheet 72-F, terrestrial habitat inventory). Regina.

Scace, Robert 1972. The management and use of a Canadian Plains oasis: the Cypress Hills public reserves. PhD thesis, University of Calgary, Calgary.

Scanlon, Michael 1981. Biogeography of forest plants in the prairie-forest ecotone in western Minnesota. In R. Burgess and D. Sharpe (eds), *Forest Island Dynamics in Man-Dominated Landscapes*. Springer-Verlag, New York, pp. 97–124.

Seymour, John and Herbert Gidardet 1986. *Far from Paradise*. British Broadcasting Corporation, London.

Shaw, James 1985. *Introduction to Wildlife Management*. McGraw-Hill, New York.

Westmacott, Richard 1983. The conservation of farmed landscapes, attitudes and problems in the United States and Britain. *Landscape Design* 8(13), 11–14.

Chapter 9

Valuation of Environmental Resources and Impacts in Developing Countries

Edward B. Barbier

Introduction

Natural resource management is crucial to the developing economies of the world. These economies, especially the lower-income countries, are highly dependent on primary production as the source of long-term, sustainable economic development. Successful exploitation of primary production – agriculture, fishing, forestry and minerals – in turn depends on efficient and sustainable management of the resource base supporting primary productive activities. Moreover, as developing countries industrialize and as their populations concentrate in urban settlements, the role of the environment in assimilating waste products and providing life-support amenities will become increasingly important. Protection and conservation of key natural systems and important ecological functions will also be essential, not only in terms of their potential value in terms of recreation and tourism but also because these systems and functions may provide invaluable support and protection for economic activity and human welfare.

Despite the essential economic role of environmental resources in developing countries, surprisingly little is known about this role. Not much is known about the economic contribution of environmental resources and ecological functions to economic development – perhaps even less about their contribution to traditional economic systems in developing countries. We also know little about the impacts of market forces and policy interventions on environmental management and the environmental effects of development projects and programmes. Research in all these areas is proceeding, but progress remains slow given the complexity of the problems and our failure to address them adequately in the past (see Chapter 3).

On the positive side, there is growing recognition that efficient and sustainable management of environmental resources is indeed critical to economic development. Any analysis of the contribution of environmental resources to development must invariably involve valuing the key economic functions performed by these resources, and in turn, the impact of economic activity on the environment. *Valuation of environmental resources and impacts* is the important starting point for the application of environmental economics in developing countries.

This chapter explores critical issues in the valuation of environmental resources and impacts in developing countries, through examining some key resource problems – tropical wetland conversion, tropical deforestation and land degradation.

Valuation issues: an overview

Economic valuation proves especially important in three areas: *accounting for the costs* of environmental damage and degradation; *analysis of market and policy failures* that contribute to natural resource degradation; and *analysis of the investments* determining land- and resource-use options.

Resource accounting

Policy-makers in developing countries are often reluctant to consider environmental degradation as a serious development problem unless they receive some indication of the aggregate costs to the economy of resource depletion and pollution. In recent years, many advances have been made in this direction through the development and application of *resource accounting*. Various methodologies have been proposed for presenting environmental impacts and values in an accounting framework (Ahmad *et al.* 1989; Repetto *et al.* 1989; Pearce *et al.* 1989, ch. 4). One is to incorporate environmental considerations into the existing system of national account (SNA), and the other is to develop the accounts for natural and environmental resources in a separate accounting framework.

The 'physical' approach involves a separate system of environmental accounts presenting the stocks and flows of environmental variables in physical units. Typical examples are energy balances, mineral stock amounts, records of emissions and measures of the state of ambient quality. This approach has been developed extensively in a number of industrialized countries, notably Norway, France and more recently Canada (Alfsen *et al.* 1987; Corniere 1986; Friend 1988). The experience of these countries indicates that the physical approach is useful in analysing the linkages between the environment and the economy and, when further developed, could aid in forecasting future levels of resource use.

However, Barbier (1988) has pointed out that the data and methodological requirements for extensive analysis of environmental variables are formidable in developing countries, necessitating rapid and expensive improvements in existing (or, in many cases, non-existent) environmental and resource data bases. Constructing such a data base would be a basic requirement for any resource accounting approach, whether involving physical or monetary accounts (see Chapter 10).

The main objective of the 'monetary' approach is to adjust the measures of national income produced by the SNA for: (i) any additional 'defensive' expenditures undertaken by households to mitigate the consequences of environmental pollution; (ii) the costs of any pollution that occurs but is not mitigated; and (iii) the depreciation that has taken place in the environmental and resource base but is not accounted for. To date, complete accounting for all the above impacts has not taken place. Approaches in industrialized countries have tended to focus on the adjustments for pollution expenditures and damage (Nordhaus and Tobin 1972; Uno 1988). In developing countries, efforts have focused more on the depreciation of key resource stocks (Repetto et al. 1989). One limitation of the latter approach is that it is currently confined to measurable, discrete stocks (e.g., petroleum, production forests, soil). Depreciation of entire natural systems, such as forest systems as a whole, coastal zones and watersheds, is not so easily accounted for. Equally, degradation of some ecological functions, such as storm prevention provided by wetlands, microclimate regulation by forests and the nutrient flows of estuaries, cannot easily be incorporated in measures of sustainable income.

Market and policy failures

Market and policy failures are at the centre of many developing country environmental problems. *Market failures* exist when markets fail to reflect fully environmental values. The presence of open access resource exploitation and public environmental goods, externalities (e.g., pollution, non-marketed environmental services), incomplete information and markets (e.g., uncertainty, lack of formal market mechanisms) and imperfect competition all contribute to market failure. Usually some form of public or collective action, involving regulation, market-based (economic) incentives or institutional measures, is required – provided that the costs of correcting market imperfections do not exceed the potential welfare benefits. *Policy failures* therefore exist when the appropriate policy interventions necessary to correct market failures are not taken, or over-correct or under-correct for the problem, *and* when government decisions or policies are themselves responsible for excessive environmental degradation (Turner 1991). For example, environmental damage may arise because economic policies and interventions in developing countries are designed primarily to promote economic growth

---

Box 9.1 The social losses from upland degradation

As a resource or natural environment is degraded, and some of its functions irreversibly disrupted or disturbed, each additional loss in these values imposes additional, or *marginal, social costs* (*MSC*).

The problem for society is that, unless individuals are made accountable for the loss in economic values from environmental degradation, the *direct costs* to these individuals, i.e. their private *marginal costs* (*MC*), from using or exploiting the environment will fall well short of the marginal social costs. That is, $MC < MSC$. The result is a loss in social welfare.

Figure 9.1 provides a simple illustration of this problem with an example of upland soil erosion. Upland farmers producing a crop, such as maize, face direct costs in terms of the amount of effort, hired labour and other inputs required for cultivation. This is represented by the upward-sloping *MC* curve. There is presumed a market for the maize, with demand *D*.

However, the production of upland maize results in soil erosion, with additional *user costs* in terms of the forgone future productivity of the soil and *external costs* in terms of the downstream impacts of sedimentation on irrigation networks, navigation, flooding and so on. User costs arise when the current removal and use of a resource precludes or reduces its availability and use in the future. These are the costs of forgoing future direct and indirect production or use benefits from environmental degradation today. External costs are costs not directly experienced by the users of the resource but inflicted on others by the consequences of environmental degradation.

The *MSC* curve indicates that society is bearing all these costs – direct, user and external – even though upland farmers only face the *MC*. The socially desirable level of upland maize production would be quantity Q_s, at price P_s, where the *MSC* curve meets the demand curve *D*. Actual production, however, is determined by the *private MC* curve and *D* – at quantity Q_p and price P_p. As indicated, the result is a *net loss in social welfare*.

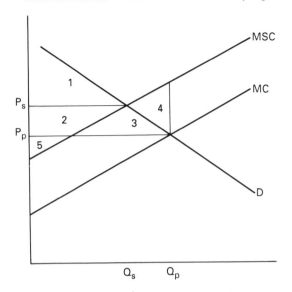

MC   = direct (private) costs of producing upland maize
MSC = direct costs + user and external costs of soil erosion
D   = market demand for maize

CS   = consumer surplus
PS   = producer surplus
NLS = net loss to society

NET LOSS TO SOCIETY =

MC = *D*		*MSC* = *D*
CS = 1 + 2 + 3	minus	*CS* = 1
PS = 5 − 3 − 4	minus	*PS* = 5 + 2

NET LOSS TO SOCIETY =

$$CS = 2 + 3$$
$$PS = -2 - 3 - 4$$

$$NLS = -4$$

Figure 9.1   Private and social costs of upland degradation

or to improve income distribution, with little regard for their consequences for the environment.

Economic valuation of the environmental impacts arising from market and policy failures is essential for determining the appropriate policy responses. Often, however, insufficient data and information exist to allow precise estimation of the economic costs arising from market and policy failures. In most cases, cost estimates as orders of magnitude and indicators of the direction of change are sufficient for policy analysis. (See, for example, the case studies in Pearce *et al.* 1990.)

Thus excessive degradation of the environment and natural resources is often the outcome of their values not being fully recognized and integrated into decision-making processes by individuals in the marketplace and by governments. The result is often a distortion in economic incentives. That is, the private costs of actions leading to environmental degradation do not reflect the full social costs of degradation, in terms of the environmental values forgone. The full social costs of environmental degradation are what economists call *opportunity costs*. Box 9.1 provides a simple illustration of this problem with an example of upland soil erosion. Other examples will be discussed in subsequent sections.

Valuing tradeoffs

Many large-scale investment projects and programmes, such as hydroelectric dams, irrigation schemes, commercial agricultural development schemes, road building, and so on, have significant environmental impacts. Some of these impacts may impose additional costs or benefits on society. In recent years, many advances have been made in applying economic valuation techniques to analysing the environmental impacts of investment projects and programmes in developing countries (Dixon *et al.* 1988; Bojö *et al.* 1990; Gregersen *et al.* 1987).

Failure to account fully for the environmental impacts of an investment project or programme means that its net economic worth is being misrepresented. Often, when significant external environmental costs are present, the result is a misallocation of resources and excessive environmental degradation. These additional costs must be included as part of the costs of the development investment.

For example, assume that there is an upstream development project on a river that is providing water for agriculture. Given *direct benefits* (e.g., irrigation water for farming), B^D, and *direct costs* (e.g., costs of constructing the dam, irrigation, channels, etc.), C^D, the the *direct net benefits* of the project are:

$$NB^D = B^D - C^{D\ 2}$$

However, by diverting water that would otherwise flow into downstream wetlands, the development project may result in losses to floodplain agriculture and other primary production activities (e.g., fishing, fuelwood, livestock grazing), less groundwater recharge and other external impacts. Given these reductions in the *net production and environmental benefits*, NB^W, of the wetlands, then the true net benefits of the development project (NB^P) are $NB^D - NB^W$. The development project can therefore only be acceptable if:

$$NB^P = NB^D - NB^W > 0$$

Thus, in the presence of significant environmental impacts, the net benefits of a development project or programme cannot be appraised in terms of its direct benefits and costs alone. The forgone net benefits of disruption to the natural environment and degradation must also be included as part of the opportunity costs of the development investment.

Tropical wetlands and irrigation development

Since 1900, over half of the world's wetlands may have disappeared. The United States alone has lost an estimated 54% (87 million hectares) of its original wetlands, of which 87% has been lost to agricultural development, 8% to urban development and 5% to other conversions (Maltby 1986). The total area and status of tropical wetlands are still unknown, but the available evidence suggests that the pattern of wetland conversion in developing countries may be similar to that of the United States – and perhaps proceeding at even a faster rate in some regions.

Natural wetlands perform many important functions for humankind – storm prevention, flood and water flow control, nutrient and waste absorption and so forth. Wetlands can also be used for recreation and water transport, and their diverse resources can be directly exploited for fishing, agriculture, wildlife products, wood products and water supply. When properly measured, the total economic value of a wetland's ecological functions, its services and its resources may exceed the economic gains of converting the area to an alternative use. Some economic studies have valued the benefits of *temperate* wetlands (for reviews, see Turner 1988; Turner and Jones 1991; and Farber and Constanza 1987). But to date, little analysis of *tropical* wetland benefits has been undertaken. The methodology for economic valuation of tropical wetland resources is relatively straightforward, usually involving

Table 9.1 Commonly applied and potential measurement and valuation techniques to tropical wetlands in developing countries

	Approach
Direct use	
Forest resources	Valuing the marginal productivity of
Wildlife resources	the resource net of any human effort;
Fisheries	marketed substitutes/alternative
Forage resources	supplies; indirect opportunity cost;
Agricultural resources	indirect substitute.
Water supply	
Energy resources	
Recreation/tourism	Travel-cost methods.
Water transport	Alternative/substitute costs.
Biological diversity	Value of genetic material, scientific and educational use.
Indirect use	
Groundwater recharge/discharge	Preventive expenditures; damage costs
Flood and flow control	avoided; alternative/substitute
Shoreline stabilisation/erosion control	costs; relocation costs; value of
Sediment retention	changes in productivity.
Nutrient retention	
Water-quality maintenance	
Storm protection/windbreak	
Micro-climate stabilisation	
External support	
Biological diversity	Value of changes in productivity.
Non-use/preservation	
Uniqueness to culture/heritage	Contingent valuation.
Biological diversity	

Source: Barbier (1989c).

the value of the production gained by directly exploiting these resources (see Table 9.1). However, measuring the indirect benefits of wetland environmental functions, such as storm prevention and flood control, through damage costs avoided and replacement costs is more problematic. More sophisticated techniques of contingent valuation, travel-cost method and hedonic pricing are more difficult to apply in most developing regions.

It is clearly important to demonstrate the economic benefits of wetland areas if appropriate land-use decisions are to be made. Given the threat posed by large-scale water development projects in many developing countries, policy-makers need convincing that wetland areas are worth preserving. Evidence from various countries is indicating that the opportunity costs of diverting water from many wetlands may be significant.

For example, an *ex-post* analysis of the large-scale Ghezela irrigation project in Tunisia took into account disruptive hydrological and other environmental impacts on the neighbouring Ichkeul National Park and surrounding areas. The analysis revealed that these costs contribute to making the project economically unviable (Thomas *et al.* 1990). The floodplain of the Park provides important fishing and grazing uses for local communities, as well as tourism/educational benefits and essential environmental functions (desalination of the water table, sewage treatment, control of birds). The benefits are being threatened by Ghezala and other upstream irrigation schemes.

However, the *ex-post* analysis showed that the net benefits of fishing and grazing alone easily exceeded the returns to irrigation. Depending on the annual water requirement of the Park's floodplain, the net benefits from fishing and grazing vary between US$0.33 and 0.55 per cubic metre of water. The net benefits of irrigation in Ghezala were actually negative, − $0.60 per cubic metre, given the high costs incurred by the scheme. It follows from this analysis that diverting water from the Park's wetlands to feed the Ghezala scheme is not beneficial.

A similar assessment was made of the net benefits from the Hadejia-Jama'are floodplain in northern Nigeria and comparing it to the net benefits of an upstream irrigation development, the Kano River Project (Barbier *et al.* 1991). As indicated in Box 9.2, when compared to the net economic benefits of the Kano River Project, the economic returns to the floodplain appear much more favourable. This is particularly the case when the relevant returns to the Project in terms of water input use are compared to that of the floodplain system. The result should cause some concern, given that the existing and planned water developments along the Hadejia-Jama'are river system, such as the Kano River Project, are currently diverting and will continue to divert water from the floodplain.

The failure to account for the economic value of tropical wetland benefits, including the benefits derived from local communities dependent on the natural wetlands, means that wetland conversion in tropical wetlands will proceed in Third World countries as long as the economic gains from conversion − mainly for agricultural purposes − exceed the direct costs of drainage, clearing and other 'reclamation' expenditures.

Box 9.2 Comparative benefits of floodplain and upstream development, Nigeria

In north-east Nigeria, an extensive floodplain has been created where the Hadejia and Jama'are rivers combine to form the Komadugu Yobe river which drains into Lake Chad. The Hadejia-Jama'are floodplain provides essential income and nutrition benefits in the form of agriculture, grazing resources, non-timber forest products, fuelwood and fishing for local populations. The wetlands also serve wider regional economic purposes, such as providing dry-season grazing for semi-nomadic pastoralists, agricultural surpluses for Kano and Borno states, groundwater recharge of the Chad Formation aquifer and 'insurance' resources in times of drought. In addition, the wetlands are a unique migratory habitat for many wildfowl and wader species from Palaearctic regions, and contain a number of forestry reserves. The region therefore has important tourism, educational and scientific potential.

Table 9.2 Comparison of present value net economic benefits, Kano River Project Phase I and Hadejia-Jama'are Floodplain, Nigeria (N7.5 = US$1, 1989–90)

	(8%, 50 yrs)	(8%, 30 yrs)	(12%, 50 yrs)	(12%, 30 yrs)
Per hectare[1]				
HJF (N/ha)	1276	1176	872	846
KRP (N/ha)	233	214	158	153
Per Water Use[2]				
HJF (N/10^3m^3)	366	337	250	242
KRP (N/10^3m^3)	0.3	0.3	0.2	0.2

Notes: [1] Based on a total production area of 730,000 ha for the Hadejia-Jama'are floodplain (HJF) and a total crop cultivated area of 19,107 ha in 1985–6 for the Kano River Project Phase I (KRP).
[2] Assumes an annual average river flow into HJF of 2549 Mm^3 and an annual water use of 15,000 m^3/ha for the KRP.

Source: Barbier *et al.* (1991).

However, in recent decades the Hadejia-Jama'are wetlands have come under increasing pressure from drought and upstream and downstream water developments. Upstream developments are affecting incoming water, either through dams altering the timing and size of flood flows or through diverting surface or groundwater for irrigation. Increased demand for water downstream for irrigated agriculture may lead to diverting water past the wetlands through construction of bypass channels. Intensified human use within the floodplain itself, notably wheat irrigation, is also putting pressure on the wetlands.

An analysis was conducted, comparing the net economic benefits of 14 agricultural crops, fuelwood and fishing in the Hadejia-Jama'are floodplain with the returns to an upstream water development, the Kano River Project (Table 9.2).

The economic importance of the wetlands suggests that the benefits it provides cannot be excluded as an opportunity cost of any scheme that diverts water away from the floodplain system.

As arable land becomes scarce in Third World countries, it is likely that subsidies and distortions to reduce the direct costs of wetland conversion may reach levels similar to those in OECD countries between the 1950s and 1970s. Efforts by IUCN, the Ramsar Convention Bureau and the International Waterfowl and Wetlands Research Bureau to cooperate on methodologies for wetland management, particularly in the hitherto neglected Third World, are welcomed. But the relevance of such efforts in affecting government development decisions and planning will depend crucially on improving and extending the economic evaluation of tropical wetlands.

Meanwhile, irrigation development continues to expand, with very little assessment of the environmental consequences. From 1950 to the mid-1980s, cropland under irrigation increased by over 3% annually, from 94 million to over 270 million hectare. Around 18% of the world's cultivated land is irrigated, producing 33% of the total harvest. The equivalent of $250 billion has already been spent to expand irrigation capacity in the Third World, and an additional $100 billion is expected to be spent between 1985 and 2000. Two-thirds of the world's irrigated lands are in Asia, where 38% of additional food production to the year 2000 is anticipated to come from existing irrigated areas and 36% from newly irrigated areas (Repetto 1986).

Virtually all irrigation developments in the Third World are through public investments, which are heavily subsidized. Distributional considerations, political concerns and common perceptions of water as a 'free good' have generally led to charges well below costs of supply.

Revenues collected from farmers in most Third World countries often cover only 10–20% of the building and operating costs of irrigation systems, usually failing to cover just operating and maintenance (O and M) costs alone. The result can be inefficient and poorly maintained irrigation networks, rent-seeking behaviour by farmers, misallocation and wasteful use of water and unnecessary investments in major surface water developments, such as dams and large-scale irrigation networks. The environmental impacts can be significant, and are not just confined to wetland loss: first, the disincentive to conserve water can lead to problems of water logging, salination and water scarcity; and second, irrigation investments and infrastructure, including dams, can lead to extensive external costs in form of displacement of local communities, loss of agricultural and forest lands, and alterations in river hydrology, in fishing and wildlife industries, in wetlands and flood recession agriculture and in erosion and sedimentation rates. For example, in India 10 million hectares of irrigated land have been lost through waterlogging and 25 million hectares are threatened by salination, and in Pakistan 12 million hectares are waterlogged and 5 million hectares saline (Repetto 1986). Agricultural irrigation in Java accounts for about 47% of the total potential water resources available and 75% of the dry season/year flow. Given the expected future demands in all uses of water, the poor cost recovery and inefficiencies in irrigation (average efficiencies are 10–35% and hardly exceed 30% in most areas for both wet and dry seasons), water scarcity is becoming a chronic problem (World Bank 1988).

Dryland degradation and soil erosion

The term 'drylands' is usually applied to all arid and semi-arid zones, plus areas in the tropical sub-humid zone subject to the same degradation processes that occur on arid lands. Accounting for about one-third of global land and supporting a population of 850 million, the world's drylands are rapidly being degraded through population growth, overgrazing, cropping on marginal lands, inappropriate irrigation and devegetation. The process of dryland degradation is often referred to as *desertification*, where the productive potential of the land is reduced to such an extent that it can neither be readily reversed by removing the cause nor easily reclaimed without substantial investment.

There are few economic studies of the costs of dryland degradation. Those that exist are mainly conducted in OECD countries. These suggest that the problem is significant. For example, the annual cost of degradation in Canada's prairie region is estimated to be US$622 million (Dixon *et al.* 1989). However, substantial work on the costs of dryland degradation in developing regions has yet to be conducted, even though the problems there are believed to be more severe than those encountered in temperate areas. Preliminary estimates of the costs of crop, livestock and fuelwood losses from dryland degradation in Burkina Faso indicate a total damage cost equivalent to about 9% of the country's GDP (Lallement 1990).

Even further behind – and more controversial – is the analysis of the effects of economic and resource management policies on dryland degradation in Third World countries. This is often attributed to the superficial identification of the causes of desertification and to the frequently poor identification of the causes of the failure of dryland projects (Nelson 1988).

Although the majority of 'causes' are attributable to population growth and natural events, dryland degradation is also symptomatic of an agricultural development bias that distorts agricultural pricing, investment flows, R & D and infrastructure towards more 'favoured' agricultural land and systems (Barbier 1989b). Where drylands 'development' is encouraged, it is usually through the introduction of large-scale commercial agricultural schemes that can conflict with more traditional farming and pastoral systems.

The complexity of social, economic and environmental relationships is formidable. Not enough is often known about dryland farming and pastoral systems; open access use and common property resource rights; land-tenure regimes and security; the distribution of wealth and income; and coping strategies in the presence of variable climatic conditions, frequent drought, market instability, political conflicts and other factors influencing risk and uncertainty. A common misperception is that the extension of private property rights, commercial agriculture and markets will 'automatically' solve dryland management problems in the long run. At the same time, not all dryland farmers and pastoralists, even in the most distant and resource-poor regions, are totally isolated from agricultural markets. Virtually all subsistence households require some regular market income for cash purchases of some agricultural inputs and basic necessities; many farmers and pastoralists provide important cash and export crops. As a result, alterations in market conditions – whether from changes in policies, climatic conditions, R & D innovations or other factors – do have a significant impact on the livelihoods of rural groups in dryland areas. Understanding the responses to these changing market conditions is a crucial aspect of the dryland management problem.

For example, a study of gum arabic production in Sudan indicates that fluctuations in the real price of gum and its price relative to those of other agricultural crops have had important impacts of farmers' cropping patterns, diversification strategies and decisions to replant gum – with important consequences for Sudan's gum belt (IIED/IES 1990). Even though it is economically profitable *and* environmentally beneficial to grow gum, only when these economic incentives are properly dealt with by the government will rehabilitation of the important gum belt of Sudan take place.

Soil erosion and land degradation are not confined just to drylands and other marginal lands; erosion of farm cropland is a pervasive problem throughout the Third World. Although some evidence has been accumulated on the costs of soil erosion to developing countries, reliable estimates are difficult.

Little long-term monitoring of soil erosion from farmers' plots has occurred; aggregation and extrapolation from the few studies that do exist are fraught with complications. Regional – not to mention national and international – comparisons should be treated with caution. Estimates of erosion based on the universal soil loss equation (USLE) and modified USLEs adapted for tropical conditions still face many difficulties, as do methods involving Geographical Information Systems (GIS). To go further and analyse the impacts of erosion on crop yields or the impacts of runoff and sedimentation on 'off-site' economic activities is even more difficult in Third World countries. Particularly frustrating is the fact that the declining trends in crop yields attributable to erosion are hard to substantiate, given that erosion impacts are often inseparable from the effects of climatic variations, relative price changes, changing cropping patterns, input mixes and labour-use strategies, and changes in technology.

Despite these methodological problems, some recent studies of the economic costs of soil erosion indicate that these costs could be substantial. For example, in Mali current net farm income forgone from soil erosion is estimated to be US$4.6 to $18.7 million annually, and current plus future forgone income due to one year's soil erosion is estimated to be US$31–123 million, or 4–16% of agricultural GDP (Bishop and Allen 1989). In Malawi, a similar analysis indicated gross annual losses from soil erosion to account for between 0.5% and 3.1% of 1988 GDP, depending on the sensitivity of crops to soil loss (Bishop 1990). In Java, the on-site costs of soil erosion in upland areas are estimated to be around US$315 million annually (3% of agricultural GDP), with additional off-site sedimentation costs of $58 million (Magrath and Arens, 1988).

Such estimates of the economic costs of soil erosion provide important indicators to policy-makers that there is a serious problem to be corrected. However, designing appropriate policy responses to control soil erosion and land degradation again requires overcoming data limitations and the lack of microeconomic analyses of farmers' responses to erosion and incentives to adopt conservation measures. The limited evidence that does exist suggests that relationships – such as the effects of agricultural input and output pricing on farm-level erosion – are complex and difficult to substantiate, but none the less critical to the analysis (Barbier 1989a, ch.7; Repetto 1988; and Pearce *et al.* 1990).

Tropical forests

Approximately 8.4 million hectares of tropical moist forest area were being cleared annually by the end of the 1980s, with over 4 million hectares deforested in Latin America alone (Schmidt 1990). Perhaps most worrying is that deforestation is occurring with little regard to long-term management of the forests. Much of the loss is the result of clearing for agriculture – both planned and unplanned – with little regard to the social opportunity costs of the disappearing forests. Even management of forests for timber production is precarious. For example, a study by Poore *et al.* (1990) concluded that, on a world scale, operational management of tropical forests for sustainable production of timber is negligible.

Yet the cost of forest conversion and degradation to developing countries can be high. For example, in Indonesia, the forgone cost in terms of timber rentals from converting primary and secondary forest land is in the order of US$625–750 million per annum. With logging damage and fire accounting for additional costs of $70 million, this would represent losses of around $800 million annually. The inclusion of forgone minor forest products would raise this cost to $1 billion per year. In addition, the loss of timber on sites used for development projects could be another $40–100 million (Pearce *et al.* 1990, ch. 5). The total cost of the depreciation of the forest stock would include not just the cost of conversion but also the cost of timber extraction and forest degradation. One study estimated this total cost for Indonesia to be around $3.1 billion in 1982, or approximately 4% of GDP (Repetto *et al.* 1989). However, this estimate must be considered a lower bound, as it does not include the value of the loss of forest protection functions (e.g., water-shed protection, micro-climate maintenance) and of biodiversity. The latter may particularly be important in terms of *option* and *existence values* – i.e. values reflecting a willingness to pay to see species conserved for future use or for their intrinsic worth – which could translate into future payments that the rest of the world might make to Indonesia to conserve forest lands.

There is now sufficient economic evidence linking the tropical deforestation problem to economic policies. Too often, the pricing and economic policies of countries with tropical forests distorts the costs of deforestation. First, the 'prices' determined for tropical timber products or the products derived from converted forest land *do not incorporate* the lost economic values in terms of forgone timber rentals, forgone minor forest products and other direct uses (e.g. tourism), disrupted forest protection and other ecological functions, and the loss of biological diversity, including any option or existence values. Second, even the direct costs of harvesting and converting tropical forests *are often subsidized and/or distorted*, thus encouraging needless destruction. For example, in the Brazilian Amazon subsidies and other policy distortions are estimated to have accounted for at least 35% of all forest area altered by 1980 through tax incentives for capital investment (e.g., industrial wood production and livestock ranching); rural credits for agricultural production (mechanized agriculture, cattle ranching and silviculture); subsidized small farmer settlement; and export subsidies (Browder 1985). In addition, government-financed investment programmes – for road-building, colonial settlement and large-scale agricultural and mining activities – may *indirectly* be contributing to deforestation by 'opening up' frontier areas that were previously inaccessible to smallholders and migrants.

Similarly, in Malaysia and Indonesia, government policies to encourage switching from the export of raw logs to processed timber products have led to substantial economic losses, the establishment of inefficient processing operations and accelerated deforestation (Repetto and Gillis 1988). Throughout South-East Asia the allocation of timber concession rights and leasing agreements on a short time-scale, coupled with the lack of incentives for reforestation, have contributed to excessive and rapid depletion of timber forests. In the Philippines, the social gains from logging old-growth forest was found to be negative (between − US$130 and − $1175 per hectare), once the social costs of timber stand replanting, of depletion and of off-site damages were included (Paris and Ruzicka 1991).

Conclusion

The above examples illustrate the important role of economic valuation and environmental impacts in developing countries. Although some important advances have been made in recent years, substantial work is required in the analysis of the costs of environmental degradation, policy and market failures and investment tradeoffs. Valuation of welfare gains and losses will remain the key to the analysis if we are to clarify the contribution of natural resource management to development, and, above all, if we are to design development policies that are environmentally 'sustainable'.

Increasingly, governments in developing countries and major donor agencies are recognizing that sound natural resource management is complementary, and not counter, to development efforts in the Third World. Continued efforts to value environmental impacts are necessary to illustrate the extent of the economic losses arising from environmental degradation and to marshall sufficient 'political will' to correct the causes. Proper economic analysis of policy options and their potential impacts will require further improvements in the appropriate data, methodology and analytical tools for economic valuation of environmental impacts. Above all, it will require the training of more economists in developing countries capable of undertaking such analysis and contributing to the formulation of economic policies.

References

Ahmad, Y.J., El Serafy, S. and Lutz, E. (eds) 1989. *Environmental Accounting for Sustainable Development: A UNEP–World Bank Symposium.* World Bank, Washington, DC.

Alfsen, K.H., Bye, T. and Lorentsen, L. 1987. *Natural Resource Accounting and Analysis: The Norwegian Experience 1978–86.* Central Bureau of Statistics, Oslo.

Barbier, E.B. 1988. Economic valuation of environmental impacts. *Project Appraisal* 3(3).

Barbier, E.B. 1989a. *Economics, Natural-Resource Scarcity and Development: Conventional and Alternative Views.* Earthscan, London.

Barbier, E.B. 1989b. Sustainable agriculture on marginal land: a policy framework. *Environment.* 31(9).

Barbier, E.B. 1989c. *Economic Evaluation of Tropical Wetland Resources: Applications in Central America.* Report to CATIE/IUCN, London.

Barbier, E.B., Adams, W.M. and Kimmage, K. 1991. *Economic Valuation of Wetland Benefits: The Hadejia-Jama'are Floodplain, Nigeria.* London Environmental Economics Centre, London.

Bishop, J. 1990. *The Costs of Soil Erosion in Malawi.* Malawi Country Operations Division, World Bank, Washington, DC.

Bishop, J. and Allen, J. 1989. *The On-Site Costs of Soil Erosion in Mali.* Environment Department Working Paper No.21. World Bank, Washington, DC.

Bojö, J., Mäler, K.-G. and Unemo, L. 1990. *Environment and Development: An Economic Approach.* Dordrecht, Kluwer Academic.

Browder, J.O. 1985. *Subsidies, Deforestation, and the Forest Sector in the Brazilian Amazon.* World Resources Institute, Washington, DC.

Corniere, P., 1986. Natural Resource (1) Accounts in France. An Example: Inland Waters, *Information and Natural Resources*, OECD, Paris.

Dixon, J.A., Carpenter, R.A., Fallon, L.A., Sherman, P.B. and Manipomoke, S. 1988. *Economic Analysis of the Environmental Impacts of Development Projects*. Earthscan, London.

Dixon, J., James, D. and Sherman, P. 1989. *The Economics of Dryland Management*. Earthscan, London.

Farber, S. and Constanza, R. 1987. The economic value of wetland systems. *Journal of Environmental Management* 24.

Friend, A. 1988. Natural resource accounting: a Canadian perspective. In Ahmad, Y.J., El Serafy, S. and Lutz, E. (eds), *Environmental and Resource Accounting and Their Relevance to the Measurement of Sustainable Development*. World Bank, Washington, DC.

Gregersen, H.M., Brooks, K.N., Dixon, J.A. and Hamilton, L.S. 1987. *Guidelines for Economic Appraisal of Watershed Management Projects*. FAO, Rome.

IIED/IES 1990. *Gum Arabic Rehabilitation Project in the Republic of Sudan: Stage I Report*. IIED, London.

Lallement, D. 1990. *Burkina Faso: Economic Issues in Renewable Natural Resource Management*. Agricultural Operations, Sahelian Department, Africa Region, World Bank, Washington, DC.

Maltby, E. 1986. *Waterlogged Wealth: Why Waste the World's Wet Places?* Earthscan, London.

Magrath, W.B. and Arens, P. 1989. *The Costs of Soil Erosion on Java – A Natural Resource Accounting Approach*. Environment Department Working Paper No. 8, World Bank, Washington, DC.

Nelson, R. 1988. *Dryland Management: The 'Desertification' Problem*. Environment Department Working Paper No. 8, World Bank, Washington, DC.

Nordhaus, W.D. and Tobin, J. 1972. *Is Growth Obsolete*. National Bureau of Economic Research, General Series 96. Columbia University Press, New York.

Paris, R. and Ruzicka, I. 1991. *Barking up the Wrong Tree: The Role of Rent Appropriation in Tropical Forest Management*. Environment Office Discussion Paper. Asian Development Bank, Manila.

Pearce, D.W., Markandya, A. and Barbier, E.B. 1989. *Blueprint for a Green Economy*. Earthscan, London.

Pearce, D.W., Barbier, E.B. and Markandya, A. 1990. *Sustainable Development: Economics and Environment in the Third World*. Edward Elgar, Aldershot.

Poore, D., Burgess, P., Palmer, J., Reitbergen, S. and Synnott, T. 1990. *No Timber without Trees: Sustainability in the Tropical Forest*. Earthscan, London.

Repetto, R. 1986. *Skimming the Water: Rent-Seeking and the Performance of Public Irrigation Systems*. World Resources Institute, Washington, DC.

Repetto, R. 1988. *Economic Policy Reform for Natural Resource Conservation*. Environment Department Working Paper No. 4. World Bank, Washington, DC.

Repetto, R. and Gillis, M. (eds) 1988. *Public Policies and the Misuse of Forest Resources*. Cambridge University Press, Cambridge.

Repetto, R., Magrath, W., Wells, M., Beer, C. and Rossini, F. 1989. *Wasting Assets: Natural Resources in National Income Accounts*. World Resources Institute, Washington, DC.

Schmidt, R. 1990. Sustainable management of tropical moist forests. Presentation for ASEAN Sub-Regional Seminar, Indonesia, Forest Resources Division, Forestry Dept, FAO, Rome.

Thomas, D., Ayache, F. and Hollis, T. 1990. Use values and non-use values in the conservation of Ichkeul National Park, Tunisia. *Environmental Conservation* 18, 120–30.

Turner, R.K. 1988. Wetland conservation: economics and ethics. In D. Collard, D.W. Pearce and D. Ulph (eds), *Economics, Growth and Sustainable Development: Essays in Memory of Richard Lecomber*. Macmillan Press, London.

Turner, R.K. 1991. Economics and wetland management. *Ambio* 20, 59–63.

Turner, R.K. and Jones T. (eds) 1991. *Wetlands: Market and Intervention Failures*. Earthscan, London.

Uno, K. 1988. Economic growth and environmental change in Japan: net national welfare and beyond. Mimeo, Institute of Socio-Economic Planning, University of Trukuba, Japan.

World Bank 1988. *Indonesia – Java: Water Resources Management Report*. World Bank, Washington, DC.

Chapter 10

Sustainable National Income and Natural Resource Degradation in Zimbabwe

Neil Adger

Introduction: Macro-indicators of sustainability

Economic activity in the primary sectors of an economy entails interaction with the natural environment. Where this stock of renewable and non-renewable assets is depleted the ability to generate welfare in the future will be decreased, in the same way that the consumption of reproducible (or man-made) capital will reduce future potential. An economy which depletes the stock of capital, natural and reproducible, risks not being able to sustain welfare in the long run, though other mitigating factors such as technological change are important. In this chapter the measurement of a country's aggregate welfare (as measured by national income) is discussed, highlighting the relationship between natural assets and this aggregate welfare. National income is only sustainable if the future potential for welfare generation is not diminished.

The most quoted principle of sustainable development is that offered by the Brundtland Commission: 'development that meets the needs of the present without compromising the ability of future generations to meet their own needs' (WCED 1987). This usual starting point for incorporating intergenerational equity hides much of the complexity which is discussed in Chapter 1. The concepts which underpin the economic definitions of sustainable development are those of *natural capital*, the stock of environmental assets from which humans and non-humans derive services, materials and aesthetic benefits; and of *intergenerational equity*, the principle of not disadvantaging those in the future through action taken in the present.

Complexity arises as natural capital is not homogeneous and because interdependencies exist between functions of environmental assets, such as pollution sinks. Economic development which assumes perfect substitution between natural capital and reproducible (or man-made) capital is not a sustainable development path, as there are certain thresholds and

Table 10.1   Sustainability rules and indicators

Assuming:	No critical natural capital	Critical natural capital
Weak sustainability	a) Savings ratio > depreciation of total capital stock b) Level of natural capital ensures ecosystem stability c) Technical change > population growth	As left + Non-declining critical natural capital
Strong sustainability	As above + Non-declining total natural capital	As left + Recognising role of cultural and ethical capital

Source: Based on Turner (1992).

interactions between these two types of capital that have potentially harmful effects. The recognition of the role of different types of capital and that these may have 'critical' functions leads to the distinction between strong and weak sustainability.

Table 10.1 sets out definitions of sustainability distinguishing between weak and strong sustainability and illustrating how the concept could be measured. Sustainability diverges from the standard economic prescriptions of efficiency by stressing equity (and in some cases cultural and ethical considerations) and in the identification of natural capital which is critical to life support (from a human perspective). Physical limits to the use of this critical capital are required for sustainability: for example, in the climate change or biodiversity loss spheres. Sustainability from an economic perspective requires the allocation of resources such that there is a non-declining capital stock consistent with intergenerational and intragenerational equity. Strong and weak variations within this central concept concentrate on the substitutability of reproducible and natural capital: strong sustainability postulates that the use of certain types of natural capital (critical natural capital) brings about irreversible losses or risk such losses and should therefore be avoided. The combined presence of uncertainty, irreversibility and risk-averse motivations among individuals provides the grounds for the strong sustainability case.

Traditional measures of aggregate income have long been recognised to reflect welfare only partially, due to their inadequate treatment of non-marketed assets, human capital and natural resources. A framework to reflect the use of natural resources at the national level is in the process of being agreed by the United Nations Statistical Office. It will be recommended as the system of national accounts (SNA) to national governments, but there is much debate as to the feasibility of the proposed alterations and potential for further amendment (Bartelmus *et al.* 1991). The adjustments include the pricing of exhaustible resources to reflect asset depletion; defensive expenditures by households and governments which maintain the stock of natural capital; and degradation or enhancement of non-marketed natural capital, such as air, soil and water, biodiversity and historical and cultural assets.

The importance of natural resources has also led to the proposed set of satellite accounts in a new SNA handbook. These satellite accounts will present both physical and monetary data on natural resources, and a section is to be included on the shortcomings of the accounts. This is a first step towards the revision of the measurement of economic performance, but is not the fundamental change or redefinition of the concept of economic growth and the scale of economic activity advocated by Daly (1991), for example.

Other indicators of the interaction of economic growth with social and environmental goals are evolving, such as the United Nations Development Programme's Human Development Index (UNDP 1992). As a first step to measuring *sustainable development*, Pearce and Atkinson (1992) estimate item (a) under weak sustainability in Table 10.1. An economy is sustainable if

$$S/Y > [(\delta_M/Y) + (\delta_N/Y)]$$

where

S = savings
Y = national income
M = reproducible capital
N = natural capital
δ = depreciation

If an economy accumulates capital at a greater rate than it depletes its present stock, then it retains the ability to generate welfare in the future. This is weak sustainability because it assumes perfect substitution between reproducible and natural capital, where strong sustainability stresses the limits to this substitutability. In conceptual terms, the identification of the extent of the substitutability and the definition of non-substitutable natural capital are major research questions.

In cases where the capital is substitutable, the economic valuation problem has to be addressed. By rearranging the sustainability rule an index can be constructed, and this is estimated by Pearce and Atkinson (1992) for 21 countries. The results are critically dependent on the availability of data on changes in the natural resource base, but can indicate that countries may be construed to be sustainable or not; this situation changes with economic profile and over time. The importance of the international economy is recognised: countries may 'import' sustainability. This is the subject of further research.

Critiques of this approach focus either on the inapplicability of a 'capital intact' rule (Nordhaus 1992) or on the consequences of this rule on the *distribution* of welfare in present and future generations (Beckerman 1992). Nordhaus (1992) presents a critique of attempts to measure Hicksian income and suggests that 'capital intact' income, as inherent in the Pearce and Atkinson (1992) sustainability approach, is inappropriate. Nordhaus's criticisms of the maintenance of natural capital approach can be summarised as follows:

(i)   Declining total capital stock can still bring about sustainable welfare in the presence of technical change.
(ii) Ecological economists exaggerate the importance of natural capital. (This is illustrated by Nordhaus's dismissal of the possibility of unpleasant surprises in future climate change scenarios.)
(iii) Natural capital constraints are inappropriate because uncertainty exists in the use of all capital and because of the possibility of large costs in sub-optimal constraints.

Nordhaus contends that Hick's definition of allowing *expected real income* to be maintained is more closely related to sustainability than keeping capital intact, as it allows for technical change. This stresses welfare changes and the role of technical change over time, and hence would lead to prescriptions for economic policy which do not emphasise the role of natural capital. Beckerman (1992) alternatively takes the view that overemphasis on intergenerational equity is inappropriate from the viewpoint of present equity.

The issues addressed in this chapter concern the modification of the standard indicator of growth and well-being in an economy: national income. The study attempts to modify net national income rather than to construct a sustainability rule. Adjustments to net product for degradation of natural capital within the agricultural sector accounts of Zimbabwe are estimated and data deficiencies highlighted. Other individual country case studies for Costa Rica, Indonesia and Mexico, for example, find reduced growth rates over the traditionally calculated rates due to the depletion of oil reserves, fish stocks and forests. Consideration of some non-marketed services of the land-use sector and the role of 'natural' greenhouse gas sinks, however, has produced *positive* net adjustments for the land-use sector of the UK, when narrowly defined (Adger and Whitby 1993). As economic growth is a central objective of macroeconomic policies, and many resource-based economies have undertaken structural adjustment and restructuring in the past decade, the indicator of 'growth' should reflect potential depletion of this resource base. Estimates of the value of the depreciation of the forest stock and economic losses due to soil erosion in Zimbabwe are given below and compared to the net product of that country's agricultural sector.

Sustainable national income

Although a range of revisions have been suggested dealing with environmental and human capital (Ahmad *et al.* 1989; Eisner 1988), there is so far no consensus as to what are correct procedures. The reasons for this stem from the inconsistencies in the underlying economic model of income generated in an economy (Norgaard 1989) and from suggested revisions requiring large capacities for data collection. Feasible revisions in national accounts have at this stage focused on environmental capital, rather than the more intractable problems tackled by social indices, such as the UN's Human Development Index (UNDP 1992). The alternative rationales of particular solutions for incorporation of environmental data into the national accounting system are to provide an indicator of growth or economic well-being, or a tool for *ex ante* planning.

The premise underlying resource accounting is, as outlined above, that natural resources are essential to production and consumption for the maintenance of life-supporting systems, as well as having intrinsic value in existence for intergenerational and other reasons. This leads to the conclusion that natural capital should be treated in a similar manner to reproducible capital in accounting terms, so that the ability to generate income in the future is reduced, if the stock falls (Victor (1991) gives a review of sustainability and capital theory). If a correct value can be placed on natural capital under an accounting system, the implication is that if stocks of natural capital are depleted to increase stocks of reproducible capital and it is assumed that there are no constraints on natural capital use (under a strong sustainability rule, for example), then the ability to generate income in the future will be maintained.

The definition of income recommended by Daly (1989), for example, is that level of consumption which can be enjoyed without jeopardising future generations of income or welfare. This, according to Daly (1989), is consistent with the classic Hicksian definition. Although gross national product (GNP) and gross domestic product (GDP) are most frequently used as proxies for economic growth, net national product (NNP) more accurately reflects the level of consumption which will not leave the economy worse off in the long run as it incorporates an allowance for the run-down of the stock of capital.

The emphasis in the literature has been on the revision of NNP for changes in reproducible capital (K_M) and natural capital (K_N). Daly's (1989) suggested modifications lead to what he describes as 'sustainable social net national product', which is found by subtracting defensive expenditures and depreciation of natural capital from traditional net product measure:

$$SSNNP = C + \dot{K}_M - C_D - \dot{K}_N$$

where

C = aggregate consumption
K_M = reproducible capital stock
C_D = defensive expenditure (or consumption) by households and governments
K_N = natural capital stock
and ($\cdot$) signifies changes in these stocks

Adjustments for defensive expenditures and for marketed natural resources both already enter the national accounting calculus, so the estimation of these, although requiring definitional conventions, does not pose the same problems as estimating changes in non-marketed K_N. Bartelmus *et al.* (1991), whose paper sets out the draft UN guidelines, suggest that *replacement* cost is the feasible way to account for environmental deterioration (i.e. changes in non-marketed K_N). This is the cost of returning the natural environment to the same level of quality as at the start of the accounting period. This is disputed by those advocating that resource rents are the correct monetary price adjustment of K_N (Hartwick 1990a, for example). With either approach the need for physical data is again large, even if the accounts are restricted to defined natural resource sectors such as water and air quality. Valuation of changes in the natural capital stock and defensive expenditures will be discussed in turn.

Marketed and non-marketed natural capital depreciation

Natural capital can be split into exhaustible and renewable resource types, but for both types the empirical estimation of the value of changes in the stock depends on whether a market for them exists directly, or some proxy can be taken. The measurement of the changes of the stock of natural capital are not discussed in Daly (1989), but solutions to this problem, based on the optimal use of exhaustible resources, have been suggested by Hartwick (1990a) and Mäler (1991), among others. Hartwick's (1990a) 'true' net national product is traditional gross product less a rent, for each type of capital (natural and reproducible). This rent is the price less marginal cost multiplied by the change in stock. This results in a formula for modified NNP, denoted NNP^*, which can be summarised as follows:

$$NNP^* = C + \dot{K}_M - (P_E - MC_E)\dot{K}_E + (P_R - MC_R)\dot{K}_R + (P_X - MC_X)\dot{X}$$

where

$\dot{K}_E$	=	$Q_E - D_E$
K_E	=	non-renewable natural capital
K_R	=	$MAI - Q_R$
K_R	=	reproducible capital
P_E	=	price of exhaustible resources
Q_E	=	extraction
D_E	=	discoveries
MC_E	=	marginal cost of extraction of exhaustible resources
MC_R	=	marginal cost of renewable
MAI	=	growth of renewables (mean annual increment)
Q_R	=	harvest
P_R	=	price of renewables
P_X	=	price of pollution
MC_X	=	marginal cost of pollution abatement
X	=	stock of pollution.

The adjustments are then for exhaustible and renewable resources and for changes in the pollution stock, valued at the difference between price and marginal cost of the physical change in quantity. The signs on the adjustments are different, for example between exhaustibles and pollution because where a change in the stock of exhaustibles is negative (extraction > discoveries), and $(P - MC) > 0$ (positive rent), then this leads to a downward adjustment in NNP. For pollution, however, according to Hartwick (1990b; 1991) increases in X, the pollution stock, have $(P - MC) < 0$. The pollution adjustment is then positive, so that an increase in the stock of pollution leads to a decrease in NNP.

The issue of discoveries of exhaustibles (D_E) needs to be highlighted as, with the adjustments set out as above, discoveries of resource stocks constitute negative depletion, so net product potentially may increase and may even exceed gross product. Hamilton (1992) proposes that 'depletion-adjusted' net product is more correct, so that discoveries would be regarded as revaluations of the stock. Hamilton also raises the issue that marginal extraction costs (MC_R) are not likely to be homogeneous across a resource extraction sector and will be difficult to estimate. Notwithstanding these empirical difficulties, this rent-based valuation of resource degradation allows estimation at the aggregate level.

Defensive expenditures

There is debate as to whether defensive expenditure by households and firms should be deducted from net product, following Daly (1989). The intuitive reason for this procedure is that if extra expenditure is incurred to maintain the level of environmental quality due to extra pollution over the period, then the individual's welfare is not enhanced, but the measured net product increases by the amount of the extra expenditure (Leipert 1989, for example). Daly (1992, 183) eloquently illustrates the logic of this position: 'When we add to GNP the costs of defending ourselves against the unwarranted consequences of growth and happily count that as further growth, we then have hyper-growthmania.' Thus, if a river is polluted through a discharge of pollutants, and public expenditure is incurred to restock the river with fish to its original level, the national income rises. A secondary argument is that this expenditure incurred by firms constitutes intermediate expenditure, unlike the same expenditure by households and governments, so the treatment of defensive-type expenditure is inconsistent.

Both of these points are rejected by Mäler (1991) – see also Bojö et al. (1990) and Börlin (1991). Defensive expenditures by firms constitute income to those selling the services and the final demand of the sector producing the intermediate goods *does* enter into GNP, so the inconsistency between the productive sector and the public and household sectors does not exist. The thrust of the argument against the deduction of defensive expenditures is not inconsistency, then, but firstly, that all expenditure increases the welfare of those who have undertaken it; and secondly, that the attribution of defensive expenditures to public sector action could lead to counter-intuitive outcomes (Bojö *et al.* 1990). If, for example, government expenditure on the maintenance of the stock of natural capital inherent in nature conservation were treated as defensive expenditure and subtracted from NNP, then the incentive for a government concerned with growth in NNP would be either to reduce this expenditure or to have it reclassified.

We reject both these points. Although the counter-intuitive outcomes do exist, they simply illustrate an anomaly in the conventions for the treatment of the public sector in national accounts. This does not disprove the more general assertion that deducting defensive expenditure results in a measure which more accurately reflects economic well-being.

The overriding problem with this modification, as with others, is the delineation of the categories of expenditure. In the case of defensive expenditure, it has been suggested that the minimum expenditures which should logically be subtracted from net product are the costs of environmental protection and expenditures for damage compensation. Additional categories suggested by Daly (1989) reflect other concerns in post-industrial societies of decreased quality of life in urban environments. These include the costs of transport congestion, such as the health service costs of accidents; and increased costs of crime prevention. To classify these as defensive expenditures clearly widens the role of aggregate income measures in reflecting quality of life, and strays from the narrow environmental modifications normally considered.

The classification problem (i.e. of government and household expenditures as 'defensive' or not) is the key constraint in the NNP adjustment process, although it remains intuitively appealing that such adjustments are required, and that Daly's formulation of net product be adopted. No attempt is made to quantify defensive expenditure, however, in the following example, which concentrates on renewable resource depletion.

Resource depletion in Zimbabwe

The relevance of the expansion of national accounts to include environmental capital is more critical in those economies with a high reliance on primary production, and development strategies based on these sectors. These are typical characteristics of many developing countries. This has been illustrated by Repetto *et al.* (1989) for Indonesia, with an analysis of the high economic growth rates experienced in the 1970s and 1980s. This growth was in part fuelled by increasing production of tropical hardwoods, hence loss of forest stock, as well as through increased oil production, and through increased agricultural output, the external costs of which include increased soil erosion.

Zimbabwe is such a natural resource dependent economy. Mining and quarrying account for 5.5 per cent of GDP and agriculture 10.9 per cent of GDP in 1987. The mining sector has decreased relative to the size of the economy in the previous 15 years, from 7.6 per cent in 1974, and agriculture has also declined, although it now makes a greater contribution to overall exports. The share of agriculture products and raw materials ranged between 53 and 61 per cent in the decade up to 1987 (Zimbabwe CSO, 1989b). More crucially, up to 80 per cent of the total population rely on agriculture as their major economic activity. The natural regions of Zimbabwe (regions I to V) are defined mainly by soil type and climate because of their influence on land use in the area (see Figure 10.1). Agriculture is organised with a large-scale commercial sector concentrated on regions I and II; a small-scale commercial sector; and communal lands which spread across all of the regions where agriculture occurs. Erosion of the physical agriculture base through soil erosion and deforestation are important considerations in any assessment of the sustainability of income generation in the Zimbabwean economy. Given the importance of the agricultural sector in the Zimbabwean economy, the distribution of income and land leads to soil erosion, in particular, being a sensitive political issue. This sensitivity is further heightened in an era when post-colonial reform of the land ownership and tenure systems is under way. Macroeconomic indicators do not reflect these distributional issues but rely on data formulated to give policy prescriptions at the sectoral level, and must therefore be used with caution. Annual changes in environmental indicators, such as soil erosion and deforestation, are also difficult to assess.

Data requirements for a meaningful analysis of the revisions to aggregate income are large, as highlighted by the comprehensive studies of the World Resources Institute. Data on energy requirements and the markets for fuelwood, for example, have been collected for many countries in energy accounting projects in the 1970s and 1980s. Developing countries have a high reliance on forests for fuel sources, with over 60% of all energy requirements in Africa being met through fuelwood.

Zimbabwe forestry sector

An energy accounting project in Zimbabwe in the 1980s (Hosier 1986) concluded that a shortfall is likely to occur between the supply and demand for fuelwood, given the then present population and relative price levels in Zimbabwe, by 2000. The reduction of the total stocks of forests each year is the difference between the mean annual increment (MAI), which is the increase due to the growth of the existing stock, and

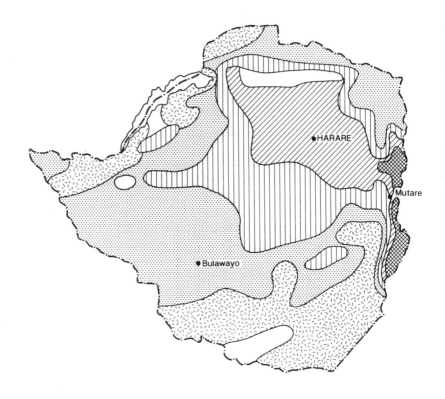

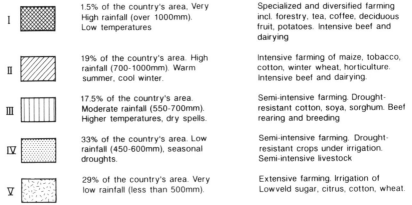

I		1.5% of the country's area, Very High rainfall (over 1000mm). Low temperatures	Specialized and diversified farming incl. forestry, tea, coffee, deciduous fruit, potatoes. Intensive beef and dairying
II		19% of the country's area. High rainfall (700-1000mm). Warm summer, cool winter.	Intensive farming of maize, tobacco, cotton, winter wheat, horticulture. Intensive beef and dairying.
III		17.5% of the country's area. Moderate rainfall (550-700mm). Higher temperatures, dry spells.	Semi-intensive farming. Drought-resistant cotton, soya, sorghum. Beef rearing and breeding
IV		33% of the country's area. Low rainfall (450-600mm), seasonal droughts.	Semi-intensive farming. Drought-resistant crops under irrigation. Semi-intensive livestock
V		29% of the country's area. Very low rainfall (less than 500mm).	Extensive farming. Irrigation of Lowveld sugar, citrus, cotton, wheat.

Figure 10.1 Natural regions of Zimbabwe and their agricultural potential

Table 10.2 Stock and changes of forests in
Zimbabwe, 1987 (million tonnes)

Opening stock	654.49
Mean annual increment (*MAI*)	6.81
Harvest (*Q*)	−9.47
Closing stock	651.83
Stock reduction (*MAI − Q*)	2.66

Source: Based on Hosier (1986).

the harvest. The estimated aggregate figures are given in Table 10.2 for
1987, showing a reduction in stock change of 2.66 million tonnes of dry
weight matter equivalent in the time period.

This reduction in the stock of natural capital would not appear in the
net product of Zimbabwe as traditionally measured, though the consequences of the use of other purchased fuels would be registered. If this
reduction were to be reflected in a modified net product figure, by
subtracting the depreciation of the physical stock valued at the rental
value (following Hartwick 1990a), then for the forestry sector the following calculation is relevant:

$$NNP = C + \dot{K}_M - (P_R - MC_R) \cdot (MAI - Q_R)$$

where

P_R = market price for fuelwood
MC_R = marginal cost of extraction

The market price of fuelwood per tonne in 1987 (P_R) was estimated
to be ZM\$68, taking a weighted average of urban and rural fuelwood
prices based on various reported surveys (Hosier 1988; Du Toit *et al.*
1984). The imputed cost of extraction of fuelwood per tonne is derived
from the estimated time to collect fuelwood in different regions, the
shadow price being the minimum agricultural wage:

$$MC_R = \frac{T_h \cdot W_{min}}{C_h}$$

where

T_h = estimated mean labour input in fuelwood collection per household
W_{min} = minimum agricultural wage
C_h = mean household fuelwood consumption.

Although the estimates are of average costs of extraction, data on marginal extraction costs are not available. It is likely that average extraction costs are rising, hence this estimate understates marginal extraction costs and overstates the value of the physical depreciation of the capital stock. Repetto *et al.* (1989) also uses the *average* rather than the (unavailable) marginal extraction cost in their estimates of depreciation in the fuelwood sector in Indonesia. The results are that net product should be reduced by the value of the physical depreciation of ZM$93.77 million in 1987. If this is borne by the agricultural sector of the Zimbabwean economy, this represents a 9 per cent reduction in the net product of the combined commercial and communal areas' agricultural net product as traditionally measured. This forms part of the K_N adjustment in Table 10.5 along with the adjustment for soil erosion now discussed.

Soil erosion

The value of soil erosion is often quantified in natural resource account studies as it is perceived as a threat to sustaining income and production in the long term, especially in agriculturally based economies. Soil erosion is amenable to the use of physical models which can be extrapolated across land-use data and is generally converted to economic accounts either through productivity loss or through replacement costs of the soil nutrients. However, development policies and land-tenure reforms often use soil erosion as an argument to emphasise the desirability of these reforms. The arguments are in terms of perceived propensity of farmers to take up farming practices which would erode soil, even on land with low erosion risk. The focus of research is therefore on the impacts of soil erosion at the micro level; land-use policy prescriptions should not be derived from extrapolated aggregate erosion cost estimates.

In the World Resources Institute's natural resource accounting study of Costa Rica (Solórzano *et al.* 1991), the national cost of soil erosion over the last two decades ranged between 6 and 11.7 per cent of the agricultural net product per annum. The estimates were, in the main, based on replacement costs of N, P and K nutrients of the physical amount of soil loss (over the tolerable levels of natural replacement),

Table 10.3 Observed and forecast soil erosion rates in Chihota Communal
Area, Zimbabwe

Plots	Slope	Observed soil loss $(t\ ha^{-1}\ yr^{-1})$	Forecast soil loss $(t\ ha^{-1}\ yr^{-1})$
Sandveld			
maize	1.8	0.71	8.3–11.0
grazing	1.8	0.27	0.4
Dambo (grassy headland)			
maize	2.3	0.36	2.0–11.9
grazing	1.9	0.14	0.3–0.6

Source: Roberts and Lambert (1990)

estimated using a physical model (the universal soil loss equation). Accounting for the physical loss requires detailed land-use data, though the results of these extrapolations tend to be overestimates of total erosion due to the coarseness of the database and the assumed management regimes of the soil loss estimators. Results in Table 10.3, for example, for one communal area site in Zimbabwe show large discrepancies between observed soil erosion rates and those predicted using a physical model, Soil Loss Estimator for Southern Africa (SLEMSA).

Uncertainty in modelled estimates is further compounded in the process of extrapolating national estimates. Accepting that aggregate estimates of erosion are uncertain a national accounting adjustment is nevertheless required. From his estimates of soil loss in Zimbabwe, Stocking (1986) provides estimates of the aggregate costs of replacing the nutrients of 15–$20\ t\ ha^{-1}\ yr^{-1}$ on arable land to $75\ t\ ha^{-1}\ yr^{-1}$ on communal grazing land. The cost of replacing N, P and K amounted to US\$1.5 billion (1985 prices) per year. Changes in carbon organic matter are also highlighted, though no attempt is made to estimate replacement cost for these, or to put this in the context of the global carbon cycle and its value as a greenhouse gas sink (Tans *et al.* 1990; Adger and Whitby 1993).

The replacement cost approach is limited, for reasons set out in Solórzano *et al.* (1991): it assumes that fertilisers can replicate lost soils which have different nutrient and structure depth profiles; that thresholds occur above which the soil cannot be recuperated; and that net erosion and natural replacement are difficult to estimate. The WRI Costa Rica study (Solórzano *et al.* 1991) uses the replacement cost method for lack of agroeconomic studies of productivity loss due to erosion. In addition to these, the productivity loss method is closely related to the capital loss as estimated for forest depreciation above. The physical resource loss is valued at the real economic costs over time.

The productivity loss approach is favoured here, though the productivity studies do not cover the range of agricultural activity and give only initial estimates. Grohs (1991; 1992) provides physical and economic assessments of soil erosion and the benefits of soil conservation policies (see also Vogel 1992; Whitlow and Campbell 1989). Given the gross margin analysis of the incremental benefits of avoiding soil erosion, and again extrapolating across Zimbabwean land use, an estimate of the soil adjustment can be made.

The relationship between soil erosion and crop productivity as outlined above is not well determined, so modelling techniques are generally used to predict the erosion–productivity relationship. From these relationships, the aggregate cost of soil erosion can be derived, extrapolating across national land-use data.

Lower yields result from the loss of topsoil, with the economic cost reflected in terms of the change in gross margin (revenue less fixed and variable cost). The change in gross margin per unit output is then:

$$\delta GM_i = (GM_i - GM_i')/Y_i$$

where

i $\quad = 1, \ldots, n$ crops
GM_i $\quad =$ gross margin on crop i without erosion
GM_i' $\quad =$ gross margin on crop i on eroded plot
Y_i $\quad =$ average yield on crop i.

To estimate the total value of the forgone income across all crops, the change in value added is multiplied by the loss of production, calculated as the total production less a range of productivity losses which were estimated through the biophysical models:

$$\sum_{i}^{n} \delta GM_i \cdot P_i$$

where

δGM_i $\quad =$ difference in gross margin per unit output (ZM\$ per tonne)
P_i $\quad =$ total production of crop i.% lost (tonnes).

The results have been estimated by Grohs (1992) for communal land

Table 10.4 Annual income loss due to soil erosion in arable agriculture in Zimbabwe (1988)

Yield loss per cm of soil loss	ZM$ million (1988)	US$ million (1988)
1%	4.52	2.13
2%	9.04	4.26
3%	13.56	6.39

Source: Grohs (1992), Zimbabwe CSO (1989b) and author's calculations.

crops, aggregated using land-use data for the whole of Zimbabwe and a range of estimates for yield loss per centimetre of top soil lost. Communal land comprises 32 per cent of the total farmed land in natural regions I to II, the remainder classified as small-scale commercial or commercial sectors. The results of Grohs (1992) are then extrapolated to cropped land of both commercial and communal lands in the applicable natural regions to give the range of aggregate estimates of the cost of soil erosion presented in Table 10.4.

These estimates are only illustrative and are likely to be underestimates as the extrapolation across all cropped land incorporates a number of factors (some of which may be offsetting) which on the whole point to the estimates in Table 10.4 being a lower bound for the soil erosion estimates for Zimbabwe:

(i)   The estimates assume that erosion loss in commercial areas is equal to that in communal areas within the same agroecological region. This critically depends on management regime, but predictions from the soil loss models would give the same erosion estimates based on topography and other factors.

(ii)   An implicit assumption is that the same cropping pattern occurs in the commercial and communal farms and that the gross margins are applicable. This may seriously underestimate the losses from commercial lands where greater proportions of cash crops are grown. The difference in gross margins between cotton and maize, for example, in communal lands is ZM$ 480 per hectare (Grohs 1991). Thus regimes with more high-value crops such as cotton and tobacco will have greater losses from yield decline due to soil erosion.

(iii)   Erosion also occurs on rangelands, which tends to be in the lower agroecological classifications IV and V, as a result of overstocking (see Abel and Blaikie 1989, for example). This is not included in the estimates taken here.

(iv)   The off-site costs of soil erosion in terms of sedimentation in rivers and irrigation dams and channels are also not considered in the estimates.

The analysis can only be taken as an indicator of the methodology required to estimate the costs of soil erosion on a basis closer to the standard accounting for depletion of renewable resources (rent on the physical asset depletion) rather than the replacement cost method which does not conform to this concept. The problems of finding an economic cost which is recreatable in each accounting period are dependent on a large database of land use and observed erosion rates.

Agricultural sector accounts

The degradation of the resource base of soil and forest stock should then affect the net product as traditionally measured. The agricultural sector accounts are adjusted here in the first instance to reflect the role of that sector as the major location for primary natural resources. The agricultural sector accounts illustrate difficulties which also occur in national accounts. The communal area account, for example, because of lack of data, imputes value to production based on estimates of yield and areas which are not updated each year. The role of subsistence and informal economic activity tends to be ignored in market-based indicators because of this paucity of data. The communal area net product in Table 10.5 is not disaggregated into returns to the factors.

Depreciation of natural capital (K_N), on the basis of the estimates of soil erosion (converting the higher-bound [3% yield decline per centimetre of topsoil lost] 1988 estimates as 1987 via a GDP deflator) and depletion of the forest stock in 1987 sums to ZM$107 million. This loss represents over 10 per cent of the net product as usually measured. The modified net product as estimated here would increase under circumstances where policies to conserve forest or soil reserves led to decreased depletion. Similarly, greater resource depletion, as a result of distorting agricultural or land-use policies, would reduce the modified net product measure, but would not be reflected in the traditional net product measure.

Table 10.5 Modified net product for commercial and communal agriculture sector of Zimbabwe, 1987 (ZM$ million)

Output		2024
Input		987
Gross product		1037
Depreciation K_M		− 62
Net product		975
Less Depreciation K_N		− 107
Modified net product		**868**
of which	Labour (commercial sector)	333
	Farming income	372
	Communal sector	332
	Depreciation K_M	− 62
	Depreciation K_N	− 107

Note: Part of K_R forms part of farming income but is not reported separately. Communal sector accounts are not broken down into disbursement of net product.

Sources: Zimbabwe CSO (1989a) and author's calculations. GDP deflator for agriculture from Masters (1990).

Conclusion: Macroeconomic indicators and economic growth policies

The need for sustainability modifications to national accounting systems has been recognised as an important contribution to better resource management as well as to the realisation that economic growth (as measured by growth in traditional economic indicators) will not necessarily reduce poverty or protect the environment. The collection of data on resource use on an annual basis, as proposed in the intermediate satellite account system, would be beneficial to planning in primary resource dependent economies, whether or not the measures of national income are revised (Repetto *et al.* 1989, 53). The initial results presented here point to the large data requirements of and analytical problems with any sustainability modifications that could be practically implemented in a revised system of national accounts.

However, measuring sustainable income is a backward-looking exercise; what is defined is *potentially* sustainable income. As an objective of economic policy, increasing the flow of sustainable income is desirable, and would serve to internalise the value of environmental assets into macroeconomic calculus. It will not, however, necessarily bring about *sustainable development*, as it does not address intragenerational equity issues or deal with exported environmental pollution.

Market liberalisation and exchange rate devaluation policies, such as those undertaken under Structural Adjustment Programmes, are perceived as having ignored the effects of the potential impacts on the natural resource base (see Mearns 1991, for example). This has occurred despite the direct relationship between macroeconomic policy, the resource base and environmental quality. Structural adjustment policies which are based on growth in sectors which cause natural resource depletion, or domestic pricing policies which encourage greater resource use, can therefore be highlighted using resource accounting techniques. Structural adjustment has wide-ranging impacts on economies and a critical effect on resource use (see, for example, Harrigan and Mosley 1991; Longhurst *et al.* 1988; Smith 1989). Smith (1989) concludes that structural adjustment has in general in sub-Saharan Africa had negative effects on income distribution but not raised agricultural production. Similarly, Cruz and Repetto (1992) conclude that macroeconomic policy in the Philippines has encouraged resource extraction and disinvestment in the primary resource sector without producing any offsetting long-term increase in the industrial sector. These policies also cause negative effects in terms of poverty and pressure on marginal natural resources.

Critically important areas in the macroeconomic restructuring of economies, such as the phasing of the implementation of policies, the role of institutional changes, food security and the urban/rural biases, would only be *indirectly* reflected in macroeconomic indicators, such as changes in GNP per capita. Other indicators are therefore required in order more comprehensively to assess the consequences of macroeconomic policy.

The degradation of the physical agricultural base through soil erosion and deforestation is an important consideration in any assessment of the sustainability of income generation in the Zimbabwean economy. The question of sustaining the level of natural resources is, however, inextricably linked to the ownership and control of these resources (especially land), and hence to the distribution of the income, in the post-colonial period in Zimbabwe. As with all macroeconomic analyses, these issues are not addressed in the estimation of sustainable income indicators. Thus the formulation and estimation of such indicators are a necessary but not sufficient condition for the achievement of sustainable development.

References

Abel, N. and Blaikie, P. (1989). Land degradation, stocking rates and conservation policies for the communal rangelands of Botswana and Zimbabwe. *Land Degradation and Rehabilitation* 1(1), 1–23.

Adger, W.N. and Whitby, M.C. (1993). Natural resource accounting in the land use sector: theory and practice. *European Review of Agricultural Economics* 20(1), 77–97.

Ahmad, Y.J., El Serafy, S. and Lutz, E. (eds) (1989). *Environmental Accounting for Sustainable Development*. World Bank, Washington, DC.

Bartelmus, P., Stahmer, C. and van Tongeren, J. (1991). Integrated environmental and economic accounting: framework for a SNA satellite system. *Review of Income and Wealth* 37(2), 111–48.

Beckerman, W. (1992). Economic growth and the environment: Whose growth? Whose environment? *World Development* 20(4), 481–96.

Bojö, J., Mäler, K.-G. and Unemo, L. (1990). *Environment and Development: An Economic Approach*. Kluwer, Dordrecht.

Börlin, M. (1991). Environmental defensive expenditures by industry do not inflate GNP? The GNP revision revisited. *The Environmentalist* 11(4), 312–13.

Cruz, W. and Repetto, R. (1992). *The Environmental Effects of Stabilization and Structural Adjustment Programs: The Philippines Case*. World Resources Institute, Washington, DC.

Daly, H.E. (1989). Toward a measure of sustainable social net national product. In Ahmad, Y.J., El Serafy, S. and Lutz, E. (eds), *Environmental Accounting for Sustainable Development*. World Bank, Washington, DC.

Daly, H.E. (1991). Elements of environmental macroeconomics. In Costanza, R. (ed.), *Ecological Economics: The Science and Management of Sustainability*. Columbia University Press, New York.

Daly, H.E. (1992). The steady-state economy: alternative to growthmania. Reprinted in *Steady-State Economics*. Second edition with new essays. Earthscan, London.

Du Toit, R.F., Campbell, B.M., Haney, R.A. and Dore, D. (1984). *Wood Usage and Tree Planting in Zimbabwe's Communal Land*. Zimbabwean Forestry Commission and World Bank, Harare.

Eisner, R. (1988). Extended accounts for national income and product. *Journal of Economic Literature* 26(4), 1611–84.

Grohs, F. (1991). *An Economic Evaluation of Soil Conservation Measures in Zvimba and Chirau Communal Lands*, Working Paper AEE 8/91. Department of Agricultural Economics and Extension, University of Zimbabwe.

Grohs, F. (1992). Monetarising environmental damages: A tool for development planning? A case study of soil erosion in Zimbabwe. Paper presented to 'Investing in Natural Capital' International Society for Ecological Economics Conference, Stockholm, August.

Hamilton, K. (1992). *Proposed treatments of the environment and natural resources in the national accounts: A critical assessment*. Statistics Canada, mimeo.

Harrigan, J. and Mosley, P. (1991). Evaluating the impact of World Bank structural adjustment lending: 1980–87. *Journal of Development Studies* 27(3), 63–94.

Hartwick, J.M. (1990a). Natural resources, national accounting and economic depreciation. *Journal of Public Economics* 43(3), 291–304.

Hartwick, J.M. (1990b). *Pollution and National Accounting*, Discussion Paper 772. Department of Economics, Queen's University, Kingston, Ontario.

Hartwick, J.M. (1991). Degradation of environmental capital and national accounting procedures. *European Economic Review* 35(2–3), 642–9.

Hosier, R.H. (ed.) (1986). *Zimbabwe: Energy Planning for National Development*. Beijer Institute and Scandinavian Institute of African Studies, Stockholm.

Hosier, R.H. (ed.) (1988). *Energy for Rural Development in Zimbabwe*. Beijer Institute and Scandinavian Institute of African Studies, Stockholm.

Leipert, C. (1989). National income and economic growth: the conceptual side of defensive expenditures. *Journal of Economic Issues* 23(3), 843–56.

Longhurst, R., Kamara, S. and Mensurah, J. (1988). Structural adjustment and vulnerable groups in Sierra Leone. *IDS Bulletin* 19(1), 25–9.

Mäler, K.-G. (1991). National accounts and environmental resources. *Environmental and Resource Economics* 1(1), 1–15.

Masters, W.A. (1990). *The Value of Foreign Exchange in Zimbabwe: Concepts and Estimates*, Working Paper AEE 2/90. Department of Agricultural Economics and Extension, University of Zimbabwe, Harare.

Mearns, R. (1991). *Environmental Implications of Structural Adjustment: Reflections on Scientific Method*, Discussion Paper 284. Institute of Development Studies, Brighton.

Nordhaus, W.D. (1992). Is growth sustainable? Reflections on the concept of sustainable economic growth. International Economic Association Conference, Varenna, Italy, October.

Norgaard, R. (1989). Three dilemmas of environmental accounting. *Ecological Economics* 1(4), 303–14.

Pearce, D. and Atkinson, G. (1992). *Are National Economies Sustainable? Measuring Sustainable Development*, Global Environmental Change Working Paper 92–11. Centre for Social and Economic Research on the Global Environment, University College London and University of East Anglia.

Repetto, R., Magrath, W., Wells, M., Beer, C. and Rossini, F. (1989). *Wasting Assets: Natural Resources in the National Income Accounts*. World Resources Institute, Washington, DC.

Roberts, N. and Lambert, R. (1990). Degradation of dambo soils and peasant agriculture in Zimbabwe. In Boardman, J., Foster, I. and Dearing, J. (eds), *Soil Erosion on Agricultural Land*. John Wiley and Sons, Chichester.

Solórzano, R., de Camino, R., Woodward, R., Tosi, J., Watson, V., Vásquez, A., Villalobos, C., Jimenez, J., Repetto., R. and Cruz, W. (1991). *Accounts Overdue: Natural Resource Depreciation in Costa Rica*. World Resources Institute, Washington, DC.

Smith, L.D. (1989). Structural adjustment, price reform and agricultural performance in sub-Saharan Africa. *Journal of Agricultural Economics* 40(1), 21–31.

Stocking, M. (1986). *The Cost of Soil Erosion in Zimbabwe in Terms of the Loss of Three Major Nutrients*, Consultants Working Paper 3. FAO, Rome.

Tans, P.P., Fung, I.Y. and Takahashi, T. (1990). Observational constraints on the global atmospheric CO_2 budget. *Science* 247(4949), 1431–8.

Turner, R.K. (1992). *Speculations on Weak and Strong Sustainability*, Global Environmental Change Working Paper 92–26. Centre for Social and Economic Research on the Global Environment, University of East Anglia and University College London.

United Nations Development Programme (1992). *Human Development Report 1992*. Oxford University Press, Oxford.

Victor, P.A. (1991). Indicators of sustainable development: some lessons from capital theory. *Ecological Economics* 4(3), 191–213.

Vogel, H. (1992). Effects of conservation tillage on sheet erosion from sandy soils at two experimental sites in Zimbabwe. *Applied Geography* 12(3), 229–42.

Whitlow, R. and Campbell, B. (1989). Factors influencing soil erosion in Zimbabwe: a statistical analysis. *Journal of Environmental Management* 29(1), 17–29.

World Commission on Environment and Development (WCED) (1987). *Our Common Future*. Oxford University Press, Oxford.

Zimbabwe Central Statistical Office (1989a). *Production Account of Agriculture, Forestry and Fisheries, 1980–1988*. Central Statistical Office, Harare.

Zimbabwe Central Statistical Office (1989b). *Statistical Yearbook of Zimbabwe*. Central Statistical Office, Harare.

Chapter 11

Controlling Water Pollution using Market Mechanisms: Results from Empirical Studies[1]

Nick Hanley and Ian Moffatt

Introduction

This chapter looks at the extent to which the theoretical advantages of 'market mechanisms' (principally tradable permits and emission taxes) can be substantiated by empirical work, with particular reference to the control of water pollution. Accordingly, we briefly review these theoretical advantages, and the theoretical problems associated with such mechanisms, before describing the methodology common to many empirical studies of water pollution control economics. We then look at three such studies: the Fox River; the Tees Estuary; and, in much more detail, the Forth Estuary. We also comment briefly on the extent to which market mechanisms are actually used to control water pollution. To preview, we will find that there are many practical problems associated with applying market mechanisms to water quality management, and that, partly because of this, it is hard to find 'pure' versions of such policies in existence. This is despite a considerable body of evidence from simulation studies that potentially large cost savings are available under taxes and/or permits. There are criteria of relevance other than efficiency, however, and we shall find that these are particularly relevant in the Forth Estuary case study.

Background

The economics of pollution control

One of the first arguments that students of environmental economics become familiar with is that uniform regulations, either of emission levels or of production technology, are inefficient as a means of achieving desired reductions in emission levels. This finding, associated primarily with Baumol and Oates (1971; 1988), holds whenever the marginal costs of pollution control or *marginal abatement costs* (MAC) vary across polluters. If the target of a control agency is specified in terms of ambient standard measures of, say, water quality, then another source of inefficiency of uniform standards is significant variations in the economic value of the environmental impact of discharges, or *marginal damage costs* (MDC). Whenever MAC and/or MDC vary across sources, then efficiency (defined here as the achievement of our target at lowest resource cost) requires flexibility in control, with most control being targeted at low-MAC/high-MDC sources. This flexibility is precisely the desirable attribute of both taxes on emissions and tradable emission permits (Montgomery 1972). Either system establishes a price on pollution, and firms adjust their emission levels to equate (for the simple case of equivalent MDCs) this price with their own MAC schedule. If there is only one price, then this results in MACs being equalized across all sources, a necessary condition for efficiency.

When MDCs vary, then for tax schemes the tax rate must vary across sources according to relative damage costs. With a perfectly differentiated tax scheme, cost-minimizing sources will act in such a way that, in equilibrium, the marginal cost of *pollution* reduction (as opposed to emission reduction) will be equalized across sources. Again, in the simple version of the story, efficiency is the result.

The many problems with tax schemes have been developed in the literature since Baumol and Oates's original paper. A full discussion of these is given in, for example, Hanley *et al.* (1990). There are three problems in particular which we should pick up here, as they relate to the empirical studies to be discussed shortly.

The first problem is the large information requirements of a tax scheme. The control agency must know all firms' MAC functions, or else attempt to iterate onto the correct tax rate (but there are problems here: see Walker and Storey 1977). Moreover, this 'correct' rate will change every time the aggregate MAC function changes in current value terms. When MDCs vary then the control agency (EPA) must at least guess how MDCs do vary across firms: imposing a single tax rate in such circumstances can be very expensive in resource cost terms, perhaps even more expensive than uniform standards (Seskin *et al.* 1983).

Secondly, simple tax schemes, where no refunds of tax revenue are made, may be cost-minimizing to society, but can be very expensive to firms in terms of their distributional impact. This makes it likely that firms will lobby against tax schemes, and in favour of standards (see Hanley *et al.* 1990).

Thirdly, unless the EPA gets the tax rate correct *and* firms actually minimize costs, then the desired improvement in environmental quality may not be forthcoming. Taxes thus, as stand-alone instruments, increase the level of uncertainty.

Tradable discharge rights (TDRs) get round all of these problems under certain conditions. For example, if they are initially freely given away ('grandfathered'), then the transfer of funds to the EPA, implicit under an uncompensated tax scheme, is avoided, the EPA does not need to know all firms' MAC functions, since permits are a quantity control, with the permit market setting the price. Finally, so long as firms do not cheat, then the desired improvement in water quality should be forthcoming if the EPA has got its sums right. However, when MDCs vary (which is very likely in many water environments), then permit trades at a one-for-one rate across the whole control area may result in violations of ambient quality objectives, as 'hotspots' occur. Given the difficulty in implementing the original solution to this problem (the ambient permit system (Tietenberg 1980)), then a permit market must be subjected to trading rules (see, for example, Atkinson and Tietenberg 1987). This reduces the number of cost-saving trades, and means that the permit system will not exactly hit the least-cost solution. Finally, permit markets may also be subject to imperfectly competitive behaviour, especially when the number of potential traders is small. This means that the cost of hitting a target rises, but most studies in this area have found this effect to be relatively weak (Maloney and Yandle 1984).

This section has summarized some of the problems and potentials of market mechanisms for controlling pollution. Before we go on to see whether these findings are borne our empirically, let us briefly look at how water quality is actually managed.

Water management practices

As recent surveys have shown, water pollution control in the West is dominated by regulation, with very few examples of market mechanisms in use (Environmental Resources Limited (ERL) 1990; Opschoor and Vos 1989). This is indeed also the case when one looks at air pollution control, with the obvious exception of the Emissions Trading Program in the USA. To take Great Britain as an example, pollution of water courses (rivers, streams, lakes, estuaries and coastal waters) is regulated by 'consents', set either by the National Rivers Authority (NRA) in England and Wales, or the River Purification Boards (RPBs) in Scotland.[2] Consents specify both the quantity and composition of discharges, and are enforced by monitoring and, if necessary, fines through the courts (although water pollution control agencies in the UK have traditionally favoured persuasion over action through the courts –

what Storey (1977) has referred to as 'arm twisting'). Process specification, used widely in air quality management, plays a much less prominent role, with regulations mostly relating to storage of hazardous substances. Both the NRA and RPBs control the composition of discharges, rather than the manner in which this composition is achieved.

The characteristics of discharges which may be controlled are many, but discharges can be divided into a number of classes:

(i)   organic discharges of a simple nature, such as domestic sewage;
(ii)   more complex organics;
(iii)   heavy metals, such as copper and lead;
(iv)   'Red List' substances, such as Atrazine (a herbicide) and DDT;
(v)   cooling water.

Water authorities use a number of parameters to determine consent conditions. These include biological oxygen demand (BOD) and chemical oxygen demand (COD), both of which measure the oxygen depletion potential of an effluent; temperature and salinity; colour; turbidity (the presence of suspended solids, giving rise to cloudiness); pH value; ammonia levels; and concentrations of trace metals and Red List substances. The quantity, as distinct from the quality, of discharge is also regulated, while combining quantity (daily flow, for example) with quality (e.g., BOD in milligrams per litre) gives a further possible control parameter, that of pollution loading.

Recently, both the NRA and the RPBs have moved towards introducing a charging scheme for discharges (NRA 1990; Scottish RPB Association 1991). However, both these schemes are intended purely to recover the costs incurred by the control agencies in regulating dischargers: they are not set at 'incentive' levels. By this, I mean that the charge levels alone would be insufficient to achieve current levels of pollution control, as they are set much too low to achieve this end. These charges will coexist with the existing regulatory scheme, with the latter being entirely responsible for bringing emission levels down to the desired level.

This setting of charges below incentive levels, and the coexistence of charges and regulation, are obviously far from the least-cost tax policy proposed by Baumol and Oates (1988). However, such schemes are becoming increasingly common in Europe. In Germany, France and the Netherlands, similar schemes exist. The German scheme imposes consent limits on dischargers, but also levies a range of charges intended to have an incentive effect, since as dischargers improve their emissions, the charge per unit of emission is lowered. Charges are calculated with regard to both the content and volume of discharges, while charge revenue is accounted for by administration costs and (the residual) the cost of environmental improvements. In the Netherlands, charges are set at a higher level than in Germany overall (ERL 1990), and a positive correlation was found between variations in charge levels across regions, and reductions in industrial pollution. Charge revenues go to administration costs, with the balance available as capital grants for the upgrading/installation of treatment works. France sets charges at the lowest level of the three, with revenues (net of administration costs) again going to fund treatment works.

Results from previous simulation studies

In trying to find out whether the theoretical cost savings of market mechanisms over uniform regulation would be likely actually to come about should such schemes be introduced, economists have engaged in simulation modelling. This work, which has been much more widespread in the field of air pollution than in water pollution, has to a large degree followed a set method. This section first outlines this method, before describing two pieces of work, one American and one British. The following section then summarises the findings of another, more recent, British study, of the Forth Estuary in central Scotland.

Methodology

Researchers trying to model alternative systems for water pollution control face two problems: first, how to simulate the reaction of dischargers to alternative policies; and second, how to simulate the impacts of these alternatives on water quality. In coping with the former of these difficulties, extensive use has been made of mathematical programming (MP) techniques, at the simplest level, by using linear programming (LP). MP methods allow an 'objective function', Z, to be minimized or maximized subject to a series of constraints. Starting from the Baumol and Oates theorem, this has usually been implemented as:

$$\min Z = \phi \; (\mathbf{C}) \tag{11.1}$$

where $\mathbf{C}$ are aggregate pollution control costs. More explicitly, if there are n dischargers, who can all engage in pollution abatement activities X_i, $i = 1, \ldots, n$, then the problem is to minimize the sum of abatement costs, but subject to a series of constraints. In the simplest possible case, these might be that the sum of emission reductions is at least as great as the target of total desired reductions in emissions (the case of uniform mixing); and that negative amounts of emission reductions are not allowed (so that no source is allowed to increase emissions: the 'non-negativity constraint'). This non-negativity constraint may be relaxed if polluters are allowed to increase emissions, so long as emissions reductions from other sources more than compensate (uniform mixing); or so long as ambient quality targets are not violated at any receptor (non-uniform mixing). If the target reduction in emissions is given as X^*, then the problem is to:

$$\min Z = \sum_{i=1}^{n} C_i (X_i) \tag{11.2}$$

subject to

$$\sum_i X_i \geqslant X^* \tag{11.3}$$

and

$$X_i \geqslant 0, \qquad \forall\, i \tag{11.4}$$

If LP is used to solve this problem, then three useful outputs are produced: the cost minimizing solution to the problem, in terms of the distribution of emission reductions across sources, and the total resource cost of this pattern; a sensitivity analysis, showing by how much the marginal abatement costs of each source can change before they either enter or leave the optimal solution; and a dual value analysis, showing by how much the value of the minimand (the resource cost of the scheme) changes if any of the constraint right-hand side values are changed by one unit. In the above problem, this means that we know the resource cost (benefit) of tightening (relaxing) the target X^* by one unit. This can be interpreted as the shadow value of the right to emit pollution, which as Baumol and Oates show (using the Kuhn–Tucker conditions for constrained optimization of a problem with inequalities such as (11.3) and (11.4), is equal to either the least-cost tax rate which would allow the target to be hit or the equilibrium permit price if a TDR market is used to hit the target.

If we move away from the uniform mixing case, we need to be able to control for the differing spatial impact of emissions. To do this requires the construction of an environmental quality model. For example, if we wish to know how sewage discharged to a lake from source i, discharging at point k, affects dissolved oxygen (DO) levels at point

j in the lake, we need a water quality model describing how the BOD levels of discharges from all points k (= 1, . . ., n) affect DO levels at all points j (= 1, . . ., p) in the lake (points j are often referred to as 'receptor points', as they are typically located at monitoring stations).

Such water quality models are complicated to construct, as they must be able to simulate, with a high degree of accuracy, a wide range of physical, chemical and biological processes going on in the particular water body. Such models may be either one-, two- or three-dimensional, with three-dimensional models being the most complex. While some of the above processes follow standard forms (such as the re-aeration of water from the atmosphere), models must be individually calibrated (by adjusting parameter values) for each separate water body. This calibration will take account of, for example, volumes and types of inflow to the system, surface area, and depth. For estuaries, the problem is complicated by the influence of the tide, and by the mixing of fresh and saline waters. We return to these points in a later section.

Suppose, however, that we have been able to obtain a calibrated and tested model of a particular lake. Using this model, we are able to predict the impact of discharges from any source i on any receptor point j in the lake. These impacts are usually formalized as 'transfer coefficients', which tell us the marginal impact of discharges from source i on point j.

Now let us reformulate our environmental target in terms of ambient water quality at each point j, instead of in terms of total reductions in emissions to the lake. One measure of ambient water quality frequently used is DO. Calling the minimum desired DO level at each receptor point DO* (which is assumed to exceed current DO levels at at least some receptor points), then we could find the necessary reduction in BOD loading that would achieve this target in each stretch. Call this necessary reduction ΔD_j. Our problem now becomes that of minimizing equation (11.2) subject to (11.4) and

$$\sum_i d_{ij} X_i \geqslant \Delta D_j, \quad \forall_j \tag{11.5}$$

where equation (11.5) now replaces equation (11.3) in the constraint set. The d_{ij} values are the transfer coefficients. Solving this problem using MP will again yield the cost-minimizing pattern of emission reductions, the resource cost involved, the sensitivity of this solution to changes in MACs, but now we get a separate shadow or dual value for each reach j. These values have similar interpretations as in the uniform mixing case, in that they constitute either the least-cost tax rates at each receptor point in a perfectly differentiated charge scheme, or the equilibrium permit prices in each ambient permit market (there will be as many markets as there are receptor points). Also, these transfer coefficients may be used to impose trading rules in simulations of emission-based TDR markets. For instance, they can be used to impose the constraint that any permit transfer must not lead to the violation of target DO levels at any receptor point.

The Fox River study

The Fox River, in Wisconsin (USA) is unique in that it has been the subject of both a simulation study of a TDR market to control BOD emissions (O'Neil *et al*. 1983) and a policy innovation to allow the trading of BOD permits. It therefore offers the opportunity of comparing predictions with actual outcomes.

The lower Fox River was classified as 'water quality limited' in 1981 by the Wisconsin Department of Natural Resources (the state water pollution control authority). The combined effects of organic discharges from ten pulp and paper mills, and four sewage works, led to DO levels falling below the water quality standard of 6.2 mg/l in conditions of low flow and high temperature (the impact of a given organic waste on DO levels is dependent, *ceteris paribus*, on both flow rates and temperature, as the former determines dilution levels, and the latter the speed at which micro-organisms reproduce). Given the location of the discharges, and the nature of the river, two 'sag' points were found, being points where DO levels reached their lowest levels. Achieving DO targets at these two sag points should achieve DO targets everywhere else on the river.

A one-dimensional dynamic model of water quality was developed ('Qual-III'), which enabled transfer coefficients to be estimated (at low flow conditions) relating DO levels at each sag to BOD discharges from all identified major point sources. These were incorporated with target DO levels (converted into target reductions in BOD loading) to give constraints of a form similar to equation (11.5) above. Data gathered on pollution control costs showed that MACs varied widely across dischargers, and that when weighted by transfer coefficient, the marginal cost of water quality improvement varied by an even greater amount. This can be seen from Table 11.1.

These MAC data were then incorporated into equation (11.2), enabling the researchers to calculate the least-cost solution for a range of DO targets. This least-cost solution was taken to be identical to the outcome of a perfectly competitive permit marker. These outcomes were then compared with the resource costs of meeting the targets with a system of uniform restrictions, whereby each source was compelled to reduce

Table 11.1 Marginal cost of oxygen increases, Fox River study

Discharger	Impact[1] coefficient h_{i1}	Marginal abatement cost 'end-of-pipe' C_i ($/lb)	Marginal cost of DO increase 'at sag' C_i/h_{i1} ($/\mu g/l$)
1	107	7.20	72
2	189	2.10	11
3	373	1.90[2]	5[2]
4	231	3.10	14
5	184	1.80	11
6	214	7.90	37
7	101	2.60	27

Source: O'Neil *et al.* (1983, 350).

[1] For flow, temperature values 950 cfs 80°F

[2] Plant is operating at maximum abatement capacity; the numbers shown in the table are costs for the last unit treated.

emissions by a uniform amount. O'Neil *et al.* found, to take a DO target of 6.2 ppm as an example, that the TDR market would involve an annualized cost of only $16.8 million (at low-flow conditions), compared with a cost of $23.6 million under uniform regulation. This finding, that there were significant cost savings for TDRs relative to uniform regulation, held up under all target DO levels/flow conditions. Powerful evidence, it would seem, in support of market mechanisms to control pollution in the Fox River.

However, when Hahn (1989) reviewed the success of the Fox River TDR scheme, he found that this had fallen far short of expectations. In fact, only one trade had occurred in the six years studied by Hahn, with the cost savings predicted by O'Neil *et al.* failing to materialize. Several explanations have been put forward for the apparent failure of the Fox River scheme:

(i)   Many of the dischargers were in product market competition with each other. This might lead to their finding it in their competitive advantage not to sell permits to each other, and so restrict rivals' output levels; while municipal dischargers may be under less of an incentive to minimize costs than private firms.

(ii)   Due to uncertainty over the impact of trading on future allocations of permits (permits were initially valid for five years), firms may also have been reluctant to sell.

(iii)   The Wisconsin Department of Natural Resources imposed a variety of trading rules which raised the transactions costs of the scheme. What is more, trades that were proposed solely to reduce operating costs were not allowed, which in one sense counters the whole point of the scheme.

(iv)   If pollution control costs are a relatively small percentage of total costs, firms will be reluctant to allocate scare managerial talent to reducing these costs through setting up potentially complex trades. For the paper industry in Wisconsin, pollution control costs were estimated at less than 1 per cent of total costs.

(v)   Finally, it appears that the scheme has never been enthusiastically received by the dischargers affected by it.

So what conclusions should we draw from the Fox River scheme? The work of O'Neil *et al.* showed that large cost savings were there to be taken advantage of. But the market circumstances of firms inside the scheme, the trading rules imposed on them, and the considerable degree of uncertainty over future allocations all combined to reduce considerably the actual number of trades. With so few permit trades, potential cost savings could not be realized.

The Tees Estuary study

In the 1970s, the Tees Estuary was one of the most polluted in the UK, with large quantities of both industrial effluents and domestic sewage inputs. Ambient levels of heavy metals, phenols, ammonia and cyanide combined with other inputs to render the estuary 'grossly polluted', and incapable of supporting fish life, from the town of Stockton to the mouth of the estuary (Rowley *et al.* 1979). In 1973, following a recommendation from the Royal Commission on Environmental Pollution, a research team from the University of Newcastle upon Tyne began to investigate the possibility of achieving satisfactory water quality levels in the Tees using emission discharges, instead of regulation by consent. To this end, they combined a dynamic programming model of nine major industrial sources of either BOD/COD or of toxic materials such as cyanide (two major municiple sources were omitted from the economic model, due to non-convexity problems). This economic model allowed for a least-cost solution to be found, when combined with transfer coefficients generated from a point-specific one-dimensional water quality model of the estuary supplied to the research team by ICI. Once the least-cost solution had been found for any particular target, then this allowed the calculation of stretch-specific emission charges. These charges could then be brought into the economic model, and their impact on water quality and on firms' costs simulated.

A range of increasingly severe water quality objectives were employed as targets. This is much more sophisticated in approach than the Fox River study, since these objectives are stipulated in terms of, for example, the migration up the estuary of coarse fish and, with a better water quality, also game fish. This means that parameters other than just DO must be controlled for. Included in the water quality objectives were maximum allowable concentrations of toxins, principally ammonia, cyanide, and copper, measured in 'fractional toxicity indices'. For these three determinants, as for BOD, charge levels were calculated at 0.3-mile intervals along the estuary, for each of the three water quality objectives.

A major finding of the Tees Estuary study was that the resource cost of meeting water quality objectives was much lower under the least-cost solution than under either uniform regulations ('cutbacks') or under a uniform tax rate. However, even the perfectly differentiated tax system modelled was unable precisely to hit the least-cost solution due to indivisibilities in MAC functions.

This 'existence problem' (Walker and Storey 1977) is shown in Figure 11.1. For simplicity, we assume just one discharger, with an MAC function of MAC*. This function is stepped, because of indivisibilities in pollution control: in moving from emissions E_4 (the no-control position) to E_3, a cleansing plant K_1 is used, with constant variable costs. In order to reduce emissions below E_3, however, new plant must be installed. The MAC function jumps, then levels off, since this new plant also has constant variable costs. Suppose the pollution control authority wishes to achieve an emissions reduction from E_4 to E_2. Suppose also that it knows the form of MAC*, and so sets the tax rate at t_1 (where MAC* = E_2). The firm, we know, will equate tax payments with control costs at the margin. This is true, however, anywhere in the range E_1–E_3. The firm minimizes *total* control costs by selecting emission level E_3, which is too great. Yet if the control agency raises the tax rate even slightly (to, say, t_2) then too much abatement is forthcoming: E_0 in this case.

The Newcastle team also found that control costs were very sensitive to the time period within which targets were to be met; and that there were very large variations in the charge rates for, say, BOD, across the different stretches. This might give rise to accusations of unfairness should such a scheme be introduced (although the tax rates partly reflect the varying marginal physical impacts of dischargers, varying according to location, and thus could be seen to be in the spirit of the polluter-pays principle.

The Tees study showed that emission taxes offered only a partial

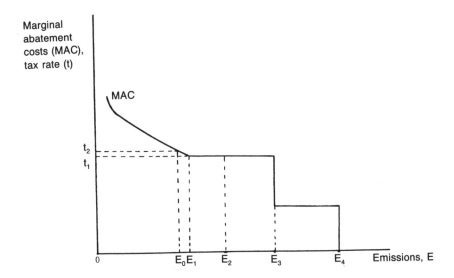

Figure 11.1   Indivisibilities in pollution control

solution to water quality management in the Tees, since the informational requirements were high, and because of the existence problem noted above. Simplified charge schemes, or a mixture of charges with regulation would retain some of the cost-saving potential of the pure charge scheme, and be more feasible.

The Forth Estuary study

Description of the case-study area

The Forth Estuary, in central Scotland, is a good example of a multiple-use water body. It is a source of cooling water to a wide range of industries located along its shoreline; a sink for the dumping of municipal and industrial effluents; a home for fish and birds; a site for water sports; and a migration route for salmon moving to and from the sea and the three salmon rivers feeding into the estuary. The estuary has been defined as 'partially mixed', with the level of salinity varying along the length of the estuary: saltwater, being heavier than freshwater, lies

Table 11.2   Environmental quality standards (EQS) versus levels for water quality determinands, Forth Estuary, 1988 (annual values)

Determinand	Unit	EQS	Range	Mean
pH		6.0–8.5	7.19–7.93	7.5
DO	mg/l	4.5	6.51–10.23	8.03
Cadmium	µg/l	5.0	0.025–0.052	0.036
Mercury	ng/l	500	0.9–19.6	8.0
Copper	µg/l	5.00	0.51–1.15	0.87
Arsenic	µg/l	25.0	2.1–4.6	3.4

Source: FRPB, various.

at the bottom of the water column, and is mixed with freshwater due to tidal effects and freshwater inflows.

Nationally, the Forth is a relatively clean estuary, as the data in Table 11.2 show. If annual means are considered, then for all measured parameters, actual levels lie within their Environmental Quality Standards (EQS). However, for DO levels, this annual mean hides a problem that occurs during conditions of low flow and warm weather, when, as Figure 11.2 shows, DO levels sag below the EQS of 4.5 mg/l for a considerable distance down the upper part of the estuary. What is more, the Forth River Purification Board (FRPB), the relevant pollution control authority, has indicated that it would like to see the EQS raised, which would mean more violations of the EQS given current discharge levels.

The major sources of BOD in the estuary are point source discharges from industrial processes and from sewage works. There is also a background BOD loading from non-point sources (e.g., runoff from farmland), but this is thought to be relatively insignificant. Table 11.3 gives data on BOD inputs to the estuary, which we define as extending from Stirling Bridge down to the Forth Road Bridge near Edinburgh. As may be seen, the background loading from the River Forth is relatively small. Major sources, which are also shown in Figure 11.3, include seven industrial sources (jointly accounting for 83 per cent of BOD loading) and 13 sewage treatment works (STWs), jointly accounting for 16 per cent of total BOD loading. Industrial discharges therefore dominate the picture, although two STWs are currently problematic, given their location at sensitive points in the estuary: these are the Alloa and Stirling works, located in the summertime sag points.

A research team at the University of Stirling looked at the potential for improving water quality in the estuary using market mechanisms. In

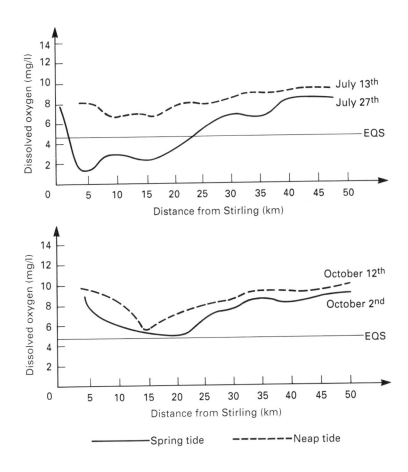

Figure 11.2   Mid-channel dissolved oxygen levels at high water in the Forth Estuary, 1989

particular, we concentrated on reducing BOD loadings in order to improve DO levels, and to remove the low flow violation of the existing EQS for DO (Hanley *et al.* 1991a). We now go on to consider how this exercise was attempted.

Table 11.3 BOD sources in the Forth Estuary

In reach	Input	BOD loading (kg/day)
Rivers		
1	R. Teith/Forth	1223.85
1	R. Allan	426.29
2	Bannockburn	50.88
4	R. Devon	1013.80
7	R. Black Devon	67.45
10	R. Carron	283.18
14	R. Avon	367.94
16	Bluthur Burn	354.24
	Sub total	**3787.63**
STWs		
1	Stirling STW	460.00
3	Fallin STW	16.65
5	Cowie STW	16.01
6	Alloa STW	3088.00
9	Kincardine STW	159.27
10	Grangemouth W. STW	561.60
10	Dalderse STW #1	280.00
14	Kinneil Kerse STW	2750.00
15	Bo'Ness STW	1250.00
16	Bluthur Burn STW	354.24
17	Ironmill Bay Sewer	172.74
18	S. Queensferry STW	820.92
19	Dunfermline STW	4034.88
	Sub total	**13964.31**
Industry		
4	United Grain #1	1650.00
4	United Grain #2	340.00
6	DCL	25000.00
8	Weir Paper	2000.00
11	ICI	20000.00
12	BP Oil	3000.00
13	BP Chemicals	10000.00
20	GP Inveresk	11374.00
	Sub total	**91,115.94**
	Total	**87933.17**

Source: FRPB, various

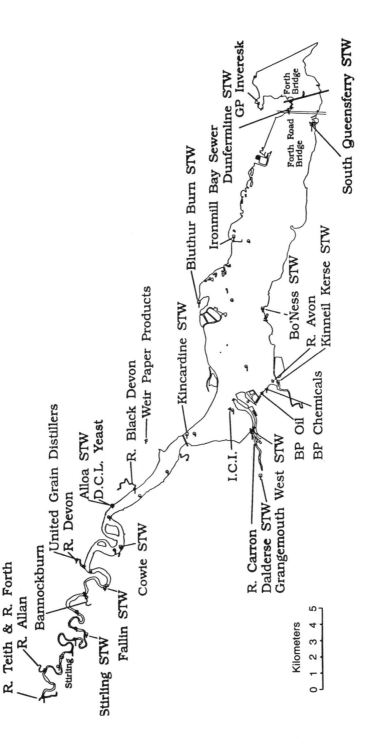

Figure 11.3 The Forth Estuary – Point Source Inputs

A model for predicting the impact of policy changes in the Forth Estuary

This model follows the general methodology described in an earlier section such that an economic model based on the minimization of control costs subject to environmental constraints was used, alongside a model of water quality in the estuary. So far, most outputs from this research programme have been in terms of target reductions in total BOD loadings, rather than target improvements in DO levels, although there is one important exception to this, the 'remove sag only' simulations described later on.

The economic model was a stepwise LP model, which explicitly incorporates the stepped nature of MAC functions, as discussed earlier. Input data for this model consist of control cost information obtained from six of the seven industrial sources, and for all STWs where improvements have been costed (this includes all major STWs). These data were supplemented with detailed information on the production and pollution control processes at each site. The activities in the model represent reductions in BOD loading from each source, specified in terms of kilograms of BOD per 24 hours, with annualized control cost data also specified in these units. The model allows for the least-cost solution of hitting water quality improvement targets to be calculated, and for either an emissions tax or TDR market to be simulated. These outcomes can be compared with the outcomes of uniform regulation, where all sources are compelled to reduce emissions by a given, equal amount.

The economic model is run alongside a one-dimensional dynamic model of water quality in the estuary ('FEDS'), which divides the estuary up into 20 reaches, and predicts levels of DO in each of these reaches. The model is calibrated at low-flow conditions, and allows for biodegradation, re-aeration, diffusivity along the mean axial length, tidal flushing and effluent inputs. These inputs can be varied to trace out the impact of one discharger's output on any other reach (Moffatt *et al.* 1991). Typically, effects outside of the reach into which a discharge is made are small, with no significant impact of, say, BOD loading reductions in reaches 12–17 (the Grangemouth area, associated with three major industrial discharges and four major STWs) on DO levels in the critical sag area (reaches 1–10). This is important for what follows.

Comparing market mechanisms with uniform regulation

Two main simulations were performed for each set of policy measures. The set of policy measures consisted of uniform restrictions, flexible regulation, TDR market, and a uniform emissions charge. These were also all compared with the least-cost solution. The two simulations consisted of targets of reducing total BOD loadings to the estuary, irrespective of source; and targets for removing the summertime DO sag. We now look at results from these simulations, before considering some practical barriers to the introduction of either TDR or emission tax systems.

First, though, we should note that existing regulation in the estuary by the FRPB does *not* correspond exactly to the textbook alternative to market mechanisms, which is uniform regulation. This is because consented discharge limits are currently set with some regard to both the varying marginal physical impacts of discharges, and their marginal control costs. It is difficult to know how much, in the past, this flexible approach has avoided the high costs traditionally associated with uniform regulation; the FRPB clearly places its main emphasis on the improvement of water quality. However, there is at least some (unofficial) cognisance of the extent to which control costs vary, and we find that the flexible regulation policy proposed for removal of the summertime DO sag gets very close to the least-cost solution.

Results for total reductions in BOD loading

We looked at three target reductions in total daily loadings discharged to the estuary, currently set at 61,940 kg/day from the six industrial sources for which MAC data were obtainable, and 12,818 kg/day from the three STWs on the estuary itself for which we had abatement cost data. The cost of meeting these reductions by uniform cuts was compared to the costs under the least-cost solution, in two time periods: where the target must be hit in 12 months; and where the target must be hit in five years. Marginal control costs were found to be very sensitive to the time-scale allowed. In all cases, the least-cost solution imposed a much lower total resource cost on dischargers than uniform cutbacks, with the least-cost solution being between 9.3 and 27.9 per cent of the uniform cutback outcome. These substantial resource cost savings are unsurprising, since we found MACs to vary dramatically across dischargers. Interestingly, the discharger with lowest marginal control costs is also located at the most sensitive part of the estuary, and dominates all efficient solutions. As the target is made more severe, then the shadow price of pollution (the marginal value of being allowed to emit) rises, as increasingly costly abatement activities are switched in to the solution. STWs are almost irrelevant in the 12-month target scenarios, as the scope for reductions in BOD without changing the capital stock is very limited.

If a perfectly competitive TDR market were introduced, with permits exchanging at a one-to-one rate across the whole estuary, then this would replicate the least-cost solution. However, there would be relatively few traders in such a market: only six firms, and three regional councils (which operate the STWs). The market would therefore be rather thin. What is more, there is a high probability that price-setting behaviour would emerge, since our simulations showed that two firms would dominate permit trades, one as a significant seller, and one as a significant buyer. Previous work on uncompetitive behaviour in permit markets has shown that, while price setting drives the resource cost above the least-cost level, the impact is relatively small, and does not erode the cost-saving property of TDRs relative to uniform regulation: Maloney and Yandle (1984), in a study for the US EPA found that even with a very high degree of cartelization, the TDR market still achieved its environmental target at 66 per cent of the uniform regulation outcome (compared with a 75 per cent saving with competitive behaviour).

However, the combined effect of a small number of traders, price-setting behaviour and likely hoarding (two of the largest sources are in product market competition with each other, as are the largest source and one medium-sized discharger: trades between these sources would be unlikely) means that a TDR market would almost certainly not achieve all the cost savings predicted by the least-cost solution.

We also simulated the introduction of a pollution tax, comparing it with uniform cutbacks to achieve the same goals (10, 25 and 50 per cent cuts in total BOD loading) as above, again with two time periods, 12 months and five years. The major finding here was that, as with the Tees study, indivisibilities in abatement meant that the targets could not be exactly hit by a single tax rate. Targets were either under- or over-achieved. However, considerable savings in resource costs were still forthcoming, if the cost of an emission cut achieved by a given tax rate was compared to the cost of meeting the same reduction using uniform regulation. The tax policy achieved the targets with savings in the order of 70–90 per cent over uniform regulations.

There is an important point here, however. So far, we have only discussed the resource costs of each policy. But dischargers will also face transfer payment costs under a simple tax scheme, and under one version of a TDR scheme. For the tax scheme, then firm i pays two sorts of costs: the abatement cost of reducing discharges to the point where $MAC_i = t$, if t is the tax rate; and the tax payments (tE_r) on their remaining (cost-effective) discharge, E_r. The 'financial burden' of the policy to the firm is the sum of resource cost and transfer payment, less any refund. Previous studies have found the transfer payment element to dominate financial burden: Palmer et al. (1980), in a study of halocarbon emission controls, found transfer payments to be ten times greater than resource costs. We found that the transfer payment element also dominated in the Forth study. For example, for the 50 per cent BOD cut target over five years, total transfer payments are five times as great as total resource costs. This factor would clearly mitigate an enthusiastic endorsement of such a scheme by dischargers.

For TDRs, then, the extent to which financial burdens are increased by transfers depends on how the permits are initially allocated. If permits are initially grandfathered, then the net transfer out of all dischargers to the FRPB would be zero. Buying and selling between dischargers creates windfall gains and losses, but these cancel out in aggregate. If, however, permits are initially auctioned, then total transfer payments of a size equal to a simple uncompensated tax scheme are extracted from dischargers.

One interesting conclusion from this part of our work was that the outcome of either variant of this simple TDR market might be viewed as equitable in the following way. We calculated the total financial burden falling on each discharger under both auctioned and grand-fathered systems. This was then divided by the BOD loading of that discharger, to give a 'financial burden per unit of discharge'. Surprisingly, these were fairly similar for all dischargers, since abatement cost expenditures are offset by permit sales for low cost-of-control sources. Whether this is viewed as equitable depends on the circumstances behind the differing MAC functions of firms: in some cases, dischargers have low MACs principally because they have been given an 'easy ride' previously, as John Pezzey has pointed out (Hanley *et al.* 1991b).

Results for removal of the DO sag only

As already pointed out, the Forth at present is a fairly clean estuary. DO levels only fall below the current EQS of 4.5 mg/l during low flow in one section of the upper estuary, corresponding approximately to the stretch between Stirling Bridge and the Kincardine Bridge, with the lowest levels of DO being recorded around Alloa. This is due to the restricted width of the estuary at this point, and the corresponding slow flushing rate. Three major BOD-causing discharges enter around Alloa (a distillery, a yeast works and Alloa STW), while an oxygen demand is still being exerted from Stirling STW at the upper points of this oxygen-deficient reach. We have also already noted that reductions in discharges from the Grangemouth area will have little impact on this reach, as the former lies down-estuary of the latter (the influence of the tide and river flow pushes the upper layer of water seawards). We looked at the possibilities of using either TDRs or taxes to remove this DO sag, ignoring all dischargers outside the relevant reaches. This reduces the economic model in size, since only four industrial discharges (two owned by one factory), and two (major) STWs then remain. We compared the least-cost solution to the FRPB's current plan to remove this EQS violation, which we take to be an example of flexible regulation (as already discussed) in action.

The required reduction in BOD discharge into the upper estuary was estimated at 17,700 kg/BOD/day by the FRPB. In the flexible regulation scheme, this was to be achieved by requiring large reductions from two sources only. This gave an annualized control cost of £1.84 million if a five-year adjustment period is allowed.

The reduced form of the economic model was then run to achieve this same total reduction in loading across all sources in the upper estuary. This produced an (annualized) resource cost of £1.01 million, a saving of only 14 per cent. It would seem that even if a TDR market or emission tax could hit the target in this instance, there will only be a small improvement over currently planned improvements. This is in fact because both abatement patterns are dominated by large reductions in emissions by one source.

It might be argued from the above that market mechanisms have little to offer over flexible regulation. However, this may be a situation unique to this particular case. Additionally, flexible regulation cannot provide the continuing incentive to reduce emissions in the most efficient manner by investing in new pollution control equipment, or new production processes, that is, provided by both TDRs and emission taxes. While, finally, flexible regulation may have a less equitable outcome than either TDRs or taxes.

Conclusions

What emerges from the three empirical studies we have looked at in this chapter, is that there are numerous problems in applying either TDRs or taxes to the control of water pollution. In both the Fox River and Forth studies, there is evidence that permit markets will not always yield the least-cost solution, due to hoarding and price-setting behaviour. The Forth study also indicated the potential size of the distributional effects of both auctioned TDRs and tax schemes which certainly mitigates against the acceptance of such schemes by industrial lobbies and public sector dischargers. The Tees study showed how complex a fully differentiated charge scheme could be, and revealed that charge or TDR schemes on their own are not able to achieve control over all the parameters that go to make up water quality indices.

However, it would be very easy to go from these findings to the position that, because of these problems, TDRs and taxes are of no practical interest for water pollution control. This is certainly not the case, since the flexibility that either concept encourages generates very significant resource cost savings over uniform regulation, and provides a continuing incentive for dischargers to find more efficient means of controlling their discharges. These potential cost savings were very evident in all three case studies examined. So while the case for market mechanisms can perhaps be more easily made for other forms of pollution (sulphur dioxide emissions from stationary sources, for instance), there is much to be gained by society if such instruments *can* be made to work in controlling water pollution.

Notes

1. Some of the research reported in this chapter was funded by the Economic and Social Research Council under the Risk and Pollution initiative. The present author would like to thank the ESRC for this support; he also recognises that much of the material contained here is in no small way due to previous joint work with the rest of the University of Stirling team (principally Ian Moffatt and Steve Hallett), and with John Pezzey, University of Bristol.
2. A full account of water quality legislation in the UK is given by Hallett *et al.* (1991).

References

Atkinson, S. and Tietenberg, T. (1987). Economic implications of emissions trading rules for local and regional pollutants. *Canadian Journal of Economics* 20, 370–86.

Baumol, W. and Oates, W. (1971). The use of standards and prices for the protection of the environment. *Swedish Journal of Economics* 73, 42–54.

Baumol, W. and Oates, W. (1988). *The Theory of Environmental Policy*. Cambridge, Cambridge University Press.

Environmental Resources Limited (1990). *Environmental Charging and Subsidy Schemes*. Report to the UK Dept. of Environment, 1990. ERL Consultants, London.

Hahn, R. (1989). Economic prescriptions for environmental problems: how the patient followed the doctor's orders. *Journal of Economic Perspectives* 3, 95–114.

Hallett, S., Hanley, N., Moffatt, I. and Taylor-Duncan, K. (1991). UK water pollution control: a review of legislation and practice. *European Environment* 1, 7–14.

Hanley, N., Moffatt, I. and Hallett, S. (1990). Why is more notice not taken of economists' prescriptions for the control of pollution? *Environment and Planning A* 22, 1421–39.

Hanley, N., Moffatt, I. and Hallett, S. (1991a). *Market Mechanisms for the Control of Water Pollution in the Forth Estuary*, Working Paper 8. River Pollution Control Unit, University of Stirling.

Hanley, N., Hallet, S., Moffatt, I., Farley, D., Taylor-Duncan, K. and Pezzey, J. (1991b). Appraisal of the potential role of market mechanisms in water quality issues. Research report to Royal Commission on Environmental Pollution.

Moffatt, I., Hanley, N. and Hallett, S. (1991). *A Dynamic Simulation Model of Water Quality in the Forth Estuary, Scotland*, Working Paper 7. River Pollution Control Unit, University of Stirling.

Maloney, M. and Yandle, B. (1984). Estimating the cost of air pollution control regulation. *Journal of Environmental Economics and Management* 11, 244–63.

Montgomery, W. (1972). Markets in licenses and efficient pollution control programmes. *Journal of Economic Theory* 5, 395–418.

National Rivers Authority (1990). *Proposed Scheme of Charging in Respect of Discharges to Controlled Waters*. NRA, London.

O'Neil, W., David, M., Moore, C. and Joeres, E. (1983). Transferable discharge permits and economic efficiency: the Fox River. *Journal of Environmental Economics and Management* 10, 346–55.

Opschoor, J. and Vos, H. (1989). *Economic Instruments for Environmental Protection*. Paris, OECD.

Palmer, A. *et al.* (1980). *Economic Implications of Regulating Chloroflourcarbon Emissions from Non-Aerosol Applications*. Rand Corporation, Santa Monica, CA.

Rowley, C., Beavis, B., Walker, M., Storey, D., Elliot, D. and McCabe, P. (1979). A study of effluent discharges to the River Tees. Report to the Departments of Environment and Transport, London.

Scottish River Purification Boards Association (1991). *Proposed Scheme of Charging for Discharges to Controlled Waters*. Aberdeen: SRPBA.

Seskin, E., Anderson, R. and Reid, R. (1983). An empirical analysis of economic strategies for controlling air pollution. *Journal of Environmental Economics and Management* 10, 112–24.

Storey, D. (1977). A socio-economic approach to water pollution law enforcement in England and Wales. *International Journal of Social Economics* 4, 207–24.

Tietenberg, T. (1980). Transferable discharge permits and the control of stationary source air pollution: a survey and a synthesis. *Land Economics* 56, 391–416.

Walker, M. and Storey, D. (1977). The standards and price approach to pollution control: problems of iteration. *Scandinavian Journal of Economics* 79, 99–109.

Chapter 12

Postscript: Future Prospects

R.K. Turner

During the 1970s environmentalist concern was primarily focused on source limits (population growth, natural resource depletion and food supply), with less emphasis on sink limits (pollution and the assimilative capacity of ecosystems). By the time the United Nations Conference on Environment and Development (UNCED) had taken place in the summer of 1992, sink limits had assumed priority status. Hence UNCED concentrated on two main issues, on which international agreements were signed: climate change and biodiversity conservation.

Thus the 'limits to growth' debate has been superseded by the 'global environmental change' debate. Nevertheless, the various competing world views, underlying ethics and related policy prescriptions, which the debates have highlighted or spawned can still be broadly categorised as neo-Malthusian and neo-Ricardian perspectives – see Figure 12.1. The concept of sustainable economic development, despite the definitional ambiguities which have surrounded it, has nevertheless proved to be a useful device for clarifying economic, political, ethical and environmental principles related to development. We do not yet have a blueprint for sustainability, but some practical rules and policy strategies can now be identified, at least in outline form (weak and strong versions).

The consensus view of the analysts who have contributed to this volume is that both the VWS and the VSS positions are probably wrong. But uncertainties are such that both the WS and SS positions can command much support (and, indeed, may not be radically different from each other in practice). The editor believes that the SS approach has much to commend it. It can accommodate the constant capital rule, but also buttresses the rule (and its equity provisions) with a precautionary approach linked to the existence of primary (and not just secondary) value in ecosystems and of critical (non-substitutable) natural capital assets. Economic growth will have to be moderated (though not stopped) in order to conserve these critical assets.

To summarise the current (1992) state of the sustainability debate, from the conventional economic perspective, the sustainability issue has

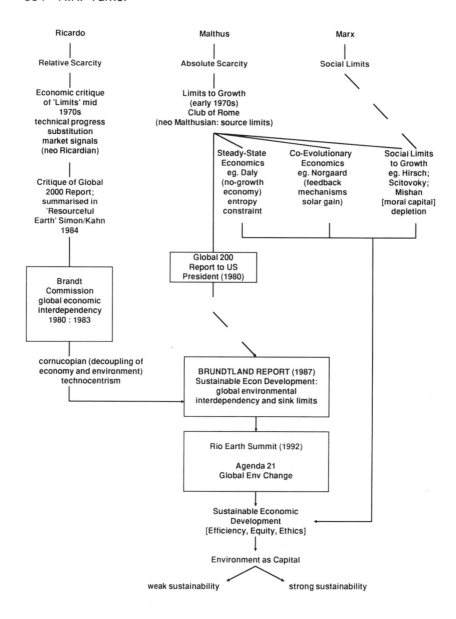

Figure 12.1   From 'limits to growth' to 'global environmental change' thinking

at its core the phenomenon of market failure and its correction via 'proper' resource pricing. What is required is an intertemporally efficient allocation of environmental resources through price corrections based on individual preference value. A vast literature has therefore grown on the various monetary valuation methods and techniques available to 'price' the range of environmental goods and services provided by the biosphere (i.e. market-adjusted, surrogate-market and stimulated-market methods).

Critics of this standard economic position, however, have put forward several arguments. The first of these is that economic sustainability is more a matter of intergenerational equity, and that incorporating environmental values *per se* into the policy-making process will not bring about sustainability unless each generation is committed to transferring to the next sufficient natural resources and capital assets to make development sustainable.

Secondly, conventional economic and ecological approaches to sustainability are largely disjoint – they address different phenomena. Intertemporal price efficiency is not a necessary condition for ecological sustainability (defined in terms of ecosystem stability and resilience and requiring the constraint that the allocation of economic resources should not result, via overall system feedbacks, in the instability of the economy-environment systems as a whole), and the pursuit of intertemporal efficiency on the basis of the individual preferences of the current generation may well be inconsistent with ecological sustainability.

Thirdly, an ethical shift away from or modification of traditional individualistic moral reasoning is required unless depletion of the moral capital stock is not to take place; in any case the normative role of individualism (consumer sovereignty) in resource allocation is no longer axiomatic.

Fourthly, cultural and biological diversity are both prerequisites for sustainable development, and therefore the conservation of cultural capital and the fostering of local sustainable livelihoods are important from a policy point of view.

According to UNCED Secretary-General, Maurice F. Strong:

the road beyond Rio will be a long and difficult one; but it will also be a journey of renewed hope, of excitement, challenge and opportunity, leading as we move into the 21st century to the dawning of a new world in which the hopes and aspirations of all the world's children for a more secure and hospitable future can be fulfilled.

Haas *et al.* (1992) examine the four central outcomes of the UNCED process – new international institutions, national reporting measures, financial mechanisms, and heightened public and non-governmental organisation participation – in order to assess the effectiveness of UNCED. They conclude that UNCED has laid a foundation with which governments and other agencies will be able to pressure each other to maintain a high level of commitment to sustainable economic development. More pessimistically, in the short to medium run lack of finance is likely to inhibit the operationalisation of the UNCED policy strategy (Agenda 21). The threat to global sustainability is not yet perceived to be severe enough for an ethic of international environmental cooperation to become established.

The list of practical requirements for sustainability is a formidably long one. We require a pragmatic package of policy instruments, deployed in a phased process: regulation (to guarantee 'environmental certainty' and perhaps to 'conserve' intrinsic ethical motivations in individual humans); economic incentive instruments, pollution taxes and tradable permits (the latter operating at the local and maybe at the global scale); environmentally compensating projects; and intragenerational and intergenerational transfers of resources (the latter via natural capital bequests). Population growth must be moderated via more education and economic opportunities for women, more birth control provisions and a more equitable distribution of income. Better valuation of natural capital and modified GNP accounting will also help, but psychological and social factors are very important and urgently require further analysis. These are very difficult matters to quantify and evaluate. Direct challenges are posed to the educational and fiscal systems that we currently operate with. Perhaps new taxes on advertising energy and transport are required, but will this be enough to alter the overly materialistic values that exist in much of contemporary society? Genetically we appear to be cooperative entities and if this is so then institutional frameworks can be devised to further encourage such ethics and behaviour. If we are both consumers and citizens then we need further encouragement to operate in the latter mode. Analysts then need to be able to elicit from citizens social preference information in order to feed this into the decision-making process, which will determine relative social values.

Reference

Haas, P.M., Levy, M.A. and Parson, E.A. (1992). Appraising the Earth Summit. *Environment* 34, 6–15 and 26–36.

Index